Advances in Deep Learning Applications for Smart Cities

Rajeev Kumar
Teerthanker Mahaveer University, India

Rakesh Kumar Dwivedi
Teerthanker Mahaveer University, India

IGI Global
PUBLISHER of TIMELY KNOWLEDGE

A volume in the Advances in Electronic Government, Digital Divide, and Regional Development (AEGDDRD) Book Series

Published in the United States of America by
IGI Global
Engineering Science Reference (an imprint of IGI Global)
701 E. Chocolate Avenue
Hershey PA, USA 17033
Tel: 717-533-8845
Fax: 717-533-8661
E-mail: cust@igi-global.com
Web site: http://www.igi-global.com

Library of Congress Cataloging-in-Publication Data

Names: Kumar, Rajeev (Professor of computer science & engineering), editor. | Dwivedi, Rakesh Kumar, 1973- editor.
Title: Advances in deep learning applications for smart cities / Rajeev Kumar and Rajesh Kumar Dwivedi, editors.
Description: Hershey PA : Engineering Science Reference, [2022] | Includes bibliographical references and index. | Summary: "This book provides both a foundational base of knowledge around deep learning and blockchain application for smart cities and explore plentiful ways that smart cities of the future must aspire to accommodate increasing technology and their expectations of, modern living, integrated societies, knowledge-based workforce, factory automation, virtual-real social behaviors and more"-- Provided by publisher.
Identifiers: LCCN 2021050158 (print) | LCCN 2021050159 (ebook) | ISBN 9781799897101 (hardcover) | ISBN 9781799897118 (paperback) | ISBN 9781799897125 (ebook)
Subjects: LCSH: Smart cities. | City planning--Data processing. | Machine learning.
Classification: LCC TD159.4 .A38 2022 (print) | LCC TD159.4 (ebook) | DDC 307.760285--dc23/eng/20211109
LC record available at https://lccn.loc.gov/2021050158
LC ebook record available at https://lccn.loc.gov/2021050159

This book is published in the IGI Global book series Advances in Electronic Government, Digital Divide, and Regional Development (AEGDDRD) (ISSN: 2326-9103; eISSN: 2326-9111)

British Cataloguing in Publication Data
A Cataloguing in Publication record for this book is available from the British Library.

For electronic access to this publication, please contact: eresources@igi-global.com.

Advances in Electronic Government, Digital Divide, and Regional Development (AEGDDRD) Book Series

ISSN:2326-9103
EISSN:2326-9111

Editor-in-Chief: Zaigham Mahmood, University of Derby, UK & North West University, South Africa

MISSION

The successful use of digital technologies (including social media and mobile technologies) to provide public services and foster economic development has become an objective for governments around the world. The development towards electronic government (or e-government) not only affects the efficiency and effectiveness of public services, but also has the potential to transform the nature of government interactions with its citizens. Current research and practice on the adoption of electronic/ digital government and the implementation in organizations around the world aims to emphasize the extensiveness of this growing field.

The Advances in Electronic Government, Digital Divide & Regional Development (AEGDDRD) book series aims to publish authored, edited and case books encompassing the current and innovative research and practice discussing all aspects of electronic government development, implementation and adoption as well the effective use of the emerging technologies (including social media and mobile technologies) for a more effective electronic governance (or e-governance).

COVERAGE

- ICT Infrastructure and Adoption for E-Government Provision
- Issues and Challenges in E-Government Adoption
- E-Governance and Use of Technology for Effective Government
- Public Information Management, Regional Planning, Rural Development
- Current Research and Emerging Trends in E-Government Development
- E-Government in Developing Countries and Technology Adoption
- Knowledge Divide, Digital Divide
- E-Citizenship, Inclusive Government, Connected Government
- Frameworks and Methodologies for E-Government Development
- Emerging Technologies within the Public Sector

IGI Global is currently accepting manuscripts for publication within this series. To submit a proposal for a volume in this series, please contact our Acquisition Editors at Acquisitions@igi-global.com or visit: http://www.igi-global.com/publish/.

The Advances in Electronic Government, Digital Divide, and Regional Development (AEGDDRD) Book Series (ISSN 2326-9103) is published by IGI Global, 701 E. Chocolate Avenue, Hershey, PA 17033-1240, USA, www.igi-global.com. This series is composed of titles available for purchase individually; each title is edited to be contextually exclusive from any other title within the series. For pricing and ordering information please visit http://www.igi-global.com/book-series/advances-electronic-government-digital-divide/37153. Postmaster: Send all address changes to above address.

Titles in this Series

For a list of additional titles in this series, please visit:
http://www.igi-global.com/book-series/advances-electronic-government-digital-divide/37153

Regional Economic Integration and Global Competition in the Post-COVID-19 Era European Union, Eurasian Economic Union, and the Belt and Road Initiative
Oxana Karnaukhova (Southern Federal University, Russia) and Dmitry Shevchenko (Southern Federal University, Russia)
Information Science Reference • © 2022 • 274pp • H/C (ISBN: 9781799892540) • US $225.00

Future Role of Sustainable Innovative Technologies in Crisis Management
Mohammed Ali (Institute of Management, University of Bolton, UK)
Information Science Reference • © 2022 • 280pp • H/C (ISBN: 9781799898153) • US $215.00

Handbook of Research on Artificial Intelligence in Government Practices and Processes
Jose Ramon Saura (Rey Juan Carlos University, Spain) and Felipe Debasa (Rey Juan Carlos University, Spain)
Information Science Reference • © 2022 • 410pp • H/C (ISBN: 9781799896098) • US $285.00

Cybersecurity Measures for E-Government Frameworks
Noor Zaman (Taylor's University, Malaysia) Imdad Ali Shah (Shah Abdul Latif University, Pakistan) and Samina Rajper (Shah Abdul Latif University, Pakistan)
Information Science Reference • © 2022 • 280pp • H/C (ISBN: 9781799896241) • US $225.00

Trends and Innovations in Urban E-Planning
Carlos Nunes Silva (University of Lisbon, Portugal)
Information Science Reference • © 2022 • 307pp • H/C (ISBN: 9781799890904) • US $195.00

For an entire list of titles in this series, please visit:
http://www.igi-global.com/book-series/advances-electronic-government-digital-divide/37153

IGI Global
PUBLISHER of TIMELY KNOWLEDGE

701 East Chocolate Avenue, Hershey, PA 17033, USA
Tel: 717-533-8845 x100 • Fax: 717-533-8661
E-Mail: cust@igi-global.com • www.igi-global.com

Table of Contents

Detailed Table of Contents

As the development of artificial intelligence continues, its impact is being felt in several areas of life, even though the method by which it is used is not always apparent. The insurance sector is shifting to incorporate artificial intelligence in response to customer demands. The impact of technology on business is growing for all industries around the world. Digital operational models have seen increasing popularity over the last decade. The concepts "AI" and "machine learning" are hardly new in the tech world and we will continue to see their rising adoption and implementation across businesses and industries.

The big players like Cisco, Hitachi, Siemens, etc. are taking part in designing and developing smart colonies across the globe. The basic applications designed so far in smart cities are for logistic, construction, maintenance, healthcare, and security. One of the prominent issues that is still to be addressed is the management of human resources and robots in smart cities. The proposed application will scan the entire city, based on the available space, green area, and number of human beings, and accordingly, it will suggest the number of buildings, residential areas, malls, parks,

In the era of IT, each and every city has become digitalized due to the high speed of internet and the world wide web. In today's scenario, each urban area has transformed to a "smart city." A smart city is a municipality that uses information and communication technologies to increase operational efficiency, share information with the public, and improve both the quality of government services and citizen welfare. The aim of smart cities is that citizens can easily access the available resources digitally. The main mission of smart cities is to promote economic growth of the nation and optimize the city services to the citizen by improving quality life for citizens. In smart cities, everything is connected to each other, and this is highly dependent on technology. Moreover, various technologies, applications, and challenges for implementing smart cities are discussed in this chapter.

In this chapter, the authors discuss advances like 5G, 6G, and 7G. They see that IoT has made our lives so natural, yet if we see it from an Indian point of view, things are completely changed. In India, there are various difficulties. The objective of smart urban communities could likewise be to advance financial development and improvement in regions to establish a perfect and stable climate, to fund-raise for all, to help the hindered, or to make city administration more straightforward. Provided that this is true, the smart metropolitan standards ought to extend and profoundly change our conventional framework by uniting new highlights, unwavering quality, versatility all through the city, and the effortlessness of day-by-day benefits in metropolitan regions. Smart city applications are presently an innovation that administrations match and a blend of numerous direct organizations for arrangements in a wide scope of conditions, for example, medical services, delivery, business and modern organization, and security.

Preface

Today's Era in this digital world is intelligent systems, followed by Deep Learning, Big Data and Artificial General Intelligence, it looks people are adapting more Smart Cities concept in their personal and professional life. This only triggers us, why not come out with a volume on *Advances in Deep Learning Applications for Smart Cities*. Prof. Rajeev Kumar was in touch with IGI Global publication and this book came into existence.

While looking into societal development, we were concerned more on the practicality of intelligence System. We preferred the term smart cities development so that it can reach up to the last mile runners. The societal structure is very typical in nature and we are more concerned for those who are living in high altitudes, rural areas, in the extreme living conditions and especially for those who are even looking for technology in their personal life.

The *Advances in Deep Learning Applications for Smart Cities* development is equally important for us, as humanity is in search of the quality of life and it is meant for every individual soul on the earth. The concept of peace and prosperity in everyone's life is the concern and if technology especially A.I. can play a vital role in it, our purpose will be served. The smart city is playing a vital role in global development. Right from the medical surgery to health diagnosis, recommendations of TV shows, suggesting friends on social media, identifying the faces on the Internet, popular voice assistants like Alexa, Siri, filtering spam mails and many more activities are there in our professional and personal life, where smart city is pawing way from back door.

While planning this book, we have a thought it will cover the basic aspects of societal development and global well-being like how technology can play a vital role in shaping societal well-being? How the smart city can make change in the life of an ordinary farmer, a small shopkeeper, household makers, working men & women, children and a common man irrespective of caste, color, creed? How technology can bring smiles for the research community and what are the aspects that smart city which can be easily implemented to attain the optimum for societal development.

There are many aspects that need to be covered as the futuristic smart city applications and we hope that the various research scholars, eminent laureates, professors and industry members who are contributing to this volume will take these challenges and deliver the same in the perspective of smart city as an effective tool for societal well-being and global development.

Rajeev Kumar
Teerthanker Mahaveer University, India

Rakesh Kumar Dwivedi
Teerthanker Mahaveer University, India

Chapter 1

A Study on the Impact of AI of Consumer Behaviour With Reference to Commercial, Industrial, and Liability Insurance in New India Assurance in Mumbai

Saumitra Sushil Sawant
Sydenham College of Commerce and Economics, Mumbai, India & University of Mumbai, Mumbai, India

Aparna Sanjay Ger
University of Mumbai, Mumbai, India

ABSTRACT

As the development of artificial intelligence continues, its impact is being felt in several areas of life, even though the method by which it is used is not always apparent. The insurance sector is shifting to incorporate artificial intelligence in response to customer demands. The impact of technology on business is growing for all industries around the world. Digital operational models have seen increasing popularity over the last decade. The concepts "AI" and "machine learning" are hardly new in the tech world and we will continue to see their rising adoption and implementation across businesses and industries.

DOI: 10.4018/978-1-7998-9710-1.ch001

INTRODUCTION

Artificial Intelligence: An Overview

As the development of artificial intelligence continues, its impact is being felt in several areas of life, even though the method by which it is used is not always apparent. The insurance sector is shifting to incorporate artificial intelligence in response to customer demands.

The impact of technology on business is growing for all industries around the world. Digital operational models have seen increasing popularity over the last decade. The concepts "AI" and "machine learning" are hardly new in the tech world and people will continue to see their rising adoption and implementation across businesses and industries (Ghosh 2021). According to a report by PwC India, India has had the greatest rise in the usage of AI during the course of COVID-19. Financially service firms are implementing it in 82% of their organisations. This year has been unlike any other, as researcher had to confront several issues that have never dealt with before. However, it was also possible to see glimmers of hope. With noteworthy exceptions, innovation and the introduction of new technologies moved at an unprecedented velocity, notably in key vertical industries. Due to the global pandemic, many organisations began searching for new solutions that would enable them to function better, smarter, and faster.

Many of these new possibilities encompass solutions that also assist them throughout the pandemic but will also assist them in the long term. Artificial intelligence is the unifying factor for these encouraging developments. Artificial intelligence (AI) is a technology that empowers computers to perform tasks that would usually need human intelligence. Gathering information, running a model, and making decisions are all things you may do to aid your decision-making process. Artificial Intelligence is a combination of massive data, sufficient computing resources, and machine learning, which can accomplish simple, repetitive tasks, or can be more sophisticated and capable of performing on their own.

In the last several years, use of AI has exploded across all industries. The access to more and better data, along with greater processing power, has created a powerful acceleration of AI research. We have come to the point where using artificial intelligence (AI) has become a permanent fixture in our daily lives. Due to this, AI is developing rapidly and is presenting significant economic options, using AI-based platforms and solutions, large companies such as IBM, Apple, Google, Face book, and Amazon are helping consumers, partners, and employees. AI is increasingly being utilised in various aspects of the insurance sector, including underwriting, customer service, claims, marketing, and fraud detection.

Definition

John McCarthy "It is the science and engineering of making intelligent machines, especially intelligent computer programs. It is related to the similar task of using computers to understand human intelligence, but AI does not have to confine itself to methods that are biologically observable."

The English Oxford Living Dictionary gives this definition: "The theory and development of computer systems able to perform tasks normally requiring human intelligence, such as visual perception, speech recognition, decision-making, and translation between languages."

BACKGROUND

Artificial Intelligence in Insurance

The integration of AI with real-time decision-making is important in today's fast-paced world, as a result of this significant change in consumer behaviour, the pace of adoption and trial of developing technologies will dramatically increase However, in the insurance company, artificial intelligence will continue to grow as a transformative instrument that will trigger great upheaval.

An insurance policy's most important function is to help its holder to secure his or her financial resources by transferring risk in exchange for an insurance premium. Carriers must adjust their business models to meet the shifting commercial landscape that will be affected by advanced artificial intelligence Shrof (2019). Incorporating AI into claims, distribution, and underwriting and pricing will significantly impact insurance industry practises. Understanding this helps them to cultivate the necessary skills and talent, as well as adopt new technology, to thrive in the insurance sector of the future.

Although many traditional insurers are experimenting with novel ways to incorporate AI into their day-to-day operations, in anticipation of further technological progress, the insurance sector as a whole has only just began this process. AI-powered Insurtech firms are using the technology to generate better underwriting models, increase customer service, and streamline operations. But while AI can help traditional insurers modernise, doing so is difficult. Many insurance companies will have to face unique obstacles like data quality, privacy, and infrastructure compatibility when using AI.

AI IN THE INSURANCE SECTOR

There is a lot of work involved in the insurance industry, both in terms of data and time. There are a number of different data categories that are used throughout the claims management process. Using unstructured or semi-structured data to fill out claim forms, incident paperwork, records and repair estimates makes it increasingly challenging to process.

This information is used in the evaluation and decision-making procedures in order to determine the authenticity of a claim, to prevent fraud, and to ultimately route the claim through all of the decision-making phases. It is possible for insurers to obtain and extract useful information from documents while also reducing processing times through the use of automation in their claim procedures.

The Beginning of the Sales Process

Train the most up-to-date artificial intelligence systems to answer the most common inquiries fast, and then maximise their capacity to utilise important client data to offer customised policy and pricing suggestions. AI can be used to discover the most responsible users or other customers and reward them with cheaper rates so that they keep returning to your agency.

Product and Marketing

Using AI solutions that are inviting, efficient, and secure, companies can provide customers with an amazing experience. The image, video, and natural language processing capabilities of AI may be used to quickly evaluate claim submissions and deliver accurate pay-out decisions in a matter of seconds. Then, using AI's capacity to identify possible hazards from scammers and hackers, companies can increase consumer confidence and loyalty while keeping their sensitive information safe. Once these critical foundations have been established, make artificial intelligence an integral component of the promotional strategy. In the future, artificial intelligence will be able to suggest tailored services based on consumer requirements and then quickly find chances for intelligent lead generation.

Taking Care of Customers

AI systems ensure that customers receive 24-hour support — they neither sleep, nor absent due to illness, or require a vacation! Cross-selling and up selling capabilities of Virtual Customer Assistants (VCAs), for example, are a valuable asset for customer support teams since they help to reduce agent turnover. Customers can be matched with the highly skilled competent agents to answer their enquiries, and when applied to huge data sets, AI systems can give analysis of general customer sentiment over time, which is very useful. Increase the effectiveness of machine learning in order to provide feedback functionality to insurance bots. Company will gain a deeper understanding of their clients' requirements, raise standards of service, and provide a highly customized experience.

Claims Processing and Reporting

Insurance companies can record, process, evaluate, and settle claims with or without the assistance of human agents, depending on how competent the AI technologies are." AI solutions, when combined with Natural Language Processing (NLP) and automatic speech recognition, can manage the First Notice of Loss (FNOL) reporting process in a more effective and efficient manner than ever before. This not only increases productivity, but it also helps to streamline operations."

There are distinct benefits to deploying conversational artificial intelligence for insurance organisations, which are: Chatbots have the potential to significantly improve the efficiency of the claim reporting process. Customers can report incidents to the chatbot from any location and at any time by interacting with it. Following that, the AI-driven chatbot can forward the information to relevant personnel for further assessment.

Improved Monitoring and Settlement of Insurance Claims

Claims evaluation and settling can be made more efficient with the help of artificial intelligence (AI) solutions. It is possible to automate the entire process from data collection and claim creation to approvals and payment tracking. An end-to-end claims processing environment that is error-free and data-driven can be created by integrating other applications, such as fraud detection. There are various advantages to implementing AI in the administration and settlement of claims in the insurance industry.

Increased Accuracy in Loss Estimation in the Face of Reduced Claims

A variety of advantages are being reaped by companies moving to digital platforms, and the insurance industry is no exception. The introduction of innovative technology such as artificial intelligence, machine learning algorithms, and face recognition systems has played a role in the transformation of the financial industry. Insurers can forecast or assess damage to an object based on a photo by utilising machine learning in the insurance industry. It has so been found that AI for claims management is the optimum technology. It aids in the estimation of prospective losses and makes repair recommendations for any necessary components. As a result, the process of estimating loss is expedited.

Operational Procedures

The use of AI-driven chatbots in customer care is already well-established. To help the insurance sector grow, these chatbots can free up human workers to focus on other important tasks. When it comes to cross-selling and up-selling products, chatbots may play to their strengths, as well. In a nutshell, AI may assist improve the overall consumer experience in a variety of ways.

Identification and Prevention of Financial Fraud

At the global level, this figure is multiplied several times over. Insurers, according to several studies, are putting their faith in technology to catch and prevent insurance fraud.

Artificial intelligence (AI) has the potential to make a significant contribution in the early detection and prevention of fraud through the identification of trends in historical data. As a result, insurance companies can conduct a comprehensive risk assessment prior to providing their services to customers.

AI brings about uniformity across various market segments, industry verticals, and service providers because of its ability to learn and adapt. As a result, procedures for getting insurance and settling claims can be more uniformly applied. Additionally, we can expect operational excellence, lower expenses, and a better client experience. Clearly, AI-driven insurance has a bright future, and the use of AI in the insurance business is set to soar in the near future.

ISSUES OF ARTIFICIAL INTELLIGENCE

Limited Technological Expertise

The insurance company must have a thorough understanding of the current advancements in AI technology and its limitations in order to integrate, develop, and implement AI applications. The absence of technological know-how is preventing most organisations from adopting this particular domain. A professional is needed to find the stumbling blocks in the enterprise's implementation process. The adoption of AI will move more quickly if the team has access to well-trained human resources.

A High-priced Affair

Small and mid-sized businesses have significant challenges when it comes to implementing artificial intelligence technologies because it is an expensive endeavour. Even insurance companies set aside a specific budget for the adoption and use of artificial intelligence technologies.

Collection and Storage of Information

The capture and maintenance of data is among the most difficult tasks in artificial intelligence. Sensor data is used as an input by artificial intelligence systems in the business world. A massive amount of sensor data is collected in order to validate artificial intelligence. Because they are difficult to store and evaluate, unimportant and messy datasets may provide an obstacle.

Personnel in High Demand but at High Cost

As previously said, the adoption and deployment of artificial intelligence technologies necessitate the involvement of specialists such as data scientists, data engineers, and other SMEs (Subject Matter Experts). These professionals are both pricey and hard to come by in today's market. Because of their limited financial resources, insurance companies are unable to bring in the necessary labour to complete the project's manpower requirements.

The Problem of Accountability

There is a massive responsibility that accompanies with the use of AI. Hardware problems are the responsibility of a single person.

Ethical Issues

The ethics and morality of artificial intelligence (AI) are among the most pressing issues yet to be solved. It's becoming increasingly difficult to tell the difference among an AI bot and a real customer support professional because of the way the designers are meticulously nurturing the AI bots to excellence.

In order for an AI algorithm to make predictions, it needs to be trained. Everything is labelled by the algorithm based on the assumptions of the data it was trained on. Private and corporate individuals will be using this technology; insurance companies need to ensure that the algorithms employed are fair.

Inability to Perform Calculations Quickly Enough

High-end processors are required for AI, machine learning, and deep learning applications since these technologies require extremely fast calculation speeds. Because of the high costs and infrastructure needs connected with these processors, they have impeded the widespread adoption of AI. In this case, a powerful alternative to meet these processing needs is the cloud computing environment with several processors working in parallel. Computation speed requirements will increase as the volume of data available for processing grows rapidly. Next-generation computational infrastructure solutions must be developed.

Legal Issues

Erroneous algorithms and data governance in an AI application can lead to legal issues for the organisation. This is once again one of the most difficult AI issues that a developer must deal with in the actual world. A faulty algorithm that was built using the wrong data can shatter an organization's profit. Predictions made by a flawed algorithm are always wrong and unfavourable. It is possible that incidents like a data breach are the result of ineffective data governance. The PII (personally identifiable information) of a user function as a feedstock for an algorithm that could be accessed by hackers. As a result, the company will find itself in the midst of a legal battle.

Myths and Expectations Regarding Artificial Intelligence

Because of this gap, our generation has unrealistic expectations for AI systems. The media claims that AI, with its cognitive powers, will replace human jobs.

However, the IT sector faces a difficult task in addressing these lofty expectations by effectively expressing that artificial intelligence is merely a tool that can only function with the cooperation of human minds. Automation of routine or common tasks, enhancements in every industrial process, and data-driven forecasts can all benefit from AI.

REVIEW OF LITERATURE

(Peiser, 2018) Pertaining to issuance, challenges in the commercial insurance sector include a large variety of covers, with non-standard policy wordings (incl. exclusions) over a wide array of industries and geographies. his implies a variation in types of processes and risks involved. Therefore, in the era of diverse and constantly evolving risks, Artificial Intelligence (AI) in combination with deep learning, big data, analytics and other emerging technologies will help commercial insurers bolster their capabilities to streamline the underwriting process; identify and price risks in an efficient fashion and thus increase profitability.

(Lloyds, 2019) AI is affecting all sectors and aspects of our world and will continue to do so in the immediate future. In insurance AI will impact all stages of the value chain, from the first enquiries to the settlement of claims through to risk prevention. As businesses increasingly incorporate AI into their systems and processes they will need insurance to protect them from a range of potential risks. AI offers huge potential; insurers must be aware of the risks associated with using this new and fast developing technology and respond to new demand by developing appropriate models and products.

(Khrais, 2020) The advent and incorporation of technology in insurance sector have reformed operations across industries. AI helps in reducing complications that may result from human errors. Thus, although AI may reduce employment opportunities, its benefits to insurance sectors are immense. The application of AI in e-commerce stands to expand in the future, as insurance sectors are appreciating their role in influencing consumer demands. The rapid development of research technology and increased access to the internet present insurance businesses with an opportunity to expand their various platforms. Notably, the influence of AI insurance sector spills over to customer retention and satisfaction. The customers are at the center of the changes and adoption of AI in insurance sector. Hence, e-commerce can further develop contact with customers and establish developed customer relationship management systems.

(Bean, 2021), March 2021 report from McKinsey, “Insurance 2030 – The Impact of AI on the future of insurance,” projects that “AI and its related technologies will have a seismic impact on all aspects of the insurance industry, from distribution to underwriting and pricing to claims. Advanced technologies and data are already affecting distribution and underwriting, with policies being priced, purchased, and bound in near real time.

THE APPLICATION OF ARTIFICIAL INTELLIGENCE IN INSURANCE: CONTINUING WITH THE CUSTOMER-CENTRIC APPROACH

Customers now expect faster and more effective service than they have in the past, because to the development and rapid growth of digital platforms, mobile apps, and big data analytics. Consumers today have great standards for a broad range of services from the time they purchase a policy until the time a claim is settled. Because of these shifting paradigms, the insurance industry's technological interventions have been enhanced, and Artificial Intelligence (AI) is one of the promising technologies that has a bright future in terms of improving customer experience, according to Dr. Ram Ramachandran. AI has the potential to completely alter the insurance industry by making it more efficient, convenient, and future proof for both insurance firms and their clients.

When it comes to optimising the customer experience, data is critical. The Internet of Things (IoT) devices have opened up new channels for insurers to collect data that can provide them with important insights on customer demands and risks associated with customer behaviour as well as external factors. The use of artificial intelligence-based analytics also increases insurance companies' abilities to make sense of their data and apply it to actuarial sciences.

It is critical to note that this increased degree of data allows insurers the ability to follow and reward positive consumer behaviour through incentives such as decreased premiums. Premiums were determined in the past based on a person's geographic location, with data only available at the zip code level. Now, owing to artificial intelligence, businesses can adopt a more detailed approach and more correctly compute premiums that are tailored to the individual needs of each consumer.

Customer interaction pain points are identified by insurers using data analytics, which helps to make the customer engagement process more fluid and tailored. Some insurers have also begun to cross-reference call centre recordings with chatbot data in order to get insight into consumer sentiment and improve the quality of service provided by agents.

The insurance industry has received a new lease on life as a result of this increasing emphasis on individualization. Customers can tailor their insurance policies to cover everything from a single item, such as a camera or smart phone, to a single event, such as a short-term renter's insurance policy for the duration of a vacation. How can we acquire access to such finely tuned information? Chatbots, virtual advisers, and speech neural interfaces, among other technologies, have made the enrolment process more easy, automated, and personalised in recent years. The use of machine learning and behavioural science by insurers is being used to expedite the application process rather than waiting for applicants to supply information and sign papers. Customers can enrol in any insurance coverage in as little as 90 seconds and have their claims processed in less than three minutes, all with the use of a Smartphone app.

Additionally, customers require an insurance system that is very transparent, simple to comprehend, and easy to get when they need it. For this purpose, insurers are rapidly using artificial intelligence-powered platforms, which can replace human input through features such as automated product recommendations and inquiry resolution by chatbots, among other things. Using cognitive Machine-Learning (ML) techniques, insurers are also better positioned to scan medical records, data on procedures, and hospitalizations in order to determine pay-outs. Furthermore, these technological tools, which are intended to expedite the claims process, are primarily focused on delivering convenience to clients while simultaneously lowering the cost of processing those claims for insurers.

The use of artificial intelligence in claims processing, as well as other new technologies such as the Internet of Things and Block chain, is broadening the reach of insurance companies' fraud detection capabilities. For example, algorithms powered by artificial intelligence and machine learning (AI and ML) can use anonymized data to develop nano-segments of actionable insurance risk behaviour based on modelling the interaction between the client and the policy. They assist providers to increase top-line income by recognising opportunities for up selling or cross-selling, increasing retention rates, reassigning orphan policies, and optimising their books of business by generating real-time visibility into their risk portfolio. Real-time visibility into risk portfolios enables executives to gain a better understanding of current lapse rates and surrenders, which in turn encourages them to reconfigure pricing and product bundling, as well as objectively and scientifically determine the company's precise reserves and capital requirements.

THE IMPORTANCE OF ARTIFICIAL INTELLIGENCE IN INCREASING THE QUALITY OF CUSTOMER SERVICE

Need to know about your customers: In order to optimise your customer experience plan, companies must leverage AI because it provides them with an abundance of real-time user data. NLP (Natural Language Processing), an AI-driven technology, allows companies to collect and analyse user data in real time, allowing you to stay abreast of changes in their behaviour and expectations.

Effectiveness and Ease of Use

Another benefit of incorporating AI into corporate operations is that it makes them more straightforward, efficient, and productive.

Automated Processes

Such as "chatbots and self-driving software", greatly reduce the time needed to execute routine operations by half. The system also collects and analyses user data in real time, allowing you to present the capabilities and concepts that they desire, in the manner in which they like to deal with them. The use of AI in quality management allows companies to develop an inventive mobile application that has a greater scope of efficacy and a more straightforward structure. In addition, these Automation chatbot and systems conduct much of the basic work, allowing the employees to focus on more productive duties while they are freed up.

Increase Decision-making Capabilities

In terms of decision-making, Artificial Intelligence is the ideal business partner. Businesses may better forecast the future with the use of technology that examines past user interactions and current industry patterns. As a result, they have a better idea of what features and functionalities to include in their business solution in order to acquire traction in the market.

The Purchase Procedure Can Be Made More Efficient

Customers are adding things to their shopping carts then abandoning them owing to a variety of factors such as delayed loading, a hard checkout procedure, and so on. Specifically,

Artificial Intelligence (AI) Assists Businesses

It helps in recognising and responding to the issues that customers face during the purchasing process, resulting in a more frictionless buying experience for customers and a lower percentage of app cart abandonment.

Recognition and Prevention of Fraud

One of the most important applications of artificial intelligence in retail, finance, and other businesses, in terms of improving the consumer experience, is that it aids in the detection of fraudulent activity. "The innovation, by utilising its ability to collect, store, and compare user data in real time, is making it easier to detect any changes in the actions of users, hence assisting in the implementation of prompt fraud prevention measures."

Customer Intelligence

Furthermore, Artificial Intelligence is becoming increasingly important in the process of collecting and analysing client data. The usage of artificial intelligence-enabled tools and platforms is making it easier to collect a large amount of user data from a variety of sources and organise it effectively according to the essential parameters.

Artificial Intelligence is making it feasible to anticipate the contexts of user experience and develop improved client communication skills by identifying the appropriate use cases for the technology and gaining knowledge from big data in a timely and precise manner.

Self Service

Nowadays, several clients would rather handle their own affairs than enlist the assistance of a representative or a technology. This is just another reason why investment in artificial intelligence is becoming increasingly necessary.

As Artificial Intelligence can give companies with useful insights on wherever clients get stopped as well as whatever issues or questions cause them to contact the customer service staff. If company are indeed utilising these analytics, company may provide users with solutions or Inquiries that give them the impression that users have already discovered the answer to their problems on their own.

Involvement in visual, audio, and written media: AI-powered systems are also enabling the chance to provide the best possible customer satisfaction to an intended customer depending on their voice or facial expressions, among other things.

It is becoming easier to gain an understanding of the consumers' emotional experiences at any given time, and to classify methods to deliver an immediate beneficial impact to them through deals or refunds, among other things, in order for businesses to make long-term profits as a result of technological advancements.

Experiences Based on Prediction and Customization: The application of artificial intelligence is enabling it simpler for starting new companies to study user interaction behaviour and forecast their next move, and to use the information gathered to provide them with the most appropriate marketing offer possible. Increased consumer involvement and earnings are realised as a result of this strategy.

FUTURE PROSPECTS AND INCREASING THREATS OF ARTIFICIAL INTELLIGENCE

To maximise the net benefits of a full introduction of artificial intelligence into society, proactive risk management measures will be required. For the long-term hazards connected with the implementation of advanced artificial intelligence applications to be managed effectively, it is necessary to overcome the five issues of concern listed above in order for the sustainable development of AI becoming less dangerous for community.

In the meanwhile, insurance will aid in the transfer and management of new risks. Protection for consumers and businesses will necessitate changes to conventional insurance policies, including those for ""liability, casualty, health, and life insurance. Insurance will need to better manage specific exposures that firms face, such as cyber-attacks, business interruption, product recalls, and reputational damage as a result of a negative incident, among other things. The necessity for solutions such as "universal basic income" and other income protection schemes would be exacerbated by the disruption of social norms, and the evolution of present income protection insurance solutions will likely serve as a counterbalance to this.

Personal data is a source of worry in the context of artificial intelligence, specifically the amount to which it can be exploited to augment the intelligence of agents. The implementation of artificial intelligence systems is already subject to significant restrictions under European data protection legislation.

Businesses will be required to mitigate, hedge, or otherwise financially protect themselves against the risks of non-compliance with new data privacy requirements in the future. Meanwhile, the assignment of obligation and the covering of liability will grow more difficult in the future. New liability insurance models will almost certainly be implemented – in areas such as autonomous driving, for example – raising the pressure on manufacturers and software vendors while simultaneously diminishing the strict liability of consumers and increasing the pressure on governments.

AI, on the other hand, will provide benefits to insurers while also introducing new hazards. Many benefits of artificial intelligence (AI) technologies are already obvious in the insurance transaction process. It is possible to better identify the wants of customers. Policies can be issued more quickly and at a lower cost, and claims can be processed more efficiently. Identifying and forecasting large-scale organisational risks such as business interruptions, cyber security threats, and macroeconomic crises is becoming easier. Customers can receive assistance from chatbots at any time of day or night.

To conclude, the insights gained from data and AI-powered analytics have the potential to broaden the boundaries of insurability, extending the reach of existing products while also spawning the development of new risk transfer solutions in areas such as non-damage business interruption and reputational damage.

RESEARCH STATEMENT

A Study on the Impact of Artificial Intelligence of Consumer Behaviour with reference to Commercial, Industrial and Liability Insurance in New India Assurance in Mumbai

Objectives of the Study

1. To identify the relationship between artificial intelligence with insurance sectors.
2. To investigate the importance of artificial intelligence for implementing the benefits of consumers towards commercial industrial and liability insurance.
3. To examine the artificial intelligence affecting the future growth of insurance sector.
4. To find out best practices that can bring trust of existing and future policy holder while using AI as a transforming tool in the insurance industry.

Hypotheses of the Study

1. H0: There is no significant difference between consumer perceptions towards the importance of artificial intelligence for implementing the benefits of consumers towards commercial industrial and liability insurance.

H1: There is a significant difference between consumer perceptions towards the importance of artificial intelligence for implementing the benefits of consumers towards commercial industrial and liability insurance.

2. H0: There is no significant effect of AI towards the future growth of the insurance sector.

H1: There is a significant effect of AI towards the future growth of the insurance sector.

RESEARCH MEHODOLOGY

1. Development of Research Variables:
   a. The dependent variable is Satisfaction of Customers
   b. The independent variables for the study are- Commercial, Industrial and Liability insurance.
2. Sample Design
   a. **Sample size-** The estimated sample size 400 respondents to collect primary data.

b. **Sampling frame -** The sampling frame for consumer data will include people between the age group 20 to 60 years who are the insurance policy holders of New India Assurance in Mumbai.
c. **Sampling method -** The sampling method will be non probability sampling in which convenience sampling will be used to collect the data through designed questionnaire.
d. **Technique of analysis -** Qualitative as well as Quantitative techniques will be used to analyse the data in order to justify the objectives and prove the hypotheses. Quantitative techniques will include various statistical tools as such as chi square, cross tabulation, factor analysis and other tools which will suit the nature of data.

DATA ANALYSIS AND INTERPRETATION

See Table 1.

Table 1. Frequency table of demographic analysis

Questions	Data Gathered	Interpretation
As a customer, do you accept Artificial Intelligence?	Yes 286 No 114	According to the data collected. Out of 400 respondents 286 said yes and rest 114 said no. Hence majority of the respondents said yes that they are highly in favour using artificial intelligence in the insurance sector.
Artificial intelligence is progressing most in	Commercial insurance 205 Industrial insurance 115 liability insurance 80	As per the table, it presents that AI is most progressing in commercial insurance as compare to the industrial and liability insurance.
Artificial intelligence practices bring trust of existing and future policyholder while using AI as a transforming tool in the insurance industry.	Yes 283 No 117	As the highest number of respondents (283) are agreed that AI practices bring trust of existing and future policyholder while using AI as a transforming tool in the insurance industry. Hence, it is observed that insurance sectors must use AI to provide customer benefits with satisfaction.

Source: Primary Data

Hypothesis 1

1. To investigate the importance of artificial intelligence for implementing the benefits of consumers towards commercial industrial and liability insurance.
2. H0: There is no significant difference between consumer perceptions towards the importance of artificial intelligence for implementing the benefits of consumers towards commercial industrial and liability insurance.

Table 2. ANOVA table

Parameters		Sum of Squares	Df	Mean Square	F	Sig.
Claims processing time could be reduced from days to hours or minutes.	Between Groups	11.126	2	5.563	1.442	.238
	Within Groups	1531.584	397	3.858		
	Total	1542.710	399			
AI Assisting insurance firms in pricing products more competitively and customizing them for each client	Between Groups	19.857	2	9.928	2.655	.072
	Within Groups	1484.581	397	3.739		
	Total	1504.438	399			
Fraud detection and risk prevention can be improved by using AI	Between Groups	44.142	2	22.071	7.786	.000
	Within Groups	1125.435	397	2.835		
	Total	1169.578	399			
It is expected that commercial, industrial and liability insurance will remain a standard with the help of AI.	Between Groups	9.754	2	4.877	1.704	.183
	Within Groups	1136.544	397	2.863		
	Total	1146.298	399			
AI empowering customers with more personalised contracts based on their specific needs.	Between Groups	.689	2	.344	1.723	.190
	Within Groups	9.391	47	.200		
	Total	10.080	49			
The potential of artificial intelligence (AI) are virtually limitless.	Between Groups	36.922	2	18.461	4.976	.007
	Within Groups	1472.828	397	3.710		
	Total	1509.750	399			
Customer possible solutions could be expanded with the use of AI.	Between Groups	27.240	2	13.620	3.273	.059
	Within Groups	1651.800	397	4.161		
	Total	1679.040	399			

Interpretation

According to the above table, the seven parameters investigated for the importance of artificial intelligence for implementing the benefits of consumers towards commercial industrial and liability insurance. As per the analysis presents, that five parameters are having greater significant values than the P value, but for the parameters the significant values are less than P value, which interprets that large number of customers are accepting that AI has the positive impact on customers by providing them benefits in all the categories of insurance sector, but for fraud detection and potential of AI for virtual limitless are not at all acceptable by the customers. Here, the selected insurers must take care of these aspects in order to provide quality services to their customers and it helps in achieving competitive advantage.

Inference

Therefore, the inference of the investigation is that customers are in favour of using AI but they want some improvements in the technology in order to get superior satisfaction related to fraud and also want to develop the potentiality of AI in the insurance sector for its better use.

Hypothesis 2

1. To examine the artificial intelligence affecting the future growth of insurance sector.
2. H0: There is no significant effect of AI towards the future growth of the insurance sector.

Table 3. Table of independent samples test

Parameters		Levene's Test for Equality of Variances		t-Test for Equality of Means						
		F	Sig.	T	Df	Sig. (2-tailed)	Mean Difference	Std. Error Difference	95% Confidence Interval of the Difference: Lower	95% Confidence Interval of the Difference: Upper
AI has the ability to gather information about a person's behavioural patterns	Equal variances assumed	.005	.945	-.814	398	.416	-.161	.197	-.549	.227
	Equal variances not assumed			-.816	389.343	.415	-.161	.197	-.548	.227
Insurers could benefit from AI implementations by increasing productivity and reducing operating costs.	Equal variances assumed	.068	.794	1.540	398	.124	.300	.195	-.083	.682
	Equal variances not assumed			1.541	387.682	.124	.300	.194	-.083	.682
Artificial Intelligence (AI) helps to speed up the exchange of information	Equal variances assumed	4.709	.031	1.828	398	.068	.356	.195	-.027	.739
	Equal variances not assumed			1.841	395.110	.066	.356	.193	-.024	.736
The use of artificial intelligence (AI) in the analysis of insurance policies	Equal variances assumed	1.718	.191	-1.066	398	.287	-.181	.170	-.516	.153
	Equal variances not assumed			-1.068	389.281	.286	-.181	.170	-.515	.153
Artificial Intelligence aids in bolstering a steady claim.	Equal variances assumed	.093	.761	.900	398	.368	.184	.204	-.218	.585
	Equal variances not assumed			.899	384.667	.369	.184	.204	-.218	.586

Interpretation

The above table examines the artificial intelligence affecting the future growth of insurance sector, the respondents gave a huge support as all the significant values are greater than P value and hence it accepts the null hypothesis. Thus, the selected insurance sector must consider AI for the better future growth as the technology is proves as boon to them.

Inference

Because of this, the present study revealed most customers feel that artificial intelligence will have an impact on the future growth of the selected insurance sectors.

Significance of the Study

AI technology has come a long way over the last several years and will continue to evolve and improve in the future. While formerly computers struggled with complex tasks such as recognising images, recognising spoken phrases, and parsing unstructured data, it has greatly improved in these areas. In addition to huge breakthroughs in technological capability, a major shift in customer expectations is also stimulating rapid AI advancement. Many millennial's expect on-demand services because of their experience with other businesses.

The significance is that, it will help to understand the various tools of artificial intelligence of consumer behaviour and how they are used by current and future policyholders, as well as the impact of artificial intelligence of consumer behaviour in relation to Commercial industrial and liability insurance which will be determined through these study.

Scope of the Study

Commercial, industrial and liability Insurance is still heavily reliant on manual, paper-based processes that are inefficient and reliant on human intervention. Even today, clients confront lengthy paperwork and bureaucracy whether filing a claim or enrolling in a new insurance policy. Additionally, customers may find up paying more for insurance as a result of policies that are not tailored to their specific needs. These studies provide a scope for multiple future studies to examine the impact of artificial intelligence on consumer behaviour among individual policyholders such as marketing, advertisements, personal selling, service, Public relations etc. An analysis of consumer behaviour can also be carried out in relation to these individual policyholders. The research focuses most prominently on the use of AI in commercial, industrial and liability insurance.

Limitations of the Study

The limitation of this study is that it will analyse the data collected from a sample of respondents belonging to a professionals and service category only with reference to commercial, industrial and liability insurance of New India Assurance. It is restricted to Mumbai region only.

FINDINGS OF THE STUDY

1. The main finding of the study is customers accept AI as it provides various benefits to them.
2. AI is most progressing in commercial insurance as compare to the industrial and liability insurance.
3. It is found that insurance sectors must use AI to provide customer benefits with satisfaction.
4. Customers are not in favour for the fraud detection and risk prevention through AI.
5. Customers are also denying the assumptions related to the potential of artificial intelligence (AI) are virtually limitless.
6. AI has the positive impact on customers by providing them benefits in all the categories of insurance sector.
7. It is also found that artificial intelligence affecting the future growth of insurance sector

Recommendations Based on Findings

1. Customers provide AI as it provides numerous benefits to them, still consumers suffers from some issues for which insurance firms must take into consideration (proper fraud detection, data security, cost effective tec.)
2. AI is most progressing in commercial insurance; therefore, other insurance sectors should also develop AI properly to take competitive advantage.
3. As the customers are highly concerns about the risk prevention, insurance companies should evaluate it properly.
4. Insurance sector should more focus towards the benefits of the AI, so that it can provide new benefits for the customer advantage.
5. Insurance companies of commercial industrial and liability insurance should develop the use of AI as it is affecting the future growth of insurance sector.

THE CHAPTER RECOMMENDS SOME INITIATIVES TO TAKE IN ORDER TO PREPARE FOR AN AI-DRIVEN FUTURE

Become Knowledgeable With Artificial Intelligence-related Technologies and Developments

Despite the fact that the seismic shifts in the sector will be technological in nature, resolving them is not the responsibility of the IT team. Instead, board members and customer-experience teams should devote the necessary time and resources to gaining a thorough understanding of artificial intelligence-related technology and applications. It will be necessary to conduct hypothesis-driven scenarios in order to identify and highlight potential areas of disruption—as well as the implications of such disruption for certain business lines—as part of this effort.

Make a Strategic Plan and Start Implementing It

Companies must select how they will employ technology to support their business plan based on the insights gained from their AI initiatives. An extensive restructuring of processes, personnel, and technologies will be required to implement the long-term strategy plan developed by the senior management. To match or outperform the market, insurance companies should establish a strategic view on the areas where they wish to invest and determine which strategic approach—for instance, starting a new business or developing strategic skills in-house—is the most appropriate for their particular organisation.

Develop and Implement an Extensive Data Strategy

Companies' ability to detect, measure, assign, and effectively manage is all dependent on the number and accuracy of services they collect during the life cycle of an insurance policy. When presented with a large volume of data from a variety of sources, artificial intelligence technologies perform at their best. Therefore, carriers must devise a well-structured and executable plan for the acquisition of both internal and external data in order to be successful.

Develop the Necessary Personnel and Technological Infrastructure

Talent with the appropriate mindsets and skills will be required by the insurance organisation of the near future. This means that the next generation of successful frontline insurance workers is going to be in increasingly high demand, and they will need to possess a unique combination of technological savvy, creativity, and the willingness to work on a dynamic mix of semi automated and machine-supported tasks that are constantly evolving. In order to generate value from future AI use cases, carriers will need to combine capabilities, technologies, and insights from across the business in order to provide unique, comprehensive client experiences. For most carriers, this will necessitate a purposeful culture shift, which will rely on CEO buy-in and leadership to be successful.

CONCLUSION

The use of artificial intelligence is getting popularity in the insurance industries, particularly in commercial, industrial, and liability insurance. According to the responses, artificial intelligence has the potential to improve the customer experience while also empowering insurance firms, customers, and staff. According to the study, the upsurge of artificial intelligence is providing new opportunities for commercial insurance, industrial insurance, and liability insurance to provide a superior customer experience. The technology is improving the way sales and services are executed, allowing for faster claims processing, and enabling more-accurate, individual risk-based underwriting processes, among other benefits. It is therefore possible for both customers and workers to provide a better customer experience; with technology improving the way, sales and services are conducted, supporting faster claims processing, and enabling more accurate, individual risk-based underwriting processes to be implemented.

REFERENCES

Bean. (2021). Transforming The Insurance Industry With Big Data, Machine Learning And AI. *Forbes*. https://www.forbes.com/sites/randybean/2021/07/06/transforming-the-insurance-industry-with-big-data-machine-learning-and-ai/?sh=542a599ff8a6

Ghosh. (2021). *AI transforming the Insurance industry in India.* https://www.expresscomputer.in/artificial-intelligence-ai/ai-transforming-the-insurance-industry-in-india/74458/

Infosys. (2019). *AI and Insurance: Everything Changes*. https://www.infosys.com/about/knowledge-institute/insights/Documents/ai-insurance.pdf

Khrais. (2020). Role of Artificial Intelligence in Shaping Consumer Demand in E-Commerce Future Internet. *MDPI*, 1-14.

Lloyds. (2019). *Taking control Artificial intelligence and insurance*. Emerging Risk Report 2019 Technology. https://www.lloyds.com/~/media/files/news-and-insight/risk-insight/2019/aireport_2019_final_pdf.pdf

NAIC. (2020). *Artificial Intelligence*. https://content.naic.org/cipr_topics/topic_artificial_intelligence.htm

Peiser, C. (2018). *Role of Artificial Intelligence in Commercial Insurance Underwriting and Renewals*. Capgemini. https://www.capgemini.com/wp-content/uploads/2018/06/Role-of-Artificial-Intelligence-in-Commercial-Insurance-Underwriting-and-Renewals.pdf

Raj. (2019). *How Are Insurance Companies Implementing Artificial Intelligence (AI)?* Towards Data Science. https://towardsdatascience.com/how-are-insurance-companies-implementing-artificial-intelligence-ai-aaf845fce6a7

Chapter 2
An Application for Human Resource and Humanoid Management for Smart Cities

Abhay Saxena
Dev Sanskriti University, India

Ashutosh Kumar Bhatt
Uttrakhand Open University, India

ABSTRACT

The big players like Cisco, Hitachi, Siemens, etc. are taking part in designing and developing smart colonies across the globe. The basic applications designed so far in smart cities are for logistic, construction, maintenance, healthcare, and security. One of the prominent issues that is still to be addressed is the management of human resources and robots in smart cities. The proposed application will scan the entire city, based on the available space, green area, and number of human beings, and accordingly, it will suggest the number of buildings, residential areas, malls, parks, free and open space, roads, and number of robots. The chapter will address the basic issues like layout and design of an infrastructure based on human beings and humanoid population. The dos and don'ts for robots will be listed to deal with the aged or young. The additional input based on human being psychology will also be formulated so that robots can handle fragile human beings, especially aged human beings, with care.

DOI: 10.4018/978-1-7998-9710-1.ch002

INTRODUCTION

The global radiance and fragrance of Smart cities is on a boom. The big players like Cisco, Hitachi, Siemens, Huawei, and Toshiba etc. are taking active part in designing and developing Smart colonies across the globe. They are addressing the major problems like congestion, ever-increasing demand of water, electricity, fresh air, climate change and increasing economic disparity. The basic applications designed so far in Smart cities are for logistics, construction, maintenance, health care, security, energy management and other major issues related with infrastructure, service and quality. Now the overall smart Cities system is important and Lots of innovation needed for smart Cities system (Johnson, 2008).

The Robots and Autonomous systems RAS are playing an important role in these prominent fields. Artificial intelligence is being used as "City Brain" for smart cities while interacting, functioning, and simulating the environment. The networks and access to the Internet is really required (Keenan & Trotter, 1999). The service robots are channeling public and industrial services. However, there is a steep rise in the aging population especially in the US, UK, Europe, Japan and many other countries whereas Asia and Africa are finding an increase in the younger generation. The smart city design, infrastructure and layout needs to address this problem effectively in order to manage human being's and humanoids effectively.

Popular Smart Cities of Prominent Companies

The smart cities are looking to be a most sellable product and now many tech giants are in this field. The different technologies are in use for the new avatara of smart cities. The technologies like smart metering, wireless sensor networks, open source or open platforms, high-speed broadband and cloud computing are emerging in new dimensions and products for the smart cities. It looks like they are the main building blocks for the smart city infrastructure. IBM and Cisco are developing smart city infrastructure along with some major competitors like Schneider Electric, Siemens, Microsoft, Hitachi, Huawei, Ericsson, Toshiba, Oracle etc (Smart City Hub, n.d.).

Figure 1. Building blocks of a Smart City

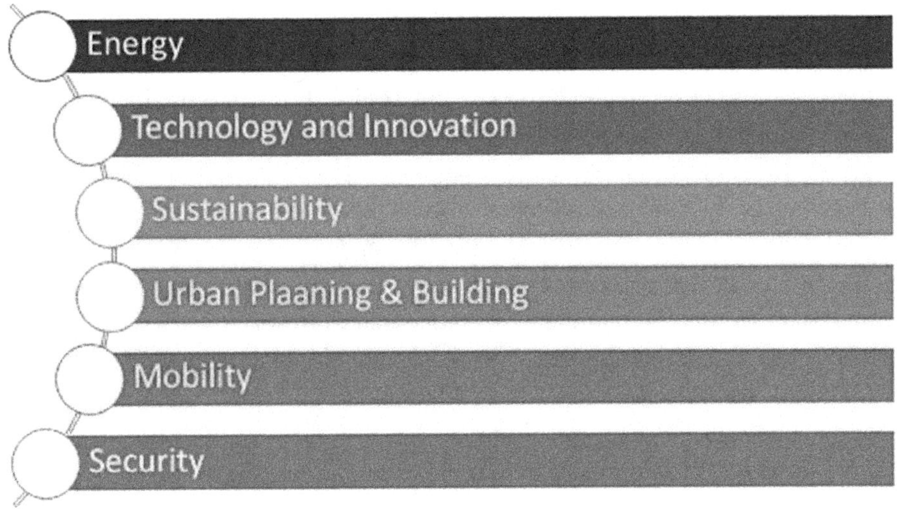

The building blocks of Smart City are Energy, Technology and innovation, Sustainability, Urban Planning and Building, Mobility, Security, which makes the societal life very smooth (United Nations, 2012). It is emerging out to be a huge market. The smart energy and technology are the most important parameters of smart cities. Transportation of the inhabitants is the biggest problem of current modern urban cities. At the same time, the environmentally friendly solutions and the electric vehicles are the need of the hour for the development of a smart city. source: https://smartcityhub.com/technology-innnovation/the-top-ten-companies-that-build-smart-cities/ (Stone, 2019).

What Is a Smart City: An Important Question to Answer

IBM defines a smart city as "one that makes optimal use of all the interconnected information available today to better understand and control its operations and optimise the use of limited resources." source:https://www.twi-global.com/technical-knowledge/faqs/what-is-a-smart, city #SmartCityDefinition (What is a Smart City?,). In brief, we can define Smart City as a framework of ICT to provide solutions for urban challenges and to create a sustainable technology-enabled infrastructure (Washburn et al., 2009).

There are many companies, who are the pioneers in their respective fields and committed to the development of smart cities. Let us quickly have a glance at them with their expertise.

- **Eliko:** A well-known technology solutions provider company. The company provides high-quality solutions based on smart innovation. Their products and the services are future-proof. Eliko helps to develop and design top-grade future technology while creating good quality of IoT products, which are helpful to develop and harness a smart city. The Eliko products are best for the indoor use and application along with in the use for sensing, signal processing, machine vision etc.
- **Terranova:** Terranova's expertise is in the software supply for smart cities. This company has been working since 2001 and now it owns 160 customers in 80+ countries. Earlier, this company was mainly involved in the gas, electric, and water sectors. The company is a pioneer in world-class software solutions. The company is expert in digital processes and efficient process management. Currently they are delivering smart grid solutions, smart metering, mobile applications and smart meter libraries.
- **Fleetonony:** This Company is helpful in improving the efficiency of fleet-based services. Fleetonony is implemented in various transit sectors and useful for automakers, mobility service providers, rental companies along with improvisation of fleet management. This company utilizes the resources efficiently. It can also help to predict demand for Car rentals and transit companies.
- **Have:** A high-rated access control system. This smart city company set its sights on the smart access sector. They are using the Li-Fi technology. It is helpful for the B2B environment. Brihtlock is its one of the important products of this company. There are many companies providing smart security-focused application. Havr is also providing access to top-notch security systems.
- **Woodoo:** This Company is the combination of technology and nature into smart materials. It helps to build more sustainable smart cities. More precisely, Woodoo takes native wood and transforms it into patented high-resistance materials. The lignin was replaced in the wood with a recycled polymer. This way the Woodoo has figured out into a more sustainable end-product. This end-product is sturdier, which is more resistant to wear and tear. It never catches fire. It is used in construction. In addition, it can be used in the smart surfaces industry.

Let us have a look at how the big and prominent smart cities are bringing technology and innovation as an integral part of their planning and successful implementation. The Internet of Things is playing a major role in designing Smart Cities. Along with IoT, the other technologies are Artificial Intelligence, Machine Learning, Deep Learning, Applications Programming Interfaces API , cloud services, green technologies etc.

Table 1. Popular smart cities and their proximity

S.No.	City	Proximity	Technology
1.	Singapore, Asia	Smart Nation Vision, Development of Virtual Singapore	IoT Driven
2.	Dubai, Asia	DubaiNow App., A.I. Driven Buses, Dubai- Abu Dhabi hyperloop	IoT, A.I.
3.	Oslo, Europe	2000 charging sections for electric vehicle, Renewable energy based Airport, Smart Grid	IoT, Green,
4.	Copenhagen, Europe	Parking systems, traffic lights, building, Smart metering in real time, Cyclist supporting App's	IoT, Green, A.I.
5.	Boston, US	Participatory Urbanism with 200 startups, Traffic snarls, Digital kiosks providing real time info. About buses, trains, bike shares, car shares.	A.I., IoT,
6.	Amsterdam	Through its IoT Living Lab, a 3,700-square-meter area fitted with IoT-enabled beacons, users can access data using Bluetooth devices.	IoT Living Lab
7.	New York	The city launched a pilot program that placed hundreds of smart sensors and a low-power wide area network throughout several business districts.	installing online kiosks for people to stay online., IoT
8.	London, UK	Mayor's Civic Innovation challenge, Connected London providing 5G services, open access wi-fi on streets,	Green IoT, 5G
9.	Barcelona	Spain's second-largest city now is saturated with LED lightpole mounted sensors that monitor traffic, air quality, pedestrian activity, and noise, and which can dim or switch off lights as needed	IoT
10.	HongKong, Asia	Mobile friendly City Dashboard Screen, lamppost fitted with sensors for 5G technology	IoT, 5G, A.I.

Source: https://www.asme.org/topics-resources/content/top-10-growing-smart-cities (Kosowatz, 2020)

Following are some companies which are involve building Smart cities:

1. Cisco
2. Schneider Electric
3. Siemens
4. Microsoft
5. Hitachi
6. Huawei

7. Ericsson
8. Toshiba
9. Oracle Source: https://smartcityhub.com/technology-innnovation/the-top-ten-companies-that-build-smart-cities/ (Smart City Hub, n.d.)

SMART CITIES PROBLEMS AND INDUCED TECHNOLOGIES

How Are the Companies Handling the Various Problems That Arise in Smart Cities?

The flying cars, innovative transportation methods, smart phones, water quality assessment (Bisht et al., n.d.), Smart Health (Bhatt et al., 2018), Energy Efficient System (Saxena et al., 2016), facial scanning and authentication technology in our smartphones are the rapid technological advancements of this era. We are expecting futuristic cities or Intelligent cities with Innovation, knowledge systems and digital spaces (Komninos, 2013). Therefore, the demand for smart technology and IT problem solving is the necessity of the recent time. The developers, IT professionals and innovators are now tackling Smart City innovations. This is a steep rise in the series of problems.

The basic challenges in various sectors of Smart Cities are as follows:

- **Infrastructure:** Utilization of sensor technology to gather and analyze information is an important aspect in Smart City. Somehow, it will improve the quality of life of its residents. The sensors are providing us various sets of data covering crime rates, water quality, soil quality, fruit chemical quality, weather data, seismic data, air quality, rain forecasting(Datt et al., 2010) but contrary to this we do require costly and complicated infrastructure for installing and maintenance of these devices.
    - Water is very important resource. It is essential need of society. Automatic Water Quality assessment and purification in the source should be done for the human in the smart city (Wikipedia contributors, 2022)(Bisht et al., n.d.).
    - To provide power supply to these variant sensors is a major issue. The varied sensors are occasionally based on hard-wiring, solar energy or sometimes on battery. Another aspect is to manage these sensors in case of power failure. The power generation companies are doing much on these aspects. Moreover, the Major metropolitan cities have replaced decades-old infrastructure, for example the underground wiring, steam pipes, and transportation tunnels can be easily seen. They have also

established high-speed internet through the Broadband wireless service. They are able established very high speed Mobile Networks and Internet of Things Infrastructures (Rosa et al., 2021). The limited internet access will not be able to connect all these sensors, actuators and grids smartly.

  - The approval processes for starting and Funding for new infrastructure projects is a very tedious and time taking process. It will not be an easy task of Installing new sensors and technologies especially, when the people are residing in these cities. This is one of the major problems every smart city developer faces. Therefore, it is the need of the hour to implement easy-to-install hardware to speed up the process of making our cities smarter. It is equally important for the Smart City developers to have prior collaboration with the residents, businesses, and academic and civic organizations to understand the common problems and to find solutions accordingly.

- **Security and Hackers:** The Internet of Things IoT and sensor technology are the base of smart cities. However, the security of these devices is a major threat. The hackers can crack the whole system of a smart city and eventually they can shut down the entire city. The cyber-terror threats can also affect the power grids, which may cause chaos and security threats to the entire smart city. It is being anticipated that the companies should create solutions with built-in mechanisms to protect against hacking and cyber-crimes. The block-chain encryption technique can be a solution to increase security in new applications.
- **Privacy Concerns:** In any Smart city, there must be a balance between quality of life and invasion of privacy. The people want to live and enjoy a healthier, peaceful, and convenient environment. The Web Cameras, installed on every nook and corner of the city, street corners obviously helps to reduce the crime. Same time one must consider the privacy concerns of people and the fear and distrust in law-abiding.
  - Another challenge of Smart city is the voluminous amount of collected data. To handle this data from smart sensors looks challenging. It is being anticipated that we should identify the types and sources of data, and notice how the collected data would be processed. One should keep an eye on how people respond to new technologies. The smart city companies need to gain trust from the people and to be taken into confidence that these solutions are intended to help them.
- **Educating & Engaging the Community:** The citizens of Smart City should also be smart in using technology. They should engage actively by taking advantage of new technologies. Before starting or implementing any technical project the biggest challenge is to educate the community about its benefits.

This can only be possible by a series of online / offline training, meeting and email campaigns etc. This will keep the citizens engaged and up-to-date.

- **Being Socially Inclusive:** The new smart city project should be socially inclusive. For example, many of the people cannot afford to take mass transit, OLA or Uber. There may be the possibility that some of the people do not use mobile devices or apps. In this situation, it will be hard for these people to cope up in a smart city. Therefore, the project's technical and social viability is a must. The smart city project is for global well-being rather than confined to technologically advanced people. Therefore, it is crucial to include or participate with everyone in the smart city. Source: https://ubidots.com/blog/the-key-challenges-for-smart-cities/ (Wikipedia contributors, 2022).

HUMAN BEING AND HUMANOID MANAGEMENT: A CHALLENGE

The professional companies and business organizations face one of the major issues is how to manage its human resources. It is also an important aspect for every management (Thite, 2011). Good management has a tremendous influence on every employee. The better management skills can help to develop productive and loyal employees and this will lead towards better financial results for an organization.

The practice of managing human resources does not come overnight. It is really challenging. Every company owns a different human resource management style. There are some general tips, which can help to improve the management of human beings and daily interactions and relationships between employees and managers.

To Understand the Psychology of the Employee

The Organization must have a clear understanding of what is being anticipated by the Individual on a personal and professional level. They should know their career goals as well as their personal interests. The lack of communication with staff members is a big hurdle. There must be a basic understanding between employees and employers. Their idea or concern should be listened to and the employer should give time full attention to them so that their ego could be satisfied. They should be motivated to work hard. The management must be vigil enough to see the positives in their employees and their work. They should also show good behavior with the employee (Wikipedia contributors, 2022). These are the segment where companies are hiring robots along with the human, in their organization.

How to Manage Humanoid: The Effectiveness of Humanoid, Their Throughput, Training, Testing, Validation, Programming Error, and Output Error

A humanoid is a robot. It resembles human beings. The Robots are interacting with human tools and environments. Example to study bipedal locomotion. The basic structure of a Humanoid has a torso, two legs and arms but in some cases, the structure varies. We can see some robots are having waist up structures. Source: https://en.wikipedia.org/wiki/Humanoid_robot(Pandyaetal.,n.d.)

During the management and planning of humanoids, one should consider the aspect of energy consumption. Humanoids performance can be calculated through a standardized, generalizable approach. The effectiveness of Humanoid is always 100%, when we calculate their throughput they always give perfect results. During training, testing, validation they always give very good accuracy and their programming error or output errors are always negligible.

The Management of Humanoids is more tedious than human beings are as the Human being management is already in the policy of various organizations but it is rare with Humanoids. It will be more typical to manage when the humanoid will be working in the offices as a coworker and start staying in the neighborhood. Same time we have to consider the human intelligence. Somehow, we should impose the human intelligence. The Articulated Human Intelligence in Perspective of Industry 4.0 should be considered during development and management of Humanoids (Saxena et al., 2019).

Humanoid Role: Caretaker, Neighbor, Coworker, Customer, Commuter

The aging people are likely to deal with the Robots as a gatekeeper, as a neighbor, as a servant or caretaker or as an assistant in the mall etc. Moreover, human beings will find Robots in the Transportation & Manufacturing sector, construction sites, distributing technologies, mobility and Warehouses as a co-worker, commuters or even customers.

Similarly, it is important to define the role of Humanoids. In the future smart societies, there will be humanoids at various places. They will be moving with human beings in the park for jogging or meditation support. We will find them as commuters in the locals, metros, trains, buses as working robots. The Humanoids will be in the shopping malls as a supporting staff or as a customer where making purchases for the aged people . We will also find the Robots in the offices, colleges, schools, shops as coworkers and even they will reside in the residential buildings as human being neighbors too.

Dos and Don'ts While Handling Aged or Young Human Beings

The needs of human beings and humanoids are different and so is their behavior and performances. There must be a certain set of policies related with the role conflicts, understanding and behavior issues among them. The dos and don'ts for humanoids needs to be clearly laid out, to deal with the aged or young human being. The Humanoid must understand the basic psychology of human beings so that they can handle Fragile Human Beings with care.

Table 2. Do's for the humanoids for handling human beings

Criteria	For Aged Citizens	For Youth	For Humanoid
Life style	Handle with Care	Give Mutual respect	Care and support
Behavior	Humble at any cost	Sober, even if they behave rashly	Sober
Support in	Housekeeping, groceries, cooking, Laundry, Entertainment	Education, Job search, infotainment	Services, Citizenship, learning skills
Covered area	Lawn, Park, House, Community Centers, Malls	Selective area and support	Selective area and support
Medical support	On top priority	On the basis of health parameters	Only on malfunctioning
Knowledge	No additional input required	Additional information can be provided	Look the service and support help

Similarly, the Don'ts can also be classified for the humanoids in order to prevent them from any sort of confusion or ambiguity.

Table 3. The don'ts for the humanoid's while handling human beings

Criteria	For Aged Citizens	For Youth	For Humanoid
Behavior	Sober, never ever Rude, soft spoken and supportive in nature	No Aggressive behavior, Depends upon the nature or respond	Sober, no irrational behavior
Language	Never abusive or attitude	Never abusive	Never abusive
Understandability issues	No Immediate actions, call relatives	Immediate response	Supportive actions
Health Issues	Consult Doctors @ database, Do not try to diagnose or treat even humanoid is a Health Robot.	Take prompt action as per programmed	Supportive, call technical head to deal with conditioning
Knowledge	No additional input required	Additional information can be provided	Look the service and support help
Knowledge	Do not try to teach or demonstrate new set of knowledge	Help them even they failed to response	No irrational shouting even if they are coworker, commuter or neighbor

Collective Management through an App

We herewith proposed an app through which one can monitor both human beings as well as humanoids in a smart city. The human beings are supposed to manage the activities and performances, whereas the humanoids will outperform human beings in repetitive things and activities. The proposed application will scan the entire city, based on the inputs such as available space, green area and number of human beings and accordingly it will suggest the number of buildings, residential areas, malls, parks, free and open space, roads and number of Robots for personal and professional space.

Figure 2. Smart city app: to develop a new smart city

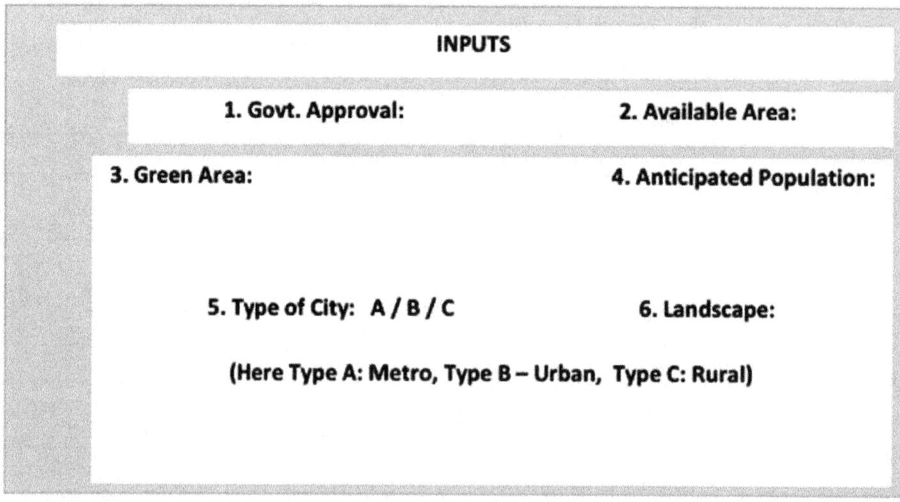

After scanning the various parametric inputs and comparing the same with the available datasets, the app will provide the basic outputs as follows:

Figure 3. The output of smart city app

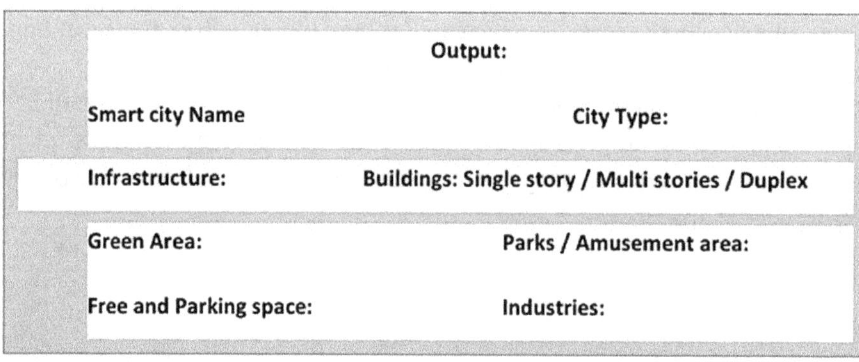

Effective Layout, Design, and Management of Both Human and Humanoid Resources

The app will also address the demand and supply of Humanoids and human beings in a Smart city especially for the professional spaces like offices, construction site, management, service facilities, maintenance, health care system and communication. A segment of it will be more like the HR department of professional companies.

Figure 4. Human and humanoid job classification

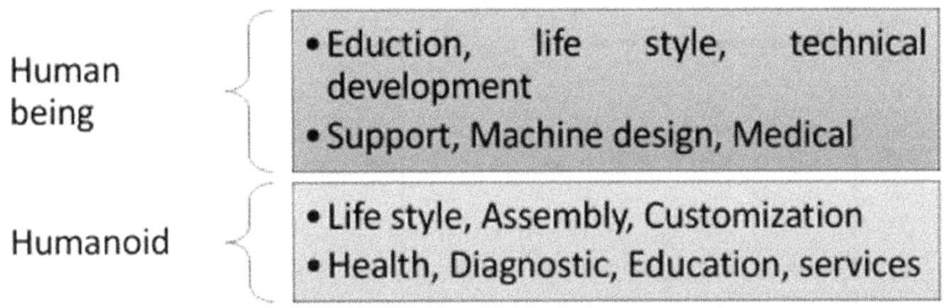

Here we are putting the Robots are not for replacing human beings rather they will be used a support factor for improvising the quality of Human beings.

Figure 5. The requirements of human and humanoids and their respective allocation

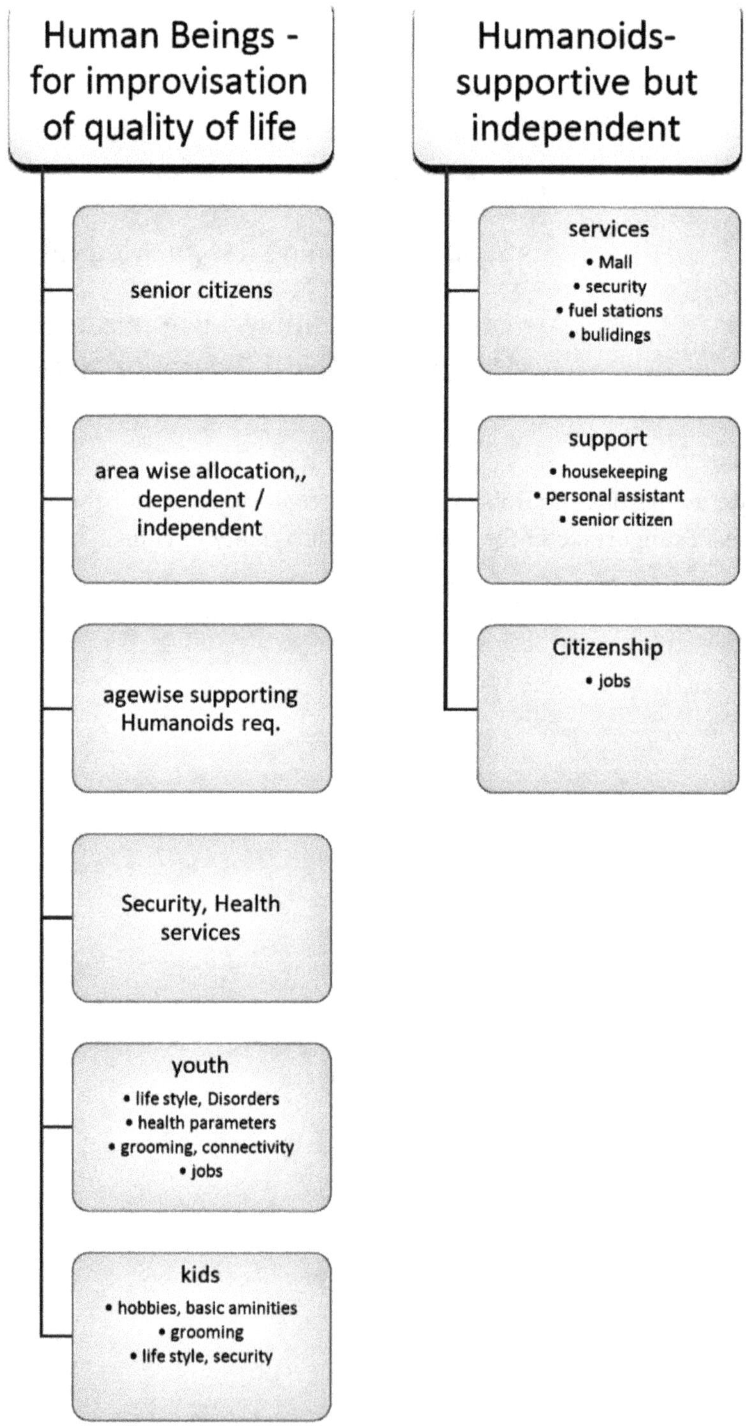

The pool of Human beings and Humanoid will be scanned, as soon as any job is available. The administrator will look into the Human Being and Humanoid databases and accordingly the job will be assigned to the respective member. The population of Human Being's and Humanoids needs to be confined in order to formulate and execute the futuristic smart city.

CONCLUSION

The Humanity future is aligned with the emergence of smart cities. Human beings and humanoids are equal stakeholders of them. The emergence of smart cities is in the development phase. The emergence of various problems related with Smart cities are development in infrastructure, connectivity, communication, information, data inflow and security, design, architecture etc. It is the need of the hour to address these problems promptly along with improving the life quality of citizens with the help of technology interventions. We do need smart cities but eventually we also need some bright smiles, warm hearted and caring people who can imbibe technology for the betterment of humanity and mankind.

REFERENCES

Bhatt, A., Dubey, S. K., & Bhatt, A. K. (2018). Analytical study on cardiovascular health issues prediction using decision model-based predictive analytic techniques. In *Soft computing: theories and applications* (pp. 289–299). Springer. doi:10.1007/978-981-10-5699-4_28

Bisht, A. K., Singh, R., Bhutiani, R., & Bhatt, A. (n.d.). *An Investigation of Assessment and Modelling the Water Quality of Rivers Based on Artificial Neural Networks-An Initiative towards the River Ganga*. Academic Press.

Datt, G., Bhatt, A. K., & Saxena, A. (2010). An Investigation of artificial neural network based prediction systems in rain forecasting. *International Journal on Recent and Innovation Trends in Computing and Communication*, *3*(8), 5338–5349.

Johnson, B. (2008). Cities, systems of innovation and economic development. *Innovation*, *10*(2-3), 146–155. doi:10.5172/impp.453.10.2-3.146

Keenan, T. P., & Trotter, D. M. (1999). The changing role of community networks in providing citizen access to the Internet. *Internet Research*, *9*(2), 100–108. doi:10.1108/10662249910264855

Komninos, N. (2013). *Intelligent cities: innovation, knowledge systems and digital spaces*. Routledge. doi:10.4324/9780203857748

Kosowatz, J. (2020, February 3). *Top 10 Smart Cities in the World*. ASME. https://www.asme.org/topics-resources/content/top-10-growing-smart-cities

Pandya, P., Saxena, A., & Bhatt, A. (n.d.). *The Human capacity assessment through time series prediction of Artificial Neural Network*. Academic Press.

Rosa, L., Silva, F., & Analide, C. (2021). Mobile Networks and Internet of Things Infrastructures to Characterize Smart Human Mobility. *Smart Cities*, *4*(2), 894–918. doi:10.3390martcities4020046

Saxena, A., Bhatt, A. K., Pandey, B., Tripathi, P., & Dutt, G. (2016). HSTL IO Standards Based Processor Specific Green Counter Design on 90nm FPGAAbhay Saxena. *International Journal of Control and Automation*, *9*(7), 331–342. doi:10.14257/ijca.2016.9.7.31

Saxena, A., Chang, A. Y., Kulathuramaiyer, N., & Bhatt, A. (2019). The Conceptual Framework of Articulated Human Intelligence in Perspective of Industry 4.0. *Indian Journal of Science and Technology*, *12*(6), 6. doi:10.17485/ijst/2019/v12i6/141147

Smart City Hub. (n.d.). https://smartcityhub.com/technology-innnovation/the-top-ten-companies-that-build-smart-cities/

Stone, S. (2019, March 6). *Key Challenges of Smart Cities & How to Overcome Them*. Ubidots Blog. https://ubidots.com/blog/the-key-challenges-for-smart-cities/

Thite, M. (2011). Smart cities: Implications of urban planning for human resource development. *Human Resource Development International*, *14*(5), 623–631. doi:10.1080/13678868.2011.618349

United Nations. (2012). *World Urbanization Prospects. United Nations, Department of Economic and Social Affairs, Population Division: the 2011 Revision*. Highlights.

Washburn, D., Sindhu, U., Balaouras, S., Dines, R. A., Hayes, N., & Nelson, L. E. (2009). Helping CIOs understand "smart city" initiatives. *Growth*, *17*(2), 1–17.

What is a Smart City? Definition and Examples. (n.d.). *TWI*. https://www.twi-global.com/technical-knowledge/faqs/what-is-a-smart-city

Wikipedia contributors. (2022, March 15). *Humanoid robot*. Wikipedia. https://en.wikipedia.org/wiki/Humanoid_robot

Chapter 3
Challenges and Opportunities for Smart Meters in Smart Cities of India Using Machine Learning and Deep Learning

Rupal Gupta
https://orcid.org/0000-0003-0905-4147
Teerthanker Mahaveer University, India

Ruchika Gupta
Whitehat Jr., India

ABSTRACT

The current era has been filled with the data and information, crawling around the web and being utilized in the best way to produce smartness. The same concept has been used for developing smart cities by increasingly deploying IoT sensor devices to collect the data from various fields. There are numerous opportunities in these fields that are related to streaming IoT data, cloud service, computer vision, and machine learning. Simultaneously, there are lots of challenges while implementing the solutions at various locations. This chapter highlights various challenges and opportunities in electricity meter consumption and billing. In this chapter, a Python-based model has been proposed for smart electricity meter for taking the reading from commercial/residential meters and producing the electricity bills smartly. Challenges for the customers will be handled by frequently monitoring the reading from the meters and reporting issues if any directly to the concerned authority. This will definitely help specially the commercial consumers to rectify their issues.

DOI: 10.4018/978-1-7998-9710-1.ch003

INTRODUCTION

As India is advancing towards being a developed nation, one of the steps in its progress is converting cities into smart cities. In converting a city into smart city there is a need of deployment of IOT devices for collection of data from various sectors such as traffic, healthcare, transportation etc. In this chapter we are focusing on electricity department and proposing a model to convert electricity meters into smart meters thus successful implementation of it will result in dispute reporting and resolution within stipulated time of encountering the problem by the electricity department. This chapter is divided into three sections. Section 1 elaborates the focus areas of smart cities along with the smart meters for Power Corporation of India, Section 2 presents the proposed models of smart meters along with the technology concerns for implementations and deployment, final section elaborates the machine learning model usage for handling disputes of consumers / Power Corporation automatically in stipulated time.

Technology Focus Areas for Building Applications of Smart Cities

For development cities into smart cities, technology advancement is the basic requirement. The focus areas of technologies are IOT, Cloud Services and Management, Machine Learning, Artificial intelligence and many more. More the technological advancement in the system, huge the changes of developing smart cities. Though this advancement has made a drastic development of smart cities in the domain of transport, health care, agriculture, security system and environment system (Weather). The outcome of the same is smart health, smart traffic, smart home, smart agriculture, smart environment etc. (Mohapatra & Panda, 2019). Further machine learning and AI field make it more interesting for predicting and classifying the results and cloud computing make the flexibility for providing the environment to accomplish the storage and further processing part.

Proposed Model for Collecting Reading and Billing From Smart Meter for Smart Cities

In this section the two frameworks for smart meters has been proposed Model-1 and Model-2 which comprises of Camera and without Camera respectively. The Models have been represented using figure 2 and figure 3. The Model-1 work with the camera and using IOT Module the captured images will be stored and processed further on cloud platform. Further using programming tools the desired information from the images will be stored and processed through machine learning / deep learning models to find fraud/meter reading issues.

Proposed Model 1 for Smart Meter

Figure 1. Proposed model 1 using camera for capturing information from smart meters

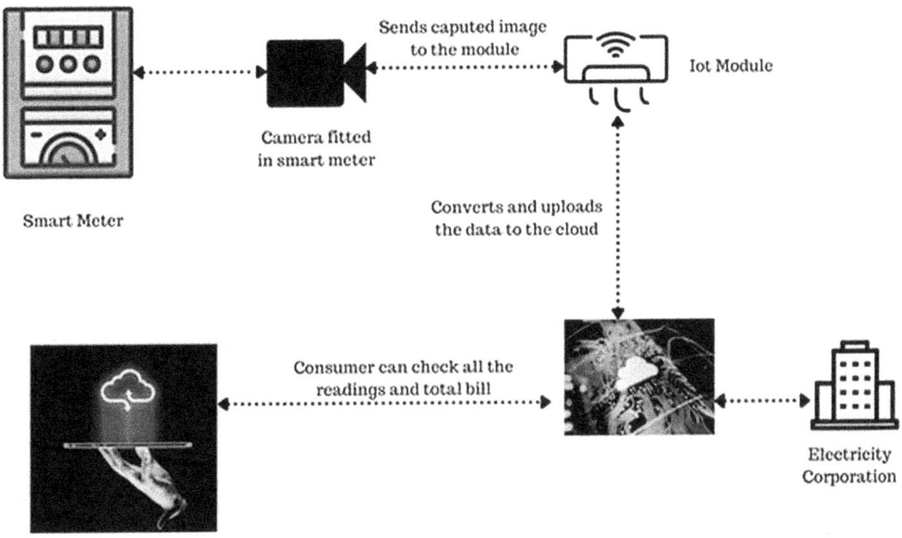

Model 1 proposes the idea of using an IOT module with a camera where the camera will capture images every day at a particular time and pass that image to the module. Then the module will convert the image to the electrical data reading required and with the help of personal network, it upload the desired information like reading, load, along with date, to the cloud interface where the interface will generate daily reports of how many units a consumer is consuming per day / per month. The consumer can check the daily records in the application provided and at the end of every month the bill will be generated automatically.

The main benefit from the proposed solution is that there will be no tampering with the meter and if the meter gets any errors than both the customers and the organization will know about the error in real time.

Proposed Model 2 for Smart Meter

Figure 2. Proposed model 2 without using camera for capturing information from smart meters

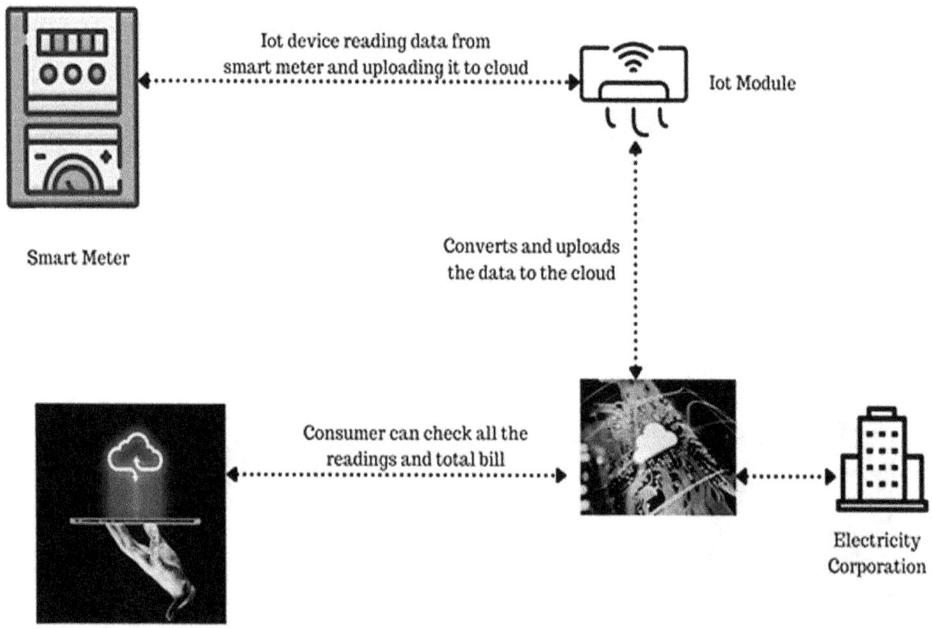

The Model 2 is same as per Model 1 with the basic functionality, further the difference is only the reading capturing part. In this model the capturing with be done directly using the chip reading system used by power corporation which already exists in the basic meters. This information will be directly sent to the cloud system for storing and further processing. This part of processing and functioning using ML/DL is same as Model 1.

FOCUSING CHALLENGES OF IMPLEMENTATIONS

There are always an implementation and maintenance challenges involved specially with the sensors devices, especially with the projects of IOT and Cloud services. As in the proposed model also there exists a huge usage of cloud and sensors along with the image data in one model, this also raises some challenges and opportunities with various domain for implementation, it also creates an opportunity for collaborating with the key players in the sector of scalable and reliable solutions to guarantee the success of massive deployment and solutions for the challenges that can come in the deployment phase, some of the challenges are discussed briefly in table 1.

Table 1. Challenges and description of the implementations

Challenge	Description
Coverage	The network coverage will play a vital role in the system as our proposed model will need an internet connection to upload the data to the cloud platform as there could be some areas which are deprived of basic internet services. The network coverage platform collaboration opportunities will bring a faster deployment in every various places for smart meter in smart cities.
Network Optimization	Network optimization is required to monitor and enhance the network requirements like traffic, bandwidth, jitter, and latency caused.
Massive Device Deployment	Replacing and deploying this device will be a great challenge as it will require the work force and huge money sums to replace all the traditional meters and placing the new devices in place of the old ones.
Device Management	Management of devices will be a hectic workload as the massive device deployment will be under way, we will have to manage all the devices placed repeatedly, also if there is any issue with the device, we must either replace or fix that issue with the device.
Security	Security will play a key role in the development of the device as the data will be stored on the cloud and it will have personal details of the customers the device needs to be most secured to the alterations and other forms of mischief.

MODEL OF MACHINE LEARNING SMART METER FOR HANDLING DISPUTES

In this section of the chapter Machine learning models has been introduced to handle the disputes which can be automatically raised using the Proposed Model-1 and Model-2. Linear regression, polynomial regression model of machine learning can be used to train the model. Further the model can be used to predict the future readings which could not be collected and on the basis of the previous reading patterns, the future date reading will be predicted. This information will be raised to all the concern parties about the error if any in the smart meters along with the deadlines to the power corporation to fix it as soon as possible. Applying machine learning can also be established as per the accuracy of the model, which will be definitely helpful for the consumers as well as to Power Corporation to avoid the hassle and especially to avoid corruption/ manipulation can be made if any. The sample linear regression model can be used on the sample dataset which has been shown in the figure3. This dataset showing the meter reading along with the dates, further machine learning model (Kukuca & Chrapciak 2016) will be used to predict the future readings along with it the model can also predict and can also check the deviation from the existing collected reading can also be used to validate the deviation of the result. This can be used using mathematical / ML model to increase the accuracy measures for predicting the future dates meter readings as per the accuracy of the models. Further deep learning models can also be introduced for better accuracy of the model (Bhattacharya, 2020).

Figure 3. Code snippet of sample dataset for smart meter using Jupyter Notebook to apply ML for handling the dispute.

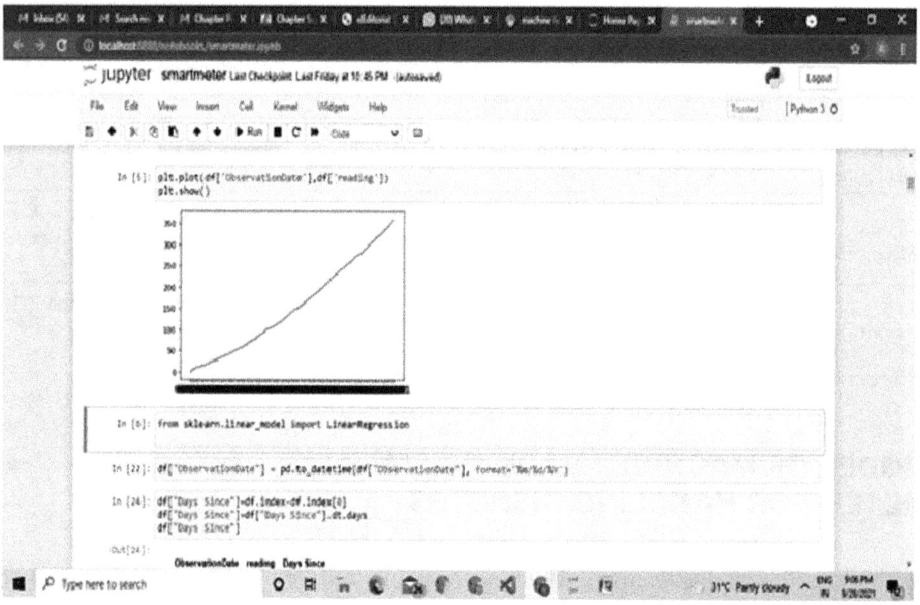

CONCLUSION

This chapter proposed two models using IOT sensor devices along with cloud computing, of smart meters for smart cities, which will helpful for consumer as well as Power Corporation of India. The proposed implementation will definitely help and provide support in smart city projects in India. Further a machine learning Model has been applied on the collected data and further if there is any dispute the model automatically will resolve the issue by predicting the future date meter reading within the stipulated time.

REFERENCES

Bhattacharya, S. (2020). *A review on Deep Learning for future smart cities*. Wiley.

Kukuca, P., & Chrapciak, I. (2016). From Smart Metering to Smart Grid. *Measurement Science Review*, (3), 142-148.

Mohapatra, B. N., & Panda, P. P. (2019). Machine learning applications to smart city. *Accents Transactions of Image Processing and Computer Vision, 5*(14).

Chapter 4
Data Analytics in the Smart City:
Dependent on Cloud Computing

Naveen Tewari
https://orcid.org/0000-0002-6104-5005
Graphic Era Hill University, India

Gopal Datt
Uttarakhand Open University, India

Arun Kumar Rai
Graphic Era Hill University, India

Mukesh Joshi
Graphic Era Hill University, India

ABSTRACT

In data analytics, cloud processing is helping make cities more powerful. The solidification of the two clouds and analytics could help cities store, decipher, and measure their enormous data to all the more possible report their citizens' issues. Organizations can spend less cash on workers and rather center on reinforcing their staff and item. Consequently, cloud drives assist more modest cities with getting the enormous information game, permitting new companies to all the more likely rival bigger associations in their industry.

DOI: 10.4018/978-1-7998-9710-1.ch004

INTRODUCTION

Bibri and Krogstie (2017) discussed in their research that Metropolitan regions have a central and portraying position in key plausible development; and thusly, they have dynamically procured a central circumstance in operationalizing and applying it.

In 2013, Pantelis and Aija (2013) says without a doubt, the main force of the great knowledge vision is the huge impact it will have on various parts of the knowledgeable city and thus on the lives of individuals. Greater knowledge is rapidly evolving, currently at a rate of 40% of advances in the global average of information produced annually compared to only 5% of advances in global IT spending. About 90 percent of digital information stored worldwide is held within just two years. As a result, many government officials have begun to use the vast amount of information available to help reform the actions and retention of knowledgeable urban communities around the world. That allowed the urban areas to conform to the norms, standards, and requirements for the use of a good city by understanding the subtle aspects of the city. These attributes include retention, power, management, improved personal satisfaction, and smart management of common assets and city offices. There are different parts of a smart city, such as portability, management, climate, and individuals just as applications and management such as medical services, transportation, intellectual education, and power says Khan et al. (2013). Working with such applications and administration is required for a large number of computers and storage areas. One way to provide these categories is to rely on Cloud Computing and utilize the many benefits of using cloud resources to help the board with great city knowledge and applications. Figure 1 shows how a distributed computer can capture large amounts of multidisciplinary information, stock collections, and testing across cloud and cloud offices.

Figure 1.
(Bibri and Krogstie, 2017).

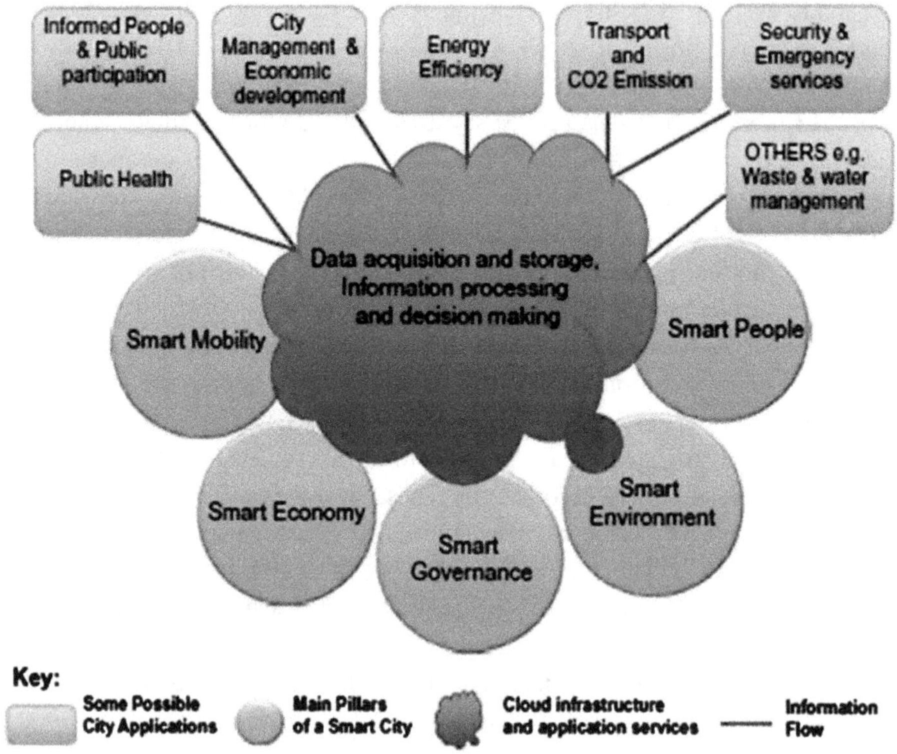

Flow work and exploration projects in this field have produced some writing that featured the significance of huge information in supporting shrewd city applications and administrations. Furthermore, some work examined by Kitchin (2013), Vilajosana et al. (2013) shows a portion of the issues of using large information in brilliant urban area. The principal commitment of this paper is inspecting the utilization of huge information in brilliant cities and investigating the changes and difficulties of using huge information in the smart city. Moreover, the paper explores the overall necessities for the plan and execution of enormous information-based applications for keen city applications and administrations.

LITERATURE REVIEW

Smart Cities

As shown by Silva et al. (2018) in their research, a smart city is a place where regular associations and organizations, city occupants, and businesses are made more useful by the usage of cutting edge and information correspondence progresses. A portion of the frameworks found in savvy urban areas, as displayed in Figure 2, are as per the following.

Savvy Industry, Smart Security, Smart Energy, Smart People, Smart Health, Smart Home, Smart Mobility, Smart Governance, Smart Retail.

Smart city plans to work on the personal satisfaction of residents by making economical systems. In this regard, the issue of the savvy city ought to be considered as an idea that incorporates distinctive monetary, helpful or legitimate components as opposed to being seen uniquely as an innovative issue. The principal objective of the keen city drive is to guarantee the maintainability of urban areas, to expand the social exercises, and to work with the day-to-day environments.

Smart Cities: Components and Characteristics

There are a large number of brilliant cities and various parts have been introduced in the figure. The parts of brilliant urban areas incorporate the accompanying: shrewd framework, savvy structures, keen transportation, keen energy, keen medical services, keen innovation, keen administration, keen training, and keen residents. A short conversation of these parts will be introduced in the ensuing areas. Distinctive keen urban areas have various levels of these keen parts, contingent upon their concentration.

The various attributes of sharp metropolitan regions fuse viability, individual fulfillment (QoL), urbanization, and splendor. The sensibility of a keen city is related to city establishment and organization, energy and ecological change, pollution and waste, and social issues, monetary issues, and prosperity. The individual fulfillment (QoL) can be assessed similar to the excitement and money related thriving of the inhabitants. The urbanization parts of the sharp city consolidate various points and pointers, similar to advancement, system, organization, and monetary issues. The savvy of a shrewd city is conceptualized as the aspiration to work on the financial, social, and natural principles of the city and its occupants. Different regularly cited parts of city insightfulness incorporate keen economy, savvy individuals, shrewd administration, brilliant portability, and keen living.

The utilization of ICT in urban areas in different structures for various city exercises has prompted the expanded adequacy of city tasks and these urban communities have been named utilizing many terms, for example, "cyberville", "advanced city", "electronic city", "flexibility", "data city", "telicity", "wired city", and "keen city". The savvy city is the biggest reflection among the marks utilized as it envelops different names utilized for urban communities. The savvy city is an idea and there is as yet not an unmistakable and predictable meaning of the idea among the scholarly community and professionals. In an oversimplified clarification, a keen city is where customary organizations and administrations are made more adaptable, productive, and supportable with the utilization of data, advanced and telecom innovations, to work on its activities to serve its occupants. In other words, in a brilliant city, the computerized advancements convert into better open administrations for occupants, and better utilization of assets while affecting the climate less. One of the conventional meanings of the savvy city is the accompanying: A city "associating the actual framework, the data innovation foundation, the social foundation, and the business foundation to use the aggregate insight of the city". Another formal and extensive definition is the accompanying: "A keen reasonable city is a creative city that utilizes data and correspondence innovations (ICTs) and different means to work on personal satisfaction, proficiency of metropolitan activities and administrations, and intensity while guaranteeing that it meets the necessities of present and people in the future as for monetary, social and natural perspectives".

Figure 2. Smart City
(Silva et al., 2018)

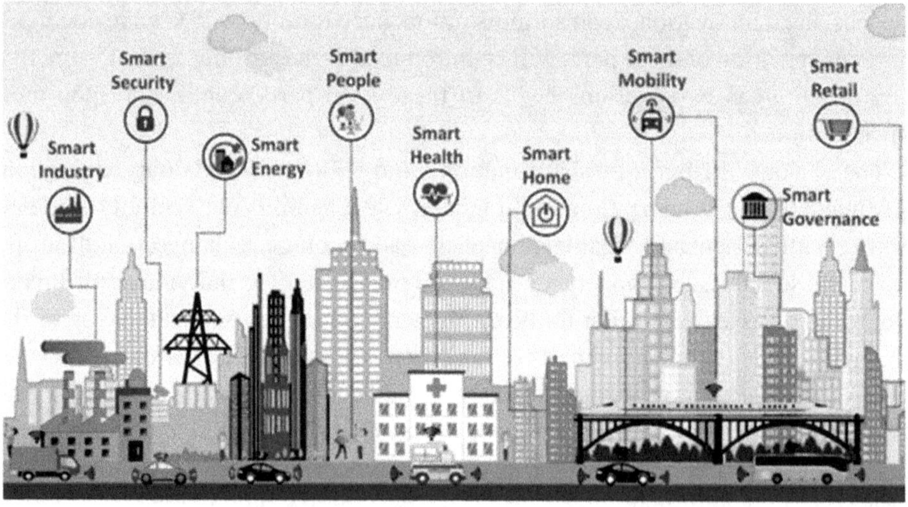

Cloud Computing

Tewari and Datt (2020) explains Cloud computing, as the Internet-based IT benefits that can be utilized by PCs and different gadgets whenever and gives divided assets between clients. With Cloud Computing, clients can get to the framework from any place on the web.

Cloud computing is turning into a piece of innovations in savvy urban areas with its achievement in cost, worldwide scale, speed, execution, productivity, security, and dependability. With distributed computing, servers, stockpiling regions, data sets, and various application administrations are effectively available over the web, making joining of the frameworks in shrewd urban communities and the trading of data between them more useful.

Also, distributed computing framework, IoT/cloud framework, and related administrations may be assigned as follows:

Infrastructure-as-a-Service (IaaS)

IoT/mists permit clients to interface with sensors and actuators in the cloud. IaaS is a huge calculation, which stores and conveys an answer upon demand. It gives a distributed computing administration. IaaS offers IoT the executives to oversee objects as an essential in the stockpile of suitable administrations.

Platform-as-a-Service (PaaS)

The IoT/cloud public framework portrayed the elite exhibition PaaS model for IoT/cloud administrations. PaaS is a full cloud-based turn of events and sending engineering, including abilities to offer types of assistance that reach from the most straightforward cloud-based administrations to complex, cloud-able organizations.

Software-as-a-Service (SaaS)

SaaS items are those that permit clients to get total programming bundles dependent on the cloud and IoT. The SaaS bundles are like standard cloud-based bundles utilizing IoT sensors and gadgets.

Considering the multiplication of sensors and the huge measures of information produced from sensors and residents, the city board must select the right answer as long as possible. Various Benefits of cloud computing-based approach on smart cities:

- Scalability - Scalability is the amount of draft power to increase or decrease performance and costs due to changes in the use and handling of draft applications.
- Agility - In a distributed computer setting, speed is often referred to as the capacity to create, test, and deploy applications that drive faster business development. Cloud Agility allows them to get involved in a variety of issues such as security, testing, and inspection, rather than providing and maintaining inventory.
- Security - Cloud security, also called distributed computer security, encompasses a wide range of methods, controls, methods, and collaborative development to ensure cloud-based frameworks, information, and infrastructure.
- Compliance - Cloud compliance is tied in with consenting to the laws and guidelines that apply to utilizing the cloud. Most associations are moving to the cloud because there are acceptable business motivations to do as such. The law doesn't forestall the reception of clouds. It has anyway had a critical effect.
- The total cost of Ownership - Complete expense of possession (TCO) is the amount of all expenses engaged with the buy, activity, and support of a given resource during its lifetime.
- Experienced partner and ecosystem

Figure 3. Cloud computing paradigm
(Tewari and Datt, 2020)

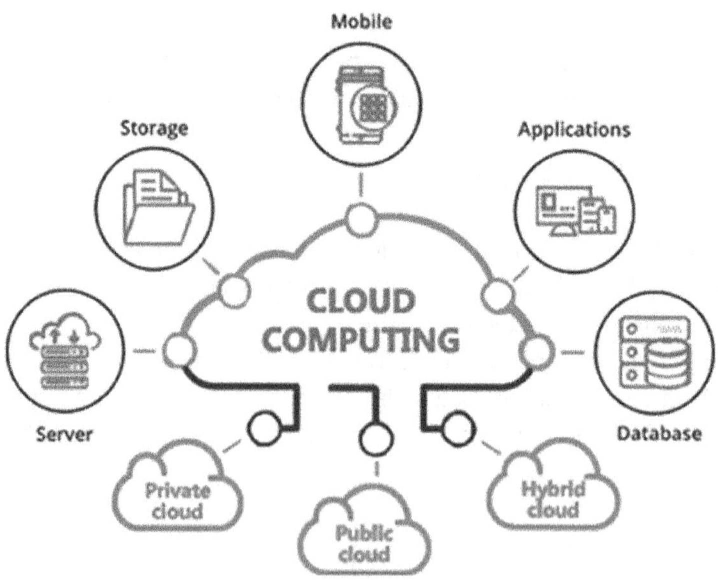

Big Data

Big Data addresses the overall domain of issues and procedures utilized for application areas that gather and keep up with gigantic volumes of crude information for space explicit information investigation. Frontiers (2013) published that present-day information concentrated innovations just as expanded computational and information stockpiling assets have subscribed intensely to the advancement of Big Data science. Development-based associations, for instance, Google, Yahoo, Microsoft, and Amazon have accumulated and stayed aware of data that is assessed in exabyte degrees or greater. Additionally, Najafabadi et al. (2015) told that electronic media relationships, for instance, Facebook, YouTube, and Twitter have billions of customers that ceaselessly make an incredibly huge measure of data.

When in doubt, Big Data suggests a variety of tremendous and complex enlightening assortments with the ultimate objective that it is difficult to deal with using standard informational collection organization mechanical assemblies or customary data taking care of utilizations. The Internet of Things (IoT), Big Data, and Smart urban areas are unequivocally related as one requirement the other two. The metropolitan information which is labeled in reality and is created in the savvy urban communities can be Big Data. The Big Data in the keen urban communities might be created from an enormous assortment of sensors, data sets, messages, sites, and web-based media the difficulties of Big Data are multifold including perception, mining, investigation, catch, stockpiling, search, and sharing. Large Data requires new methodologies of handling to empower upgraded dynamics, knowledge disclosure, and cycle enhancement. Complex information examination instruments are important to look at and concentrate significant examples and information from the Big Data of the IoT and shrewd urban areas.

Mining and separating significant examples from huge info information for dynamic, expectation, what's more, other inferencing is at the focal point of Big Data Analytics. Just as inspecting gigantic sizes of data, Big Data Analytics presents other novel hardships for AI and data examination, including plan assortment of the unrefined data, fast streaming information, constancy of the data assessment, significantly scattered data sources, rambunctious and bad quality information, high dimensionality, the flexibility of estimations, imbalanced data, solo, and un-organized data, confined coordinated/named data, etc. Satisfactory data storing, data requesting/naming, and speedy information recuperation are other major questions in Big Data Analytics. Along these lines, inventive data examination and data the board plans are advocated when working with Big Data. For example, in another work, we examined the high-dimensionality of bioinformatics area information and explored highlight determination procedures to resolve the issue (Hasanin et al., 2019).

Data Analytics

One of the essential ideas of information science is the computerized extraction of helpful information from enormous masses of metropolitan information to tackle metropolitan supportability issues (physical, ecological, social, and monetary) identified with different metropolitan spaces, which can be dealt with methodically by following a bunch of sensibly clear cut stages, for example, a few codifications of the course of information mining, most remarkably the cross business-standard interaction for information mining (CRISP-DM). This interaction gives a system to structure metropolitan contemplating information investigation issues identified with various components of manageability, as in savvy feasible metropolitan arranging practice, devise logical arrangements dependent on cautious examination and assessment of the significant issues utilizing powerful scientific and evaluative apparatuses, just as imagination, sound judgment, and particular information. Thusly, organized pondering metropolitan information examination is of importance to supporting dynamic cycles concerning various elements of manageability inside practically all metropolitan spaces.

Fan and Bifet (2013), Shearer (2020) gives a codification of the information mining process that includes the accompanying:

1. Metropolitan manageability issue understanding
2. Information understanding
3. Information readiness
4. Model structure
5. Results assessment
6. Results arrangements.

Data is processed from a variety of sources to improve what is now known as big data. Data sources are all around us, advanced cell phones, PCs, biological sensors, cameras, GPS (Positioning Systems), and even people. Various applications such as online-based media circuits, machine tools, and accounts, business trading, app development, games, and much more accelerated data for the last two or three years.

ADVANTAGES OF USING CLOUD SERVICES IN SMART CITIES

At this moment, various metropolitan networks battle to be adroit metropolitan regions to get a piece of their prizes fiscally, normally, and socially. Therefore, many are looking toward the possibilities made possible by using enormous data assessment in sharp city applications. As needs are, we will discuss in this section a part of the benefits and openings that may help in deciding to change or upgrade a city to transform into a splendid city.

Proficient Asset Use

In their research Tewari and Joshi (2020) suggested that with various resources ending up being either meager or expensive, organizing deals with any consequences regarding having better and more controlled utilization of these resources. Starting with imaginative structures, for instance, Enterprise resource organizing (ERP) and Geographic Information systems (GIS).

Better Personal Satisfaction

With better associations, more valuable work, and living models, and less waste (on schedule and assets), sharp city occupants will have a dominating individual satisfaction. This is the inevitable result of better getting sorted out of living/work areas and locales, seriously convincing transportation structures, better and speedier benefits, and the accessibility of enough data to settle on an educated choice.

More Significant Levels of Straightforwardness and Receptiveness

The need for better association and control of the different speedy city perspectives and applications will drive the interoperability and responsiveness to more raised levels. Information and asset sharing will be the standard. In addition, this will expand data straightforwardness for each and every contributed person. This will invigorate interest and correspondence among substances and make more associations and applications that further overhaul the marvelous city. One model is the US government that gathered and passed on a wide degree of information, dispersal, and content for straightforwardness and responsiveness. These offered the occupants and the public position substances the opportunity to trade and utilize the information plausibly.

Effective Administration of Information

Cloud arrangements assist with smoothing out the most common way of overseeing information. At the point when information is sent employing cloud applications to their middle hub, it is put away in the cloud till utilized for future references.

Lift in the Economy

When imaginative advancements are utilized to help brilliant urban communities, it brings about an increase in the economy. The common collaboration among urban communities and organizations can draw in tremendous benefits. Alongside the city, even organizations can use the information being assembled. They can fathom their objective populace better and offer applicable assistance.

Cost Decrease

Reducing cost levels in the city becomes significant over the long haul. Distributed computing conveys better outcomes, at decreased functional expenses. Moving to cloud applications or capacity can likewise prompt essentially bring down work and support costs while amplifying space and cutting energy consumption.

Maintainable Living Choices

Sustainable living is a pivotal component in the present society. As it were, Globally Cloud arrangements suppliers are embracing cleaner and greener energy to drive gigantic server farms and different other energy the board apparatuses which at the Individual association level are not reasonable.

VARIOUS COMPUTING SERVICES IN SMART CITIES

Clever urban communities require numerous information causes and dependable models to create choice-supporting data. It turns out to be particularly difficult when a monstrous number of savvy gadgets and sensors are locked in. This part presents five common savvy city applications, the information drawn in, comparing models, and their prerequisites for registering.

Transport and Traffic Management

Conveyance is one of the most significant angles for metropolitan living exercises. Different wellsprings of conveyance information are identified with individuals' movement and driving, which is a muddled and essential piece of brilliant urban communities. For example, traffic data are delivered and assembled by sensors in busy time gridlock automobiles (e.g., cabs, transports, metros, trains, vessels, and planes) or screens presented along the roads (e.g., circle sensors and perception cameras). Driving data insinuate data that record people's standard improvement in metropolitan regions. Geo-labeled informal community information gathers posts (e.g., sites, tweets) through interpersonal organizations which are labeled with geoinformation. Street network information addresses street fragments and crossing points, separately. The transportation network is displayed as a coordinated chart that incorporates travel courses and stops offices of transports and metro organizations. Sheikhi et al. (2015) says that focal point (POI) information portrays related data for offices, for example, eateries, shopping centers, parks, air terminals, schools, and emergency clinics in the city, which assists in directing individuals with discovering their objections.

Utilities and Energy Management

The utilities and energy the executive's frameworks ought to be green, manageable, and with high functional speed and effectiveness. Future economical energy frameworks will be savvy and incorporated with shrewd networks, sustainable sources, stockpiling, and energy the board and checking frameworks. Song et al. (2015) shows that the energy and utility frameworks of urban areas are convoluted because they need to fulfill countless necessities with equivalently restricted inventory. The computational frameworks need not exclusively incorporate discontinuous force sources productively and adequately, yet in addition to anticipate gear disappointments and blackouts, permitting utilities to enhance their upkeep financial plans.

Ecological Protection and Sustainability

Natural insurance and manageability additionally assume significant parts in keen urban areas. The natural assets allude to minerals, timberlands and prairies, wetlands, streams, lakes, and the sea. These regular assets have been taken advantage of unduly, and the unseemly administration of normal assets has caused extreme natural corruption. The information that metropolitan natural security and maintainability the board frameworks are managing incorporate hydrogeological information, ecological reconnaissance information, environmental insights, and meteorological information. The information amount and measurements are huge as per the attributes of enormous information. According to Al-Hader et al. (2009) the elements of this information are not exclusively to precisely introduce the current circumstance of the climate yet additionally to adequately anticipate the future and manageability. Consequently, incredible computational capacity is expected to help states and individual clients to forestall and settle natural difficulties.

Smart Waste Management Systems

With waste creation in urban communities expanding, regions are searching for ways of making their assortment processes more effective. Rather than utilizing predefined courses and a decent assortment plan, squander the board laborers depend on sensors set in squander repositories to quantify fill levels and advise them when receptacles are fit to be discharged. The city of Seoul in South Korea diminished its assortment costs by 83% with a keen waste administration framework and disposed of burn through flood simultaneously. San Francisco likewise as of late furnished 1,000 garbage cans all through the city; regional government trusts it will lessen rubbish pickup time and kill calls from general society about spilling over receptacles.

Gunshot Detection Technology

A few urban areas, similar to Denver, are going to acoustic discharge recognition frameworks (AGDS) to recognize when a firearm has been shot and the specific area of the shot(s). Acoustic sensors sent at key regions (for instance, air terminals, huge settings, lodgings, and so forth) identify a shot and send the area data to a nearby crisis authority. This permits police to show up on the scene quicker than they could have in any case; they likewise know precisely where to go when shot episodes happen. Denver police gauge that AGDS has added to 134 captures since it was executed. Gunfire-discovery frameworks are additionally being carried out in some U.S. workplaces and plants too.

Smart Air Quality Sensors

With air contamination tormenting numerous enormous urban areas, checking air quality consistently is a need. Chicago (the eighteenth most dirtied city in the U.S.) is one of a few urban communities that presently utilize IoT sensors to gather an assortment of air quality information, including levels of particulate matter, carbon monoxide, ozone, nitrogen dioxide, and then some, as a feature of its more extensive Array of Things (AoT) drive. Government authorities intend to utilize this information to make the city better and more bearable.

Smart Street Lighting

Various metropolitan networks are upgrading their street lighting in a bigger number of ways than one—they're both changing to LEDs and executing related lighting courses of action. This splendid city IoT use case enjoys different benefits, from extending energy capability to diminishing energy and upkeep costs. (Chicago wants to save $10 million consistently in energy costs with its street lighting drive.)

Insightful lights can normally change their quality ward on occasions of idleness; they can similarly convey upkeep information for quicker response times. The city of Miami has more related streetlights than another city in the world; it saves 44% on energy yearly stood out from the aggregate paid ahead of time for regular streetlights.

The brilliant energy framework comprises a wise combination of decentralized manageable energy sources, effective appropriation, and advanced power utilization. Brilliant energy in this way comprises of three autonomous building blocks that should be sewed together and adequately speak with one another to frame a brought together with brilliant energy framework. Low-carbon age, otherwise called efficient power energy, photograph voltaic, sun-oriented warm, biogas, and wind energy can be a significant piece of a brilliant energy framework. Productive dissemination in the brilliant energy framework is made conceivable by the utilization of shrewd foundation, savvy network, keen meters also as a proper degree of use of the data and correspondence innovation (ICT). The center of a brilliant energy framework is the data foundation which is answerable for gathering the energy utilization data just as sharing the supplier rate data.

Smart Data Management

The immense proportions of data assembled by a splendid city ought to be examined quickly to make it significant. Open data doorways are one decision that a couple of metropolitan regions have picked to circulate city data on the web, so anyone can get to it and use judicious examination to assess future models. Associations, for instance, Community Logic are working with metropolitan regions to help them with separating data, and they're in the Startup in Residence (STiR) program for the city of San Francisco.

Smart IoT Devices

Sensors are crucial in a clever city, a wide extent of reporting contraptions like sensors, penetrability devices, and far edge centers that make the data that makes a splendid city work. In an adroit city, information will logically be procured directly from deliberately passed on sensors or by a suggestion from sensors sent for another explanation yet what gather and offer supportive information. With this information, uninhibitedly exchanged, complex city systems can be directed consistently and, with satisfactory joining, to restrict concealed aftereffects. As dependence on sensors grows, so too will the need that they are reliable and that the systems to which they are related will need to bear the unavoidable dissatisfactions.

The savvy urban areas need to have three key highlights: knowledge, interconnection, and instrumentation which the IoT can give. One might say that the utilization of the IoT can make shrewd urban areas doable. The utilization of PDAs, savvy meters, shrewd sensors, and radio-recurrence recognizable proof (RFID) shapes the IoT structure in brilliant urban communities. The IoT structure comprises different parts including gadgets, sensors, organizations, firmware, and programming. IoT is the organization of interconnected actual items (called "things") including PCs, advanced mobile phones, sensors, actuators, wearable gadgets, homes, structures, constructions, vehicles, and energy frameworks. The IoT guarantees the correspondence of numerous assortment sorts of frameworks and applications for giving progressively brilliant, dependable, and secure administrations. An enormous assortment of sensors including RFID, IR, and GPS, interface the structures, foundation, transport, organizations, and utilities through ICT.

Smart Healthcare

Because of the rapid development of the population, the general clinical benefits are outweighed. There are health professionals who are in short supply to solve the population issue. Institutions often transmit errors in dealing with coercive diseases. Often, patients receive certain medications that are not acceptable. In various remote parts of the world getting good health care at this point is a long-term dream. Consequently, with restricted assets and consistently expanding requests, customary medical care should be canny, effective, and maintainable; that is the place where brilliant medical services come in. Shrewd medical services can be conceptualized as a mix of different substances including customary medical care, keen biosensors, wearable gadgets, data and correspondence innovation (ICTs), and savvy rescue vehicle frameworks. The different parts of shrewd medical services incorporate arising on-body sensors, savvy clinics, and keen crisis reactions. In brilliant medical clinics, different systems including ICTs, distributed computing, PDA applications, and progressed information investigation procedures, are utilized for their activity. Patient information can be continuously unlocked in a variety of work environments in a problematic environment or even in intelligent centers in different networks in the same city or city. Clinical topic specialists, planners, and specialists can pass on test data without wasting any time in conveying comparable information starting with one office and moving on to the next. Basically, various specialists can see details in order to make decisions about a patient's condition. Therefore, endless decisions to understand the conditions of prosperity and conflicting drugs can be made possible.

VARIOUS ISSUES RELATED TO CLOUD SERVICES IN SMART CITIES

Albeit the information examination, currently may be incompetent for enormous data achieved by the situation, contraptions, structures, and issues that are extremely novel corresponding to ordinary mining issues, considering the way that a couple of properties of tremendous data in like manner exist in the standard data examination. A few open issues brought about by the large information will be tended to as the stage/system and information mining viewpoints in this part to clarify what situations we might stand up to as a result of huge information. Here is a portion of the open issues:

Correspondence Between Frameworks

In this manner, most tremendous information evaluation associations will be normal for identical selecting, and they, as a rule, will chip away at different constructions (e.g., cloud stage) or work with different frameworks (e.g., web searcher or data set), the correspondence between the immense information assessment and different frameworks will truly impact the presentation of the entire course of KDD. The main assessment issue for the correspondence is that the correspondence cost will cause between frameworks of information assessment. Bearings to reduce the correspondence cost will be certainly the primary concern that information researchers need to mind. One more appraisal issue for the correspondence is the way the colossal information assessment chats with different frameworks. The consistency of data between different structures, modules, and chairmen is in like manner a huge open issue on the correspondence between systems. Since the correspondence will appear all the more regularly between systems of huge data assessment, how to diminish the cost of correspondence and how to cause the correspondence between these structures as strong as possible will be the two huge open issues for colossal data examination.

Bottlenecks on Information Examination Framework

The bottlenecks will be shown up in better places of the information examination for large information because the conditions, frameworks, and information have changed which are unique concerning the conventional information investigation. The information storm of enormous information will top off the "input" arrangement of information examination, and it will likewise build the calculation heap of the information "investigation" framework. The present circumstance is similar to the downpour of water (i.e., information storm) hurried dejected the foothill (i.e., information examination), how to part it and how to stay away from it streaming into a restricted spot (e.g., the administrator can't deal with the info information) will be the main things to keep away from the bottlenecks in information investigation framework. One of the current answers for the evasion of bottlenecks on an information investigation framework is to add more calculation assets while the other is to part the examination works to various calculation hubs. A total thought for the entire information investigation to stay away from the bottlenecks of that sort of examination framework is as yet required for enormous information.

Security Issues

With so much information on climate change and human behavior to be collected in a major information investigation, how can you ensure that that will be an open problem because, without a secure data management system, the most comprehensive data testing would not be reliable framework? Despite the security, we need to improve to investigate more information before it can gather more information from all over the world, the fact of the matter is that since recently, there have been very few tests dealing with security issues. In our view, security issues for major information investigations can be divided into four categories: input, information testing, yield, and inter-sectoral communication. For information, it can be viewed as a collection of sensory-related information, mobile gadgets, and surprisingly web-based gadgets. One of the most important security concerns in the information section of a major information investigation is to ensure that the nerves will not be subjected to attacks. Investigations and information, it is often seen as a security matter for such a context. In communication with the alternative framework, the issue of security is still divided among large-scale evidence testing and other external bodies. Because of these obsolete issues, security has developed one of the most exposed areas of information intelligence.

The most dangerous of digitizing our framework is the danger of cybercrime that could unleash ruin on a great many lives. Brilliant urban communities are helpful focuses for coordinated cybercriminal gatherings and can prompt enormous misfortunes of significant data. Cybercrime exercises are consistently developing, bringing about the chance of a break anytime.

Commotion, Anomalies, Deficient and Conflicting Information

Although big data research is another era of knowledge exploration since few arrangements include old-fashioned methods of disseminating data in big data investigations, open issues of traditional mining statistics are equally present in these new frameworks. Open problems of chaos, diversity, inconsistencies, and conflicting information in the generalization of mining information will also arise in mining accounting. Many fragmented and contradictory details will easily emerge because the information is captured or processed from a variety of sensors and structures. The effect of commotion, exceptions, deficient and conflicting information will be developed for enormous information investigation. Accordingly, how to alleviate the effect will be an open issue for huge information investigation.

SECURING SMART CITIES

Safety efforts additionally should be carried out as far as the particular abilities of savvy urban communities like web-based banking. Monetary organizations should establish approaches and measures to oppose extortion and record takeovers. This could incorporate measures like biometric confirmation and multifaceted verification for versatile applications to forestall the laundering of cash. These actions will assist with building purchaser trust.

For a city to turn into "brilliant," put resources into modernizing data innovation and security framework. Execution of brilliant networks and man-made consciousness is vital, similar to the need to take on ISO 27001, a global standard that sets out the best practice for data security in the executive's frameworks (ISMS).

The more designers and other realtors can see shrewd city arranging as brilliant homes anticipating a more terrific scope, the smoother the advances can turn into. One of the critical parts of land engineers' selling is accommodation. Brilliant urban communities are the encapsulation of the IoT. The Internet of Things in a shrewd home permits an inhabitant to pass through their day from bed to the front entryway and back again in the most helpful way. The best savvy urban communities permit that equivalent occupant an equivalent degree of comfort from the second they set foot across the line of the city.

CONCLUSION

A bright city with great details is current and important idea; therefore, many are starting to apply themselves to promote smart city requests that will benefit achieving control, improved flexibility, operative management, improved personal satisfaction, and smart city asset management. Our review investigated these two perspectives and their various interpretations and came to distinguish some common debts each. Despite the changing definitions, all ideas have different personalities. In light of these common features, we have had the option to diversify the overall paybacks of using extensive knowledge to plan and back critical city submissions.

A city is a game plan of structures with a stand-out history and set in a specific social and regular setting. For a city to succeed, all the key city systems need to collaborate, by utilizing every one of their resources to rout the troubles the city faces. The "insight" of a city portrays its ability to join all of its resources, to reasonably work with the most outrageous possible capability to fulfill the reasons it has set itself. The adroit city is a thought and an arrangement of definitions exist among the academic local area and experts. An adroit city can have no less than one splendid part, including insightful transportation, sharp organization, sharp clinical consideration, and sharp organization.

Creating and deploying large-scale operational information intelligent city applications will need to address complex and open problems, following comprehensive strategies and development models, having a well-organized HR, using recreational and well-organized models, and relying heavily on managerial elements. With all the elements of established success and a better understanding of ideas, making the city more vibrant will be thoughtful, and continuously developing it to get the most intelligent representations and management will be viable and manageable targets.

REFERENCES

Al-Hader, M., Rodzi, A., Sharif, A. R., & Ahmad, N. (2009). Smart city components Architicture. *2009 International Conference on Computational Intelligence, Modelling and Simulation*. 10.1109/CSSim.2009.34

Al Nuaimi, E., Al Neyadi, H., Mohamed, N., & Al-Jaroodi, J. (2015). Applications of big data to smart cities. *Journal of Internet Services and Applications*, *6*(1), 25. Advance online publication. doi:10.118613174-015-0041-5

Bibri, S. E., & Krogstie, J. (2017). The core enabling technologies of big data analytics and context-aware computing for smart sustainable cities: A review and synthesis. *Journal of Big Data*, *4*(1), 38. Advance online publication. doi:10.118640537-017-0091-6

Cloud computing trends for 2019. (2019, February 22). *Networks Unlimited*. https://www.networksunlimited.com/cloud-computing-trends-for-2019/

Fan, W., & Bifet, A. (2013). Mining big data. *SIGKDD Explorations*, *14*(2), 1–5. doi:10.1145/2481244.2481246

Frontiers in massive data analysis. (2013). doi:10.17226/18374

Hasanin, T., Khoshgoftaar, T. M., Leevy, J. L., & Bauder, R. A. (2019). Severely imbalanced big data challenges: Investigating data sampling approaches. *Journal of Big Data*, *6*(1), 107. Advance online publication. doi:10.118640537-019-0274-4

Khan, Z., Anjum, A., & Kiani, S. L. (2013). Cloud based big data analytics for smart future cities. *2013 IEEE/ACM 6th International Conference on Utility and Cloud Computing*. 10.1109/UCC.2013.77

Khan, Z., Anjum, A., & Kiani, S. L. (2013). Cloud based big data analytics for smart future cities. *2013 IEEE/ACM 6th International Conference on Utility and Cloud Computing*. 10.1109/UCC.2013.77

Kitchin, R. (2013). The real-time city? Big data and smart urbanism. *GeoJournal*, *79*(1), 1–14. doi:10.100710708-013-9516-8

Liu, Y., Shi, J., & Jian, M. (2017). Understanding visitors' responses to intelligent transportation system in a tourist city with a mixed ranked Logit model. *Journal of Advanced Transportation*, *2017*, 1–13. doi:10.1155/2017/2854895

Najafabadi, M. M., Villanustre, F., Khoshgoftaar, T. M., Seliya, N., Wald, R., & Muharemagic, E. (2015). Deep learning applications and challenges in big data analytics. *Journal of Big Data*, *2*(1), 1. Advance online publication. doi:10.118640537-014-0007-7

Pantelis, K., & Aija, L. (2013). Understanding the value of (big) data. *2013 IEEE International Conference on Big Data*. 10.1109/BigData.2013.6691691

Shearer, C. (2000). The CRISP-DM Model: The New Blueprint for Data Mining. *Journal of Data Warehousing*, *5*, 13–22.

Sheikhi, A., Rayati, M., Bahrami, S., Ranjbar, A. M., & Sattari, S. (2015). A cloud computing framework on demand side management game in smart energy hubs. *International Journal of Electrical Power & Energy Systems*, *64*, 1007–1016. doi:10.1016/j.ijepes.2014.08.020

Silva, B., Khan, M., Jung, C., Seo, J., Muhammad, D., Han, J., Yoon, Y., & Han, K. (2018). Urban planning and smart city decision management empowered by real-time data processing using big data analytics. *Sensors (Basel)*, *18*(9), 2994. doi:10.339018092994 PMID:30205499

Song, M., Cen, L., Zheng, Z., Fisher, R., Liang, X., Wang, Y., & Huisingh, D. (2017). How would big data support societal development and environmental sustainability? Insights and practices. *Journal of Cleaner Production*, *142*, 489–500. doi:10.1016/j.jclepro.2016.10.091

Tewari, N., & Datt, G. (2020). Towards FoT (fog-of-Things) enabled architecture in governance: Transforming E-Governance to smart governance. *2020 International Conference on Intelligent Engineering and Management (ICIEM)*. 10.1109/ICIEM48762.2020.9160037

Tewari, N., & Joshi, M. (2020). Water quality prediction system (WQPS) and method using fog of thing (FoT). *2020 9th International Conference System Modeling and Advancement in Research Trends (SMART)*. 10.1109/SMART50582.2020.9337109

Vilajosana, I., Llosa, J., Martinez, B., Domingo-Prieto, M., Angles, A., & Vilajosana, X. (2013). Bootstrapping smart cities through a self-sustainable model based on big data flows. *IEEE Communications Magazine*, *51*(6), 128–134. doi:10.1109/MCOM.2013.6525605

Chapter 5
Hindrance Detection and Avoidance in Driverless Cars Through Deep Learning Techniques

Priyanka S.
Bannari Amman Institue of Technology, India

Saran Kumar A.
Bannari Amman Institue of Technology, India

Praveen V.
Bannari Amman Institue of Technology, India

Sivapriya G.
Kongu Engineering College, India

Meenakshi Dhanalakshmi M.
Bannari Amman Institue of Technology, India

ABSTRACT

Generally, a technique to detect each shifting and static item concurrently is needed due to the fact the static items inclusive of containers can fall on the street in the front of a vehicle and they may be risky. Obstacles can be detected and avoided by using neural networks. Obstacle detection algorithms have the capability to detect the items that can be out of doors. Irregularities on the street floor that aren't affecting the riding aren't considered. In this case, the detection may be primarily based on looking for unique patterns. In the chapter, a hybrid network based on the combination of stacked auto encoder-based deep belief network is built to detect the obstacles on the road. The system achieves accuracy greater than 90% when compared with existing literature.

DOI: 10.4018/978-1-7998-9710-1.ch005

INTRODUCTION

Nowadays many city roads have developed with the growth of the transportation environment (we can't control the development of the technology) and improving accidents making the safety of the transport an important issue. For hindrance discovery, many researchers decided to focuses on robots and independent or automatic vehicles (Yin, 2017).

Automatic cars need an accurate barrier detection methodology, which means it needs to safely find the way for abandoned circumstances. Number of several technologies came into existence that scans the surroundings for possible problems, mainly depending on gathering the reflection of light or sound waves (Zhang, 2015).

The discovery of material substance through releasing the wave patterns and computing the response time awaiting for the reflection captured, is called time of flight. Microsoft Kinect uses this method, for the emission of infrared waves. Kinect is only errand for indoors discovery of objects, since natural light tends to flood its receivers (Sanil, 2020). Numerous successful research institutes such as Mobil-eye, Daimler, and KIT Autonomous Vision Group (AVG) have proposed preferable research approaches in this area.

Driverless cars are categorized on the foundation of their skill to work systematically and are classified under various levels. There are 5 various levels, first one is LEVEL1 where the driver has to have his direct action, which means that driver's help is compulsory. Next, LEVEL2, the driver can have his/her hands outside as it is like semi automation where the methodology can govern the steering and velocity at the same time, without driver stimulation for little periods. Next, LEVEL 3, is the level of automatic-enabled level, in this, the driver would hand over his full control to the vehicle and operate during select parts of a journey when certain operating conditions are met. In LEVEL 4, it is the highly automated level, does not need human intervention, which means he/she could not have at mind on driving. The vehicle can do needed driving, but the driver can intervene in the driving when it is required. Finally, in the last level, LEVEL 5 it is the fully automatic level, no driver allowed and humans are only just passengers (Du, 2019). The evolution of driverless cars is shown in figure 1.

In vision of the international automobile trend and the reward and the growth of devoted sensor methodology and publish of a more reliability of the car level to sense the vehicle must be the first task for establishing the self-driving methodology. Offer consistent visibility of self-driving inadequate finding information. Unmanned driving can develop rapidly (Lai, 2018). More companies are depending on the research and development of superior driver help systems. Many companies have already invested a lot of money in developing the smart vehicle technology, the advanced auxiliary system can not only reduce the occurrence of car accidents, but also promote the driverless industry developing rapidly (Zhao, 2020).

Figure 1. Evolution of driverless cars

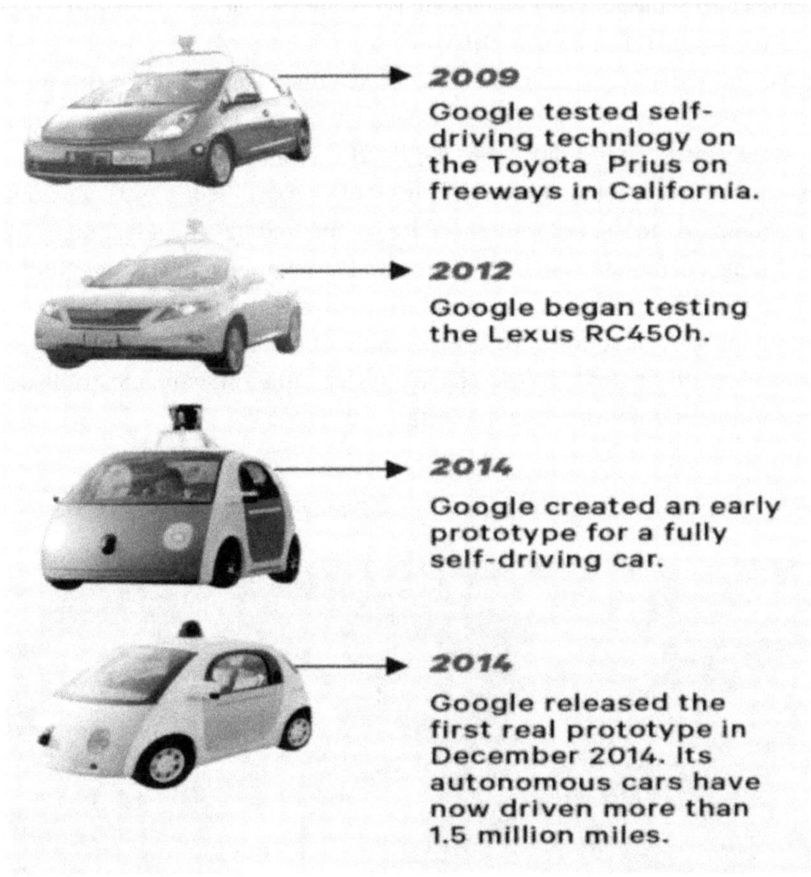

Vehicle circumstance barrier recognition is a viral matter in the domain of bright vehicle research. It is a requirement for intelligent vehicle to realize usual driving, automatic navigation and other functions. Among the system for vehicle surroundings obstacles recognition, lidar, millimeter wave radar (MMW radar) and ultrasonic radar have been mainly used (Prabhakar, 2017).

The lidar finds the obstacles by emitting the infrared beam to the object and measures the response time delay to reflect the waves and calculate the distance between the hindrance and vehicle as shown in figure 2. The high measurement accuracy and has wide discovery distance, but it does not calculate the distance in the bad weather changes (such as heavy rain and fog) and heavy and expensive and can interfere with the laser signals emitted by other vehicles.

MMW radar used to detect the distance by the reflected wave, and it can also accomplish enough accuracy under cruel situations. However, MMW radar has a narrow view field (usually has the field with milli-radians level) and low lateral information correctness, which means that if vehicle is far away it is difficult to get the reflected waves (Hernández, 2016).

Ultrasound radar is also a general hindrance discovery method. But ultrasonic radar could not cover all the distances because of the poor long-range detection ability, weakness to sound intrusion, deprived surrounding awareness correctness and robustness. In computer vision, it is used to support the three-dimensional line of target compared to the radar method. Radar method has the advantages of wide detection range, complete road information and remote sensing. This method is very vital in car hindrance discovery method. But some virtual sensors are very subtle to weather and radiant conditions. With parallel of this method, the calculation and analysis and rebuilding of 3D sights have large calculation, which restricts the uses of stereo vision in omnidirectional hindrance discovery.

Figure 2. Radar monitoring perimeter in vehicles for safety

With current growth in artificial intelligence (AI), machine learning (ML) and deep learning (DL), various uses of these methodologies have increased fame and come to front. One like application is self-driving cars that are expected to have a reflective and innovative influence on society and the method people travel. Though the approval and taming of new methodology could face the early and lengthy rejection, yet these innovative cars will be the first to reach the new connection of personal robots into the human society. The last period has seen growth of research attention in applying AI to drive cars.

Due to quick development in AI and related technologies, cars are ultimately composed to grow into independent robots delegated with human lives, and bring about a diverse socio-economic impact. But, for these cars to become a functional possibility, they need to be armed with awareness and reasoning to handle the strenuous real-life situations, arrive at appropriate conclusions, and take suitable and harmless exploit at all times. Entrenched in the self-driving vehicles' AI are visual recognition systems (VRS) that encompass image classification, object detection, segmentation, and localization for common visual performance (Gupta, 2021).

Object detection is developing as a sub-domain of computer vision (CV) that benefits from DL, especially convolutional neural networks (CNNs). This chapter discusses the self-driving cars visualization systems, part of DL to understand complex vision, improve awareness, and motivate kinematic maneuvers in self-driving cars (Karmakar, 2021).

This book gives the basic idea of having DL to perform the object detection and scene perception in self-driving cars. We can also achieve the contribution of DL in these areas. Using these idea we can answer the question which appreciates contribution of DS in these areas (Lai, 2018):

1. What are the equally strengthening and important working necessities for fully functional self-driving cars?
2. What breakthroughs and growths have been achieved in self-driving cars in the last 20 years and what are certain hopeful research paths for the next era?
3. What is DL and how does DL create artificial perception?
4. Why is DL a promising technique for solving object detection and scene perception in self-driving cars?

Many deep learning architecture develop image recognition algorithm accompanied by image depth information, it is helpful to identify the objects in the path on different weather conditions and to improve the image recognition on the path way objects through deep learning methodology to improve the identification accuracy and recognition speed. The identification of deep learning method needs an original path way database to develop the research on thing detection methodology. It is essential to collect many dataset for training purposes, including sunny, rainy, and foggy and path way under various weather conditions.

The image sensing technology development for domestic road object sensing industry is to develop and use, this to stimulate the domestic imaging sub-industry manufactures for research and development for quality updates.

The process of data collection for research is mainly conducted by a manual approach. That manually driven vehicle is fully organized with the set of sensors and other automotive devices such as camera, LiDAR(Light detection and ranging), radar, GPU and IMU(inertial measurement unit). The sensor/device configuration (e.g., selected sensors, number and placement of each sensor, parameter configuration of each sensor) might change depending on the data collection method. For instance, a camera can be monocular, stereo, or omni-directional, grayscale, color, and HDR (high dynamic range).

The viewpoint, image resolution, and frame rates are the main limitations of a camera that is to be configured before a data collection. At the time of data collection, the raw data from different sensors are collected with the timestamp and time synchronization. Data composed on civic roads is of specific interest for automated driving by asset of accurate traffic situations imitated from the data. LiDARs and vision-based automotive sensors produce immense amount of data with high frequency. Hence, the resultant data size can reach the greatness of tens or even hundreds of gigabytes (Leng, 2019). In the Drivers Assistance System, we usually only focus on the obstacles on the road. It means that we should extract the road first, and then locate the candidate obstacle regions on the road. The quality of road extraction seriously affects the detection results. Most of the existing road detection methods have achieved good performance using monocular cameras. However, the computational complexities of these algorithms are too high to satisfy real-time requirements (Sreedhar, 2020).

LITERATURE SURVEY

Zutao Zhang (2015) proposed a novel methodology to control the speed of vehicles based on obstacle detection and avoidance. The proposed methodology has three steps namely detection of obstacle and segmentation, tracking and recognition of the obstacle and an algorithm to control the speed of the vehicle. The obstacle is detected by using a binocular camera system and then its distance information is computed using disparity computation and triangulation. Next particle filter framework and sparse representation mechanisms are utilized to classify the different types of obstacles like animal, human and vehicles. Atlast, an algorithm to control the speed of the vehicle is used to avoid the accident. The main drawback of the proposed system involves the high cost of building the object tracking and hindrance system.

Zhao (2020) proposed a hindrance detection system using depth cameras in-order to detect the obstacle in Omni directions. The proposed system uses a high definition camera in the vehicle to capture the obstacle image and then those images are processed using image segmentation, iterative normalization based cut process and three dimensional visualization methods to detect the image of obstacles correctly. The proposed system out performs the traditional methods like K-means clustering, traditional threshold and few deep learning algorithms. The detection rate of the proposed method depends on the use of good quality cameras which is a major issue of the framework.

Leng (2019) proposed an obstacle detection system for driverless cars which can effectively detect the obstacles in the road. The proposed method uses a U-V disparity map to detect the on-road obstacles effectively. The U disparity map makes an accurate detection of the images on the road and then uses the V disparity map to identify the obstacles on the road accurately. The end results of the U-V disparity map are fed into faster RCNN (Region-Based Convolutional Neural Network) in-order to process and efficiently identify the correct obstacles on the road. The proposed system is tested on two publicly available datasets which gives an accuracy of 72% and 76% respectively. The main drawback of the proposed system is the non-detection of overlapping objects on the road.

Hernández (2016) proposed a kinect sensor based detection system to detect the obstacles on the road in various autonomous vehicles. Kinect sensor helps to detect the vehicles accurately on the road by applying multiple filtering algorithms when compared with normal camera systems in the vehicle. A testing prototype was made to detect the performance of the proposed system which detected the obstacles on the road by performing multiple layers of segmentation and filtering mechanisms and gives a high accuracy of 82%.

Lai (2018) proposed a real-time object detection system which helps to find the vehicles on the road in unmanned vehicles. The proposed framework uses Mobile net network, which is a type of neural network with multiple layers of filtering techniques to accurately detect the obstacles in the road. The proposed system was trained on a coco dataset which has a large number of road traffic images that helps to detect the vehicles on the road and gives an accuracy of 90% on GPU. The speed of obstacle detection is quite slow which a major drawback of the proposed system is.

Prabhakar (2017) proposed a novel deep learning algorithm using faster region based CNN (RCNN) for detection of obstacles on the road. The proposed algorithm detects obstacles of different types like person, vehicle or animal in the road. The region based CNN framework is trained with pascal voc image dataset which detects and classifies the obstacles in the road faster. The pascal voc dataset was tuned in-order to detect the 20 different type object classes. Also, RCNN based algorithm archives frame rate of greater than 10 fps so that obstacles can be detected effectively. The precision of the proposed algorithm is greater than 95% for each different type of obstacle.

Kohl (2018) proposed a novel deep learning based artificial intelligence framework for detection of obstacles in the road for driverless self-driving cars. The deep neural network detects the obstacle in the road based on the images captured by the camera fitted in the car. The AI technique helps to give alerts or warnings to the driver in-order to avoid accidents once an obstacle is detected. The proposed system gives accuracy greater than 80% when implemented in real time scenarios.

Sanil (2020) proposed a convolutional neural network based object detection system for self-driving vehicles with the help of raspberry pi kit to control the speed of the car when an obstacle is detected. The attached camera in the car collects the image data with the help of raspberry pi and then transfers the collected data to the CNN in-order to detect the obstacles in the track of the car. The proposed model is implemented in a track of length 12 to 14 meters. During the training phase, the built model gives an accuracy of 88.6% and the accuracy decreases when the track path has sharp turns.

Krishna Chaitanya (2021) proposed a deep learning based CNN architecture for detecting the obstacles in driverless cars. Googlenet, a type of CNN architecture, is used to detect the obstacles in the road. The captured image data is first fed into the googlenet architecture which then segregates the objects into different types and finally detects the obstacles in the path of driverless cars in-order to avoid accidents. The proposed framework achieves an accuracy of 80% when tested with training dataset.

Nguyen (2016) proposed a deep learning based obstacle detection framework which can identify the obstacles in the road effectively. The proposed framework consists of three stages namely obstacle tracking, obstacle recognition and obstacle detection. Obstacle tracking is done with the help of adaptive U-V disparity algorithm at the initial stages. In the next step, results of the obstacle tracking step are fed as an input to the auto encoder framework which is a type of deep learning framework to detect the depth of obstacle images. Finally, the output of the recognition stage is utilized to detect the different types of obstacles in the road. The proposed framework is implemented using different datasets like EDISAT dataset, HCI stereo dataset and Daimler dataset which gives accuracy greater than 90%.

Haris (2020) proposed a hybrid framework for hindrance detection in roads using a combination of Markov random field (MRF) and Convolutional neural network (CNN) for autonomous cars. MRF is used to split the obstacles on the road into hazardous and non-hazardous objects for drivers. MRF also calculates the wheel steering angle for the vehicle in order to find the obstacles in the road. Finally CNN model is used to make the cars drive safely without any hindrance or to make the persons drive safely in case of non- autonomous vehicles. The proposed model gives high accuracy of 78% when executed with the newly created ZED stereo device dataset which has more than 5000 images.

PROPOSED SYSTEM

The proposed framework for obstacle detection and avoidance consists of utilizing deep learning (DL) algorithms. The main reason behind the use of DL algorithms is the fast convergence and obstacle detection capability which are identified from the literature survey. Also deep learning algorithms have high processing capability and self-learning capability in which the algorithm itself learns from the original raw data (Kumar, 2022). In this work, a deep stacked auto encoder is used in combination with deep belief networks in-order to predict the different types of obstacles in the road for self-driving autonomous vehicles. The detailed architecture of deep stacked auto encoder, deep belief networks and techniques involved in the prediction of obstacles for self-driving vehicles are explained in the rest of this chapter.

Deep Stacked Auto Encoders

Auto encoder is a type of unsupervised learning methodology that is used for the feature extraction process. Auto encoder consists of several layers of encoder and decoder modules. It is a type of deep learning model which attempts to perform data preparation and selection processes. The encoder module in auto encoder performs the data preparation process and then finally performs feature extraction so that the results of auto encoder can be used to train a machine learning model for classification. Auto encoders are currently used in many applications like face recognition, speech recognition and self-driving autonomous vehicles. The structure of the auto encoder is shown in figure 3.

Figure 3. Architecture of auto encoder

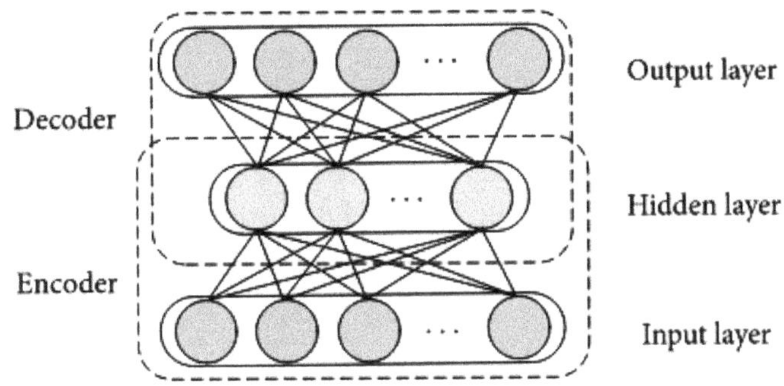

It has three layers of processing namely input layer, hidden layer (bottle neck) and output layer. The training of an auto encoder involves only encoder and decoder modules. The encoder converts the input data into hidden representation format while the decoder converts the hidden data format into correct input data. In this way, auto encoders will decrease the size of the dataset thereby reducing the dimensions of the dataset. The training procedure of an auto encoder is same as that of feed forward neural networks as the structure of auto encoder resembles the architecture of feed forward neural network. Also, an auto encoder can perform principal component analysis (PCA) if there is only one dense layer with one activation function in the encoder and decoder module (Dairi, 2018). The encoding process is given as:

$$h_n = f\left(W_1 x_n + b_1\right)$$

where

h_n denotes the hidden encoder vector value
x_n is given unlabelled dataset
f denotes encoding function,
W_1 denotes the weight matrix of the encoder and
b_1 represents the bias vector.

Also the decoding process is given as:

$$\hat{x}_n = g\left(W_2 h_n + b_2\right)$$

where

g denotes decoding process function,
W_2 denotes decoder's weight matrix and
b_2 denotes bias vector.

Stacked auto encoder is the combination of multiple encoders which are built on one another. In this modern era, every dataset has many dimensions with complex relationships among them. In order to reduce the complexity of such datasets, a single auto encoder is not sufficient. So stacked auto encoders are evolved in order to reduce the dimensionality of such datasets. The structure of stacked auto encoder is given below in figure

Figure 4. Block diagram of deep stacked auto encoder

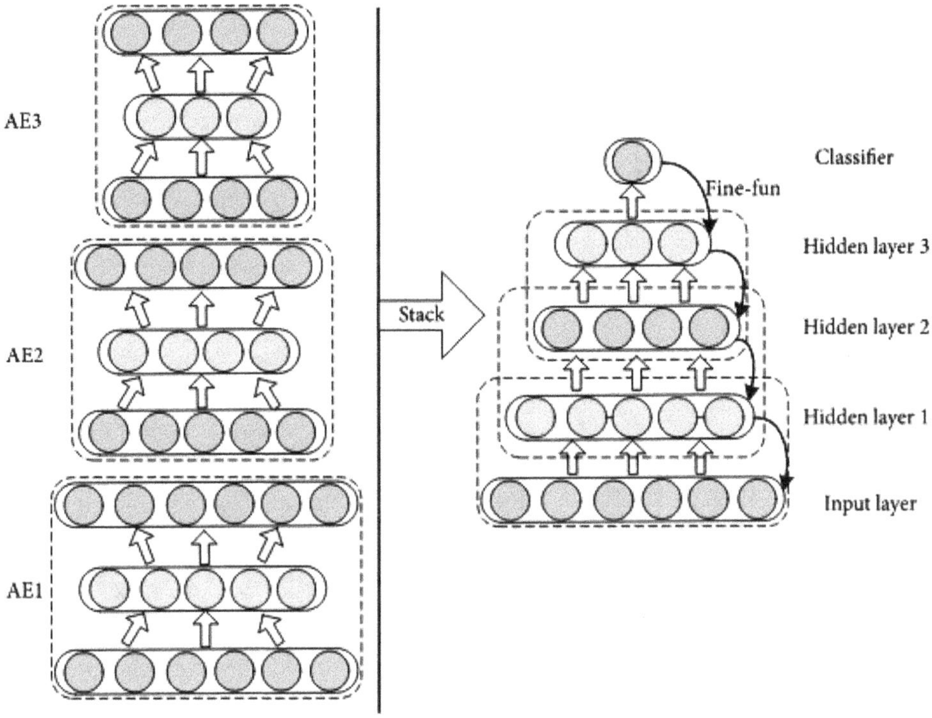

In the figure 4, the input dataset is first given to the auto encoder (AE1) and the output of AE1 is given as input to AE2 and then the output of AE2 is given as input to AE3. Atlast the output of AE3 is given as input to the classification model(in our case DBN) (Liu, 2018). In this way, the dimensionality of the dataset can be reduced in order to use the dataset as input to the classifier.

Deep Belief Networks

Deep belief network (DBN) is one of the most effective probabilistic generative models. It is constructed by stacking Multiple Restricted Boltzmann Machines (RBM). In the stacking process, the hidden layer of the first RBM acts as a visible layer for the next RBM. In this way, DBN is constructed as a combination of multiple RBM for prediction. The structure of DBN is shown in figure 5.

Figure 5. Architecture of DBN

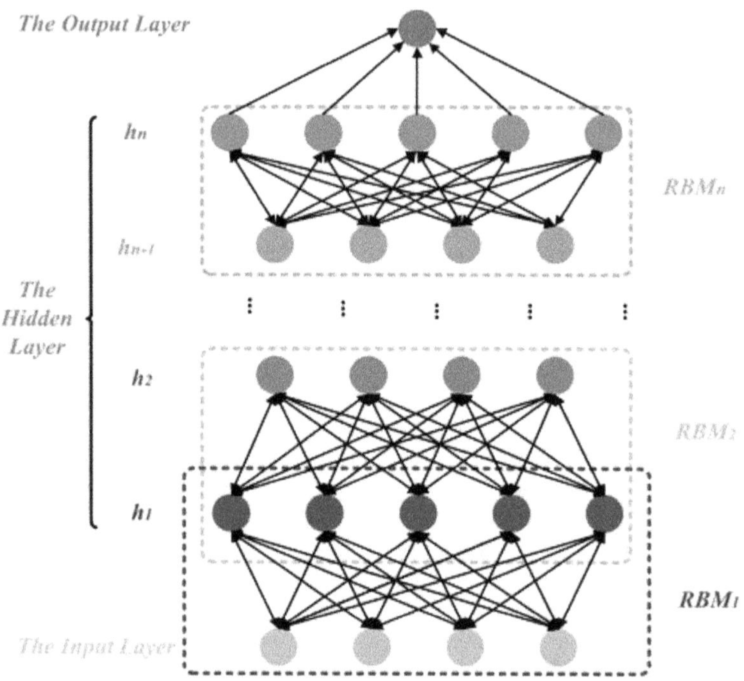

DBN has two different types of layers namely hidden layer and visible layer. In the visible layer, input nodes and output nodes are present. In the hidden layer, hidden nodes are present. A deep greedy strategy is followed in DBN in sequence with RBM to extract the correct statistical structure of the input data (Hinton, 2006).

A RBM consists of two layers namely: hidden layer and visible layer with nodes (units) and weighted connections between the neurons. The structure of RBM is shown in figure 6.

Figure 6. Structure of RBM

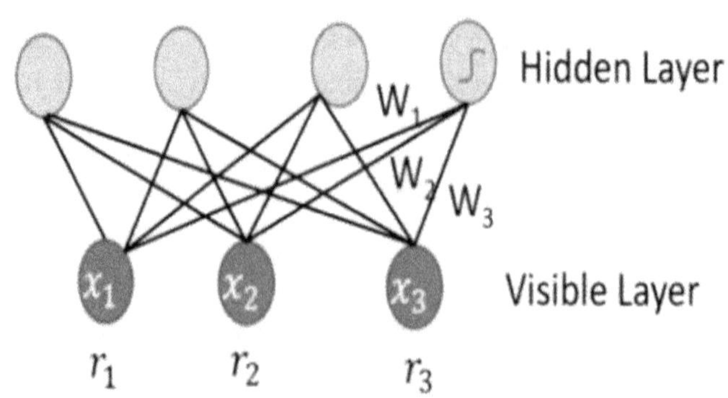

In RBM, the nodes in hidden and visible layers have no connection with each other and the units in one layer are connected to units in the other layer (Alzubaidi, 2021). The input data is passed from one node in the visible layer to nodes in the hidden layer. The input values are multiplied with weight of the corresponding node and then bias values are added and at last the activation function is applied to get the end results. The commonly used activation function includes softmax and sigmoid functions. The forward propagation and back propagation are applied in order to calculate the end probability values of the output based on the given input, weight and activation function. In the end, Kullback Leibler (KL) Divergence is used to compare the given input values and reconstructed input values.
The proposed system architecture is given in figure 7.

Figure 7. Block diagram of proposed model

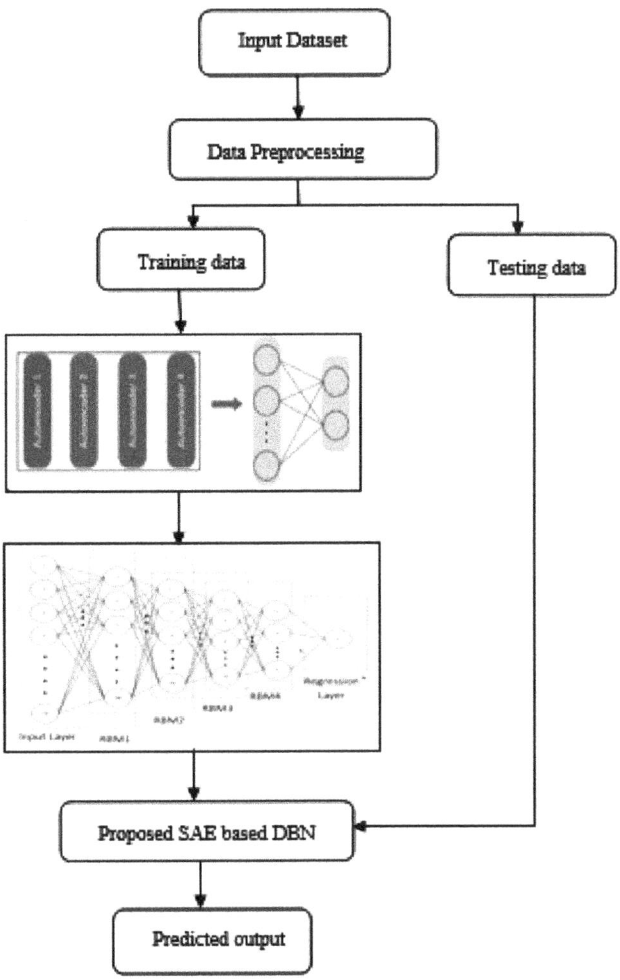

From the figure, first the input dataset is preprocessed in order to remove the missing values and also to remove the noisy values. Then, the processed dataset is split into training set and testing set and the training dataset is given to the deep stacked auto encoder and the output values are given to the deep belief networks and the trained model is tested with testing dataset in order to get the final classified output values.

RESULTS AND DISCUSSION

Dataset Collection

Nowadays, enormous data are generated by people. Data generated can be in the form of any type such as numeric (name,age,weight,etc.),categorical data can be classified into groups such as gender,race,highest degree attained, etc. Data collection (Gupta, 2021) is the process of gathering the information from different type of source that can be used to build the deep learning solutions. Auto mobile which is utilized for discovering the data and gathering are integrated with sensors, camcorder, light detection and ranging (LiDAR) devices. The gathered data can be stored in local system or it can be send to cloud for further processing. Data can be related holder, passenger information, location data, sensor data, etc. These collected data will be further utilized for training and testing the various algorithms involved in driverless cars such as transporting passengers on fixed routes, Autonomous Emergency Braking, Lane-Keeping Assist and Traffic Jam Assist.

As self-driving cars grows exponentially, leads to consideration of public roads data collection (Yin, 2017). In general, sensors are equipped over the vehicle to carry the information. During manual driving, the raw fact will be entered into a system which can be eventually utilized for training and testing the different algorithms. Since data collection involves more time consumption process and this can be avoided by using already available datasets. These datasets may differ in terms of data format,size,sensor setup,tool,etc. The dataset should hold attribute like camera, LiDAR, or radar data.Most of datasets are freely accessible .In earlier days, only small amount of dataset is available to find the obstacle. Hindrance may be ston e,paper,animals,pedestrian,speed bumps,dirt,etc with different heights. Generally our dataset can be classified into two: almost 80% is for training and other 20% is for testing. If the parameter found in dataset is not trained well leads to two major problems: overfitting and under fitting (Lai, 2018). Under fitting may arise while learning rate is too large or too low, duration of training time is not enough and lack of model complexity is shown in the figure 8.

Figure 8. Underfitting

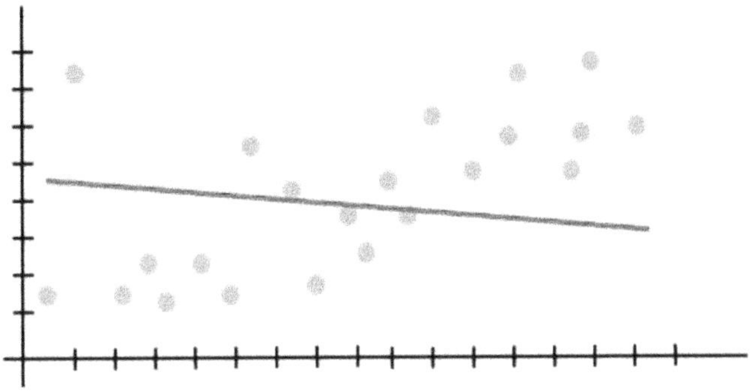

Over fitting may happen while training the data. The factors to be considered for over fitting (as shown in Figure 9) are training materials are too small in scale,prolonged training time and complexity of model is very high. Stopping training at earlier or overdue stage leads to training failure.

Figure 9. Overfitting

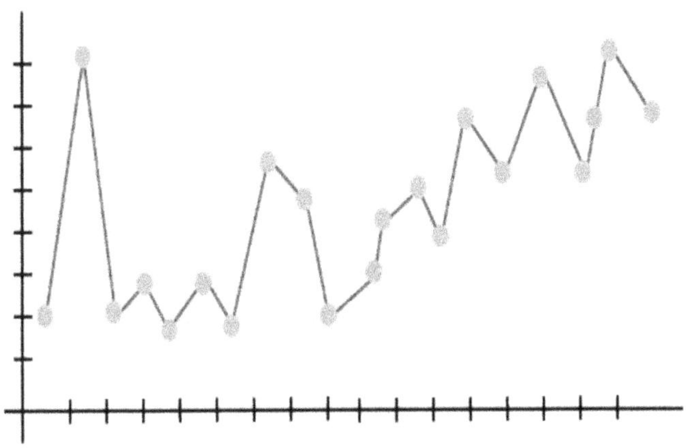

Dataset used in the proposed study is Pascal Visual Object Class 2012(shown in table 1). The main objective of this dataset is to find out the object class in our circumstances. Supervised learning approach is used here for training the labelled data. Twenty subjects are taken into account for classification and detecting the obstacle such as Vehicles, household, animals, aero plane, bicycle, boat, bus, car, motorbike, train, bottle, chair, dining table, potted plant, sofa, TV/monitor, bird, cat, cow, dog, horse, sheep, and person.

Table 1. Pascal visual object class 2012 dataset

Images 11,530
ROI annotated objects 27,450
Segmentations 6,929
Training set1,464 images
Validation 1,449 images
Classes20
Remaining Private testing set

Dataset Preprocessing

Data preprocessing is the key step to create a machine learning model. It is the process of cleaning the raw data to make into suitable data for building the deep leaning model. In general, raw data consist noise, unstructured, missing values and errors, outliers, inconsistent and inaccurate. By doing preprocessing, model accuracy and efficiency can be increased. Steps to be followed in preprocessing are:

1. Cleaning the data
2. Integrating the data
3. Data reduction
4. Transforming the data

A fast adaptive median filter is carried forwarded here in order to remove the noise present in the images. Filter evaluates each pixel available in the region of interest (ROI) space contains noise or not. The below figure 10 represents the region of interest and non-ROI area during different movement by vehicle.

Figure 10. ROI and Non-ROI prediction

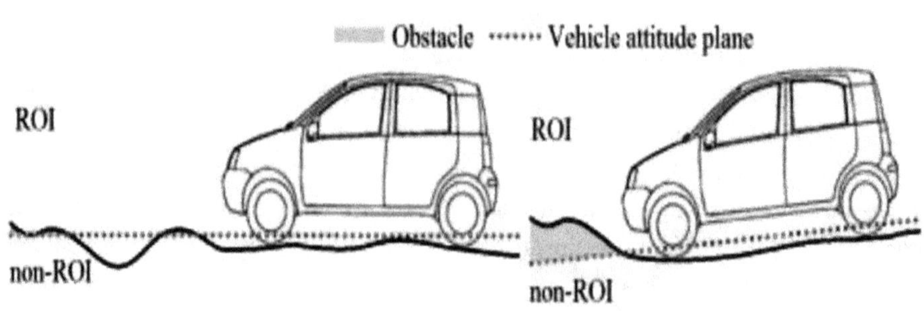

A window with initial size will be provided, then calculating the neighborhood average value and determining this neighborhood pixel value lies between minimum and maximum. While these criteria are satisfied then it is not a noise point. Therefore, median filtering is not required. If not, then enlarging the neighborhood range and repeating the process. Obviously, this enlarging the range will be continued till the size of neighborhood reaches threshold. In this way we can eliminate the holes and other noises present in the image as shown in figure 11.

Figure 11. Filtering results: a. Original image b. Filtered image

Obstacle Detection Using Deep Belief Network

Deep belief network comprises of multi-layer neural networks designed for two dimensional data. In every layer, extracting the useful features and output of this layer will be encoded as input to other layer. End layer will be fed as input to the classifier. Finally, presence of obstacle and non-obstacle cases will be figured out. Network gets raw image as an input and feature are extracted by itself (Zhao, 2020). During the process, feature map is attained. To detect video on real-time, every individual image frame is put into the model. This picture frame will be refined by deep belief network for the bounding boxes of diverse magnificence-specific items. This system is designed to detect the hindrance on the road. The processed picture is then annotated with bounding packing containers tagged with the respective class. The bounding field colors are indexed in Table 2 where each on-road obstacle object will be indicated by colors.These bounding packing containers are fed to a monitoring module for motionmaking plans of a self-sustaining vehicle(Prabhakar, 2017).

Table 2. Bounding field colors

Name of the Class	Bounding Box Color
Bicycle	Red
Bus	Yellow
Car	Cyan
Train	Cyan
Motorbike	Red
Animal	Blue
Person	Green

Discovering the Obstacle Using Disparity Map

Existing method available for obstacle detection using deep learning approaches are time consuming to predict the region of interest. Faster RCNN is successful in figuring out the objects available in the images but it is found unsuccessful to predict the minor obstacles (Prabhakar, 2017). In driver assistance system will be focusing the on-road obstacle. Hence, first step to detect obstacle is on road extraction. It is the foremost process to affect our detection rate. It is achieved mainly by monocular camera. Computational complexity for this algorithm is very high. U-V disparity map can be used for faster extraction of road and obstacle detection. It is calculated by using the pixel count in rows and columns from map with same disparity value.

$$V_d\left(j,k\right)_{j\in H,K\in\Delta}=\sum_{l\in W}\varphi\left(j,k,l\right)$$

$$\varphi(j,k,l) = \begin{cases} 1 & if\, I_d(j,l) = k \\ 0 & else \end{cases} \tag{1}$$

$$U_d(j,k)_{j\in\Delta,K\in W} = \sum_{l\in H}\varphi(j,k,l)$$

$$\varphi(j,k,l) = \begin{cases} 1 & if\, I_d(l,k) = j \\ 0 & else \end{cases} \tag{2}$$

where $V_d(j,k)$ and $U_d(j,k)$ represents intensity values of(j,k) in disparity map. Real image width and height is represented using H and W. Δ is the maximum disparity value. $\varphi(j,k,l)$ are the binary functions used for the intensity value threshold. Our system consists of two portions: extraction of road and locating the obstacle.

Extraction of Road

During flat road portion it is very much easier way for extraction in V-disparity using the Hough transform. Here road will be viewed as oblique line. Suppose if road with ups and downs, it should be represented by a rectangular area. We can extend the quality of road extraction facility by combining the least square with Hough transform. Oblique line can be expressed in the form of u=ad+b, where k refers to slope and be refers to intercept of line. Collecting the pixel $(\overline{u},\overline{d})$ which is present around the blue line which satisfies the below condition.

$$\left|(\overline{u} - (a\overline{d} + b)\right| < \varepsilon \tag{3}$$

where ε represents the maximum threshold of distance exists between $(\overline{u},\overline{d})$ and detection of oblique line by Hough transform. Most noise can be filtered out in this manner and it is helpful for extracting the road. Assigning the different weights to the pixel $(\overline{u},\overline{d})$ by using adaptive weight method. Figure 12 represents the road region extraction from red and blue line respectively.

Figure 12. Detection of red and blue line in v-disparity map

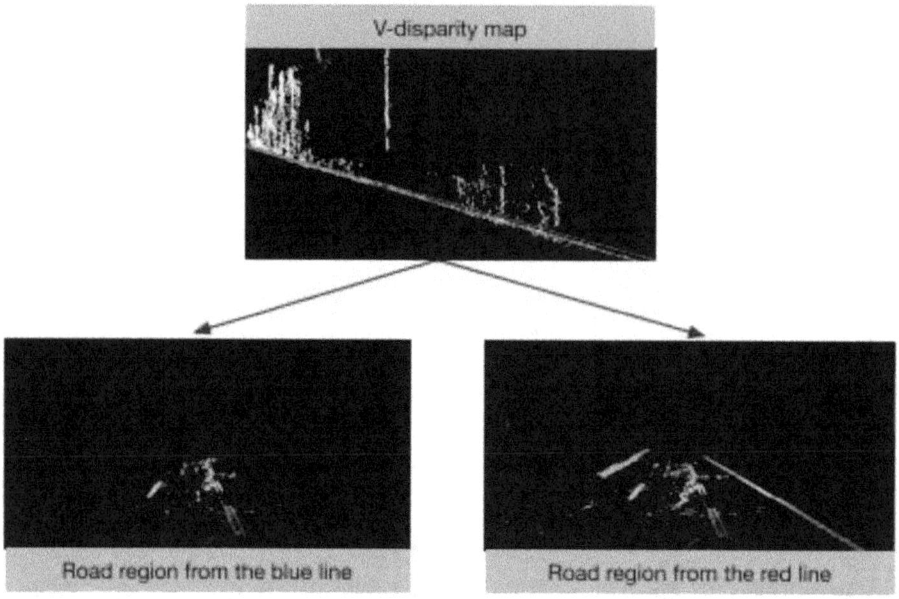

Location of Obstacle

In this method, our system is going to identify the obstacle which is available on the road by the mapping the U-disparity with its actual three dimensional picture. It can be calculated using the below mentioned formula,

$$\begin{cases} X = ac / d \\ Y = \dfrac{\left(b - b_0\right)X}{C} + \dfrac{1}{2}a \end{cases} \tag{4}$$

Where X and Y are the U-disparity map horizontal coordinate and vertical coordinates of the scene. a and c are the shutter parameters stating the distance and focal length in the form dimensions of pixel. Equation(4) clearly states all the obstacles original position which is generally viewed in form of 3D scene. Figure 13 represents the obstacle detection mechanism using U-disparity and V-disparity map.

Figure 13. Obstacle detection mechanism using U-disparity and V-disparity map.

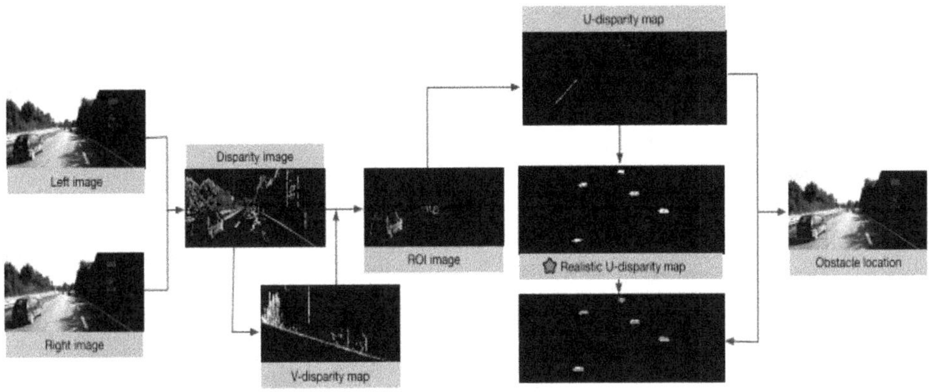

Obstacle Detection Algorithm

Input: Color image with corresponding disparity map.

Output: Surrounding the on-road obstacles with boundary boxes in the colored image.

1. Produce U disparity and V disparity map form the disparity map.
2. Oblique line extraction by combination of Hough transform and least square.
3. Realistic U-disparity map creation for accurate detection of obstacle.
4. Figure out the horizontal lines P_n in U-disparity map.
5. for n¬ 1 to P_n
6. Calculate the obstacle(n) width W which corresponds to detected line width in U-disparity map
7. Compute the obstacle height H in the U-disparity map
8. Calculate IoU with two different obstacle a,b which is having the closer disparity value.
9. If IoU value is gereater than 0.5 then
10. Combine obstacle a and b
11. Save the obstacle.

Performance Metrics

Intersection Over Union

Intersection over Union is commonly preferred assessment metric for the object revelation and localization task. It calculates the intersection of coordinates of divinated and ground accuracy bounding stands. IoU=1 specifies that the two bounding boxes perfectly overlays. A threshold point can be fixed to check whether the object is correctly detected as shown in figure 14.

Figure 14. Calculation of intersection over union

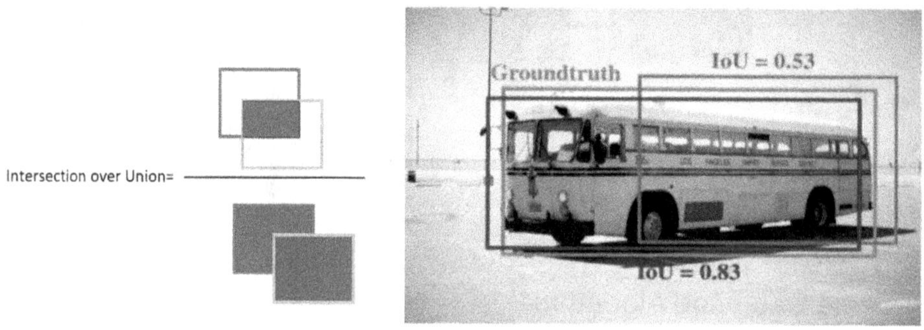

If the threshold of 0.6 is set, then IoU>0.6 is considered as True positive (TP) else it is taken as False Negative (FP). When the model fails to predict the image where the ground truth is present then it is False Negative (FN). True Negative is not considered as it represents the part of image where the object is not located. Multiple bounding boxes may overlap one object, in such cases the bounding box with highest value is takes as true positive.

Precision and Recall

Precision and Recall are used to evaluate the model performance with the help of TP, FP and FN values. It is usually measured for all objects in the captured image as shown in figure 15.

Figure 15. Calculation of precision and recall

$$\text{Precision} = \frac{\text{True Positive}}{\text{True Positive} + \text{False Positive}}$$

$$\text{Recall} = \frac{\text{True Positive}}{\text{True Positive} + \text{False Negative}}$$

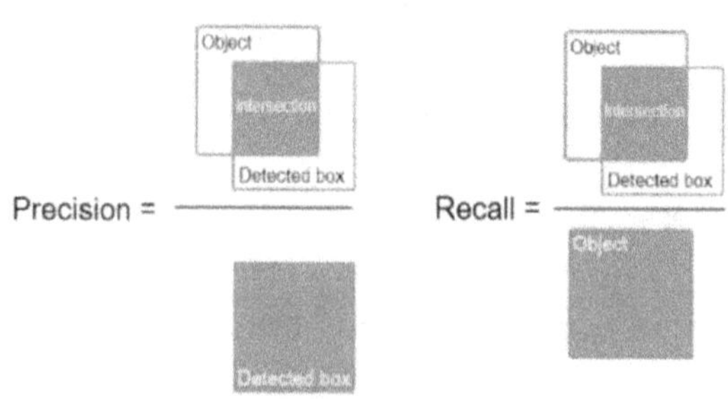

F-Measure

It combine the properties of both precision and recall and can be represented with a single measure. It is also commonly referred as F_β.

$$F_\beta = \left(1+\beta^2\right) * \frac{Precision * Recall}{(\beta^2 * Precision) + Recall}$$

If β=1, then it is F_1 score which gives equal importance to both precision and recall. This balance should be broken in some cases by assigning weights. F_2 Measure is an example by taking β=2, it concentrates on minimizing the FN than the FP. Accuracy can be considered as a best evaluation metric in case of multiclass classification problem but it will not be effective in case if the dataset in unbalanced (Number of datasets in one class is higher than the other). In such cases F-Score can be a better solution.

Average Precision

The obtained recall and precision values are combined to form a Precision recall curve. The area under the curve is generally referred as average precision. PR curve follows a zig zag like pattern instead of a monotonically decrease. Interpolation methods are used to decrease the curve monotonically.

11-point digression method: 11-point digression method can be done by taking 11 equally spaced recall values followed by taking mean of those values. Eg. [0, 0.1, 0.2, …, 1] the values with a step size of 0.1 is taken.

$$AP = \frac{1}{11} \sum_{r \in \left(0,0.1,0.2,..1\right)} Pinterp\left(r\right) \text{and } Pinterp\left(r\right) = \max_{r':r' \geq r} P\left(r'\right)$$

Mean Average Precision (mAP)

Average precision is estimated individually for every objects. It means that the number of AP would be equal to the number of objects. Average of the AP values are taken to determine mAP.

$$mAP = \frac{1}{n}\sum_{i=1}^{n} AP_i$$

Segmentation accuracy can also be considered as one of the evaluation criteria which can be calculated with the formula

$$Segmentation\ accuracy = \frac{TP}{TP + FP + FN}$$

COCO

A prediction of bounding box above certain threshold is taken as true positive (e.g. 0.5). Thus prediction values for two bounding boxes with values 0.6 and 0.95 have equal weightage. A way to solve this problem is to have few ranges of IoU values and the mAP is calculated for every value and a final mAP is taken.

$$mAP_{COCO} = \frac{mAP_{0.5} + mAP_{0.55} + \ldots\ldots + mAP_{0.95}}{10}$$

10 IoU inception values from 0.50 to 0.95 with a stride capacity of 0.05 is taken. The average of an IoU provides an object detectors with best localization results.

Comparison of DL With ML

Deep Learning (DL) is a subgroup of Machine Learning (ML). DL network needs very less human intervention compared to ML algorithms where the features extraction process is carried out manually. Best results can be expected with the deep learning algorithms rather than traditional ML methods. ML and DL differ each other in dataset consumption, extraction of features, hardware interfacing and uses cases. DL requires very large collection of datasets with high quality and should be properly annotated. Gathering the large datasets and labeling them results in high cost, more time and domain expertise is needed for proper labeling. Whereas ML algorithm works fine with smaller datasets as shown in figure 16. As larger datasets are used it takes longer time for training and GPU systems are preferred for it and simple setups are enough for training the machine learning algorithms.

Figure 16. Difference between ML and DL methods

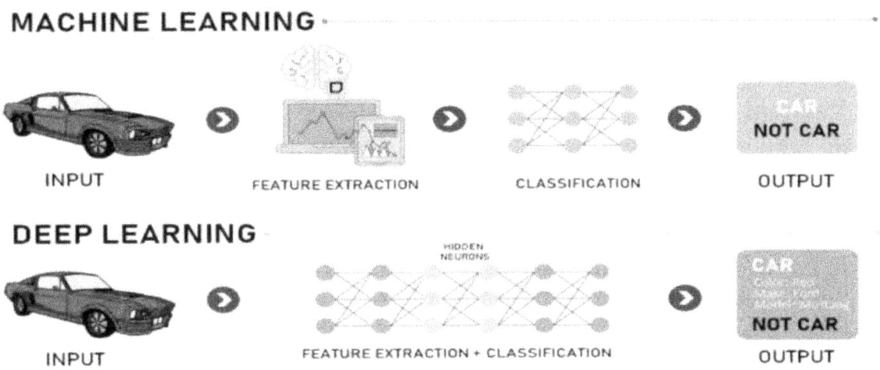

In traditional ML, the algorithm is trained with the set of features to detect the presence of an object or to classify it (e.g. Car). To make it possible human intervention is required to obtain the feature set from the available dataset. The features usually help the algorithms to detect the presence of particular indication of an object (e.g. Wheel, headlamps etc.). In contrast the DL algorithms attempts to detect the feature set on its own with the available data volume.

Different ML algorithms like Support Vector Machine (SVM), AdaBoost, Random Forest, Naïve Bayes and decision tree have produced better results. Many researchers have introduced ensemble of different machine learning approaches to increase the overall accuracy. DL approaches are playing a significant role in image segmentation, it is proven that it outperforms the ML algorithms. In the field of image and video analysis, DL methods made a benchmark by surpassing the traditional ML technologies. At 2012 ImageNet challenge, CNN and AlexNet had beaten the ML algorithms with a difference of about 10%.

Commonly Used ML and DL Algorithms

- **Logistic Regression:** It is used for classification of two class objects where the target output should be categorical in nature. A threshold value is normally set to determine the value at which a class can be put into.
- **Naive Bayes:** Another classification algorithm which works on the principle of bayes theorem. Bayes theorem can be used to identify the probability of occurrence of A when the occurrence of the other event B is given. Navie bayes can be opted when there are numerous data points and only very few training variables are available.

- **kNN:** A supervised learning algorithm mostly preferred for classification problems which can be more effective for larger training datasets. kNN can be used when we have a two categories A and B and if a new data arrives then it can be easily categorized.
- **Random Forest:** It is relevant for regression & classification problems and the random forest is an ensemble strategy which is formed by combining many classifiers for solving a highly complex problem. It can solve the problem of overfitting and it can work effectively than decision tree

Many more machine learning algorithms like decision tree, k-means, Adaboost etc., are available. Basic Deep Learning Architecture which outperforms all the ML algorithms is CNN. DL does not require human invented rules, whereas it maps the large volume of dataset with the input labels. Many variants of CNN have evolved each of which produced outstanding performance in object detection and localization than all the traditional machine learning algorithms.

Stating from AlexNet to all high-resolution networks have been designed with CNN. Achieving the desired outcome from ML requires various steps like pre-processing, feature extraction, selection of features and finally classification and segmentation. Biased selection of features may sometimes lead to improper classification of datasets. DL exhibits a fully automatic selection of feature sets from the datasets for several applications.

Figure 17. Performance of DL algorithms

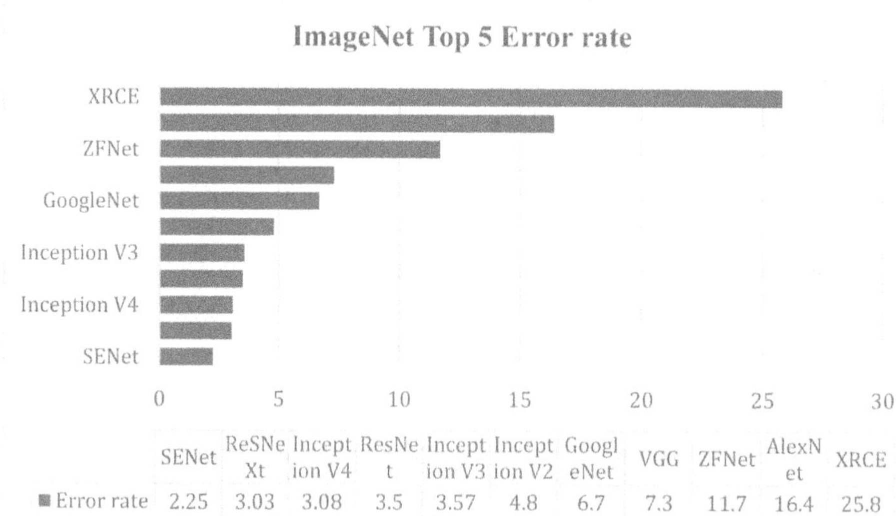

From the above figure 17, it can be seen that the error rate in the prediction is drastically reduced upon the evolution of different algorithms which can even works well than human performance. These DL algorithms can be applied when human experts are not available or to make a highly automated device.

Deep learning has been used for solving complex and non-linear problems. Pascal Visual Object Class2012 dataset comprising of 11530 images. The dataset was collected in various city street areas, highways and residential areas in different lightening conditions.

Location of objects is a primary task in self-driving cars and this dataset provides bounding box annotations for each object in Pascal Visual Object Class in 11350 images. The dataset is divided into training and validation datasets each containing 1464 images and 1449 images. Finding the architecture size and its hyper parameters that will be capable of learning the representations of the training dataset for object localization. Sequence of model selection task were performed to determine the optimal set of hyper parameters such as hidden layers size, training epochs, learning rate and batch size as shown in table 3.

Table 3. Different hyper parameter chosen for training DBN model

Model	No. of Hidden Layers			Learning Rate		No. of Epochs		Size of the Input
	I	II	III	Pretraining	Training	Pretraining	Training	
I	1100	700	150	7.80E-05	2.45E-03	15	101	9475
II	1300	500	100	7.80E-05	2.45E-03	15	92	9475
III	1300	700	150	7.80E-05	2.45E-03	15	87	9475
IV	1300	700	200	7.80E-05	2.45E-03	15	63	9475
V	1400	1000	200	3.50E-05	3.26E-03	15	40	7161
VI	1300	700	150	7.80E-05	2.45E-03	15	98	7161

Features extracted are multiplied with the chosen weights and these feature vectors are given as input to the DBN. The weights multiplied with the features are tuned to produce the optimum performance where the error between the predicted and actual outputs are reduced. Modified dragonfly optimization is used for the weight's optimization. To carry out the proposed experiment, we used the system configurations of Windows 10 OS with 16GB RAM and implemented using MATLAB 2020a.

Table 4. Comparison to the state of art performance with different models

Model Name	Backbone	mAP (%)
U-Net	VGG-16	82
U-Net	ResNet-50	78
U-Net	Mobile Net 16	82
SSD	VGG – 16	72.95
YOLO	CNN	67.99
YOLO V2	CNN	77.12
Faster R-CNN	VGG - 16	73.2
SSD 300	VGG - 16	77.5
MR-CNN	VGG - 16	78.2
LED 300	VGG - 16	78.7
SSD	VGG - 16	79.5
Faster R-CNN	ResNet – 101	76.4
R-FCN	ResNet – 101	79.5
Attention Network	ResNet – 101	80.1
DBN	**Deep SAE**	**85.0**

The proposed method is evaluated with various metrics like precision, Recall, F1 score and mean average precision (mAP). For a fair comparison the state of art deep learning architecture with ResNet, VGG-16 and CNN as a backbone is used and all the architectures were trained with same dataset. The proposed system provides a better mean average precision as mentioned in table 4 and DBN produces optimal precision, recall and F1 score compared with previous works on deep learning architectures as shown in table 5 and figure 18.

Table 5. Comparison of state of art works with various performance metrics

Model Name	Precision Rate	Recall Rate	F1 Score
Faster R-CNN	58.36	88.38	0.7030
DFL-CNN	64.61	89.44	0.7502
AVPN_basic	85.93	75.59	0.80
AVPN_large	87.81	77.02	0.82
H-Fast	86.2	74	0.80
H-RPN	89.2	78.3	0.83
Improved faster R-CNN	88.1	91.6	0.918
Deep SAE based DBN	**89.2**	**90.0**	**0.928**

Figure 18. Comparison of state of art works with various performance metrics

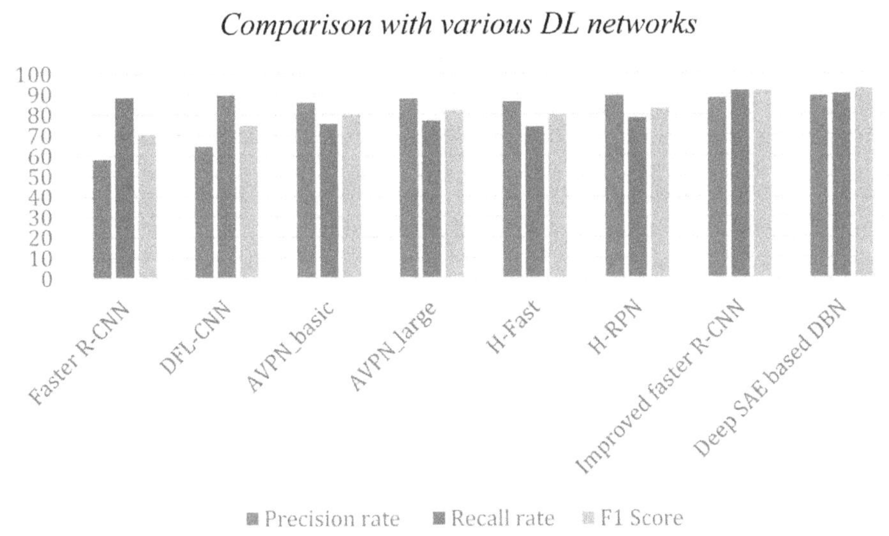

CONCLUSION

A smart city is a metropolitan area that utilizes various types of electronic methods and sensors to collect data. Transportation plays a significant role in urban area and development to accommodate an ever growing number of vehicles. This crucial application has drawn in numerous scientists in both scholarly world and industry to explore autonomous vehicles. In this article, we figured out an obstacle detection and avoidance mechanism in self-driving vehicle by using a deep belief networks. Here the method makes use of disparity map to predict the hindrance. The proposed methodology follows the deep belief neural network for the betterment of network structure to identify the hindrance. Followed by experiments were conducted with Pascal Visual Object Class 2012 Dataset for twenty subjects where fast adaptive median filter is used. Filter smoothen the noise found in the region of interest. Hindrance detection method utilizes the U-V disparity map for accurate prediction of objects. By means of analysis, roads are extracted and obstacles are located in self-driving vehicle. The results show that proposed method attained faster detection ratio compared to existing method. The trained model has achieved accuracy and is in good consent with expected performance. In an automated car system, accuracy and efficiency are directly proportional to the training data amount and object recognition model quality. Finally, the proposed vehicle safety method is more appropriate for real time environment.

REFERENCES

Alzubaidi, L., Zhang, J., Humaidi, A. J., Al-Dujaili, A., Duan, Y., Al-Shamma, O., Santamaría, J., Fadhel, M. A., Al-Amidie, M., & Farhan, L. (2021). Review of deep learning: Concepts, CNN architectures, challenges, applications, future directions. *Journal of Big Data*, *8*(1), 1–74. doi:10.118640537-021-00444-8 PMID:33816053

Dairi, A., Harrou, F., Sun, Y., & Senouci, M. (2018). Obstacle detection for intelligent transportation systems using deep stacked autoencoder and $ k $-nearest neighbor scheme. *IEEE Sensors Journal*, *18*(12), 5122–5132. doi:10.1109/JSEN.2018.2831082

Du, W., Rao, N., Liu, D., Jiang, H., Luo, C., Li, Z., Gan, T., & Zeng, B. (2019). Review on the applications of deep learning in the analysis of gastrointestinal endoscopy images. *IEEE Access: Practical Innovations, Open Solutions*, *7*, 142053–142069. doi:10.1109/ACCESS.2019.2944676

Gupta, A., Anpalagan, A., Guan, L., & Khwaja, A. S. (2021). Deep learning for object detection and scene perception in self-driving cars: Survey, challenges, and open issues. *Array*, *10*, 100057. doi:10.1016/j.array.2021.100057

Haris, M., & Hou, J. (2020). Obstacle detection and safely navigate the autonomous vehicle from unexpected obstacles on the driving lane. *Sensors (Basel)*, *20*(17), 4719. doi:10.339020174719 PMID:32825601

Hernandez-Aceituno, J., Arnay, R., Toledo, J., & Acosta, L. (2016). Using kinect on an autonomous vehicle for outdoors obstacle detection. *IEEE Sensors Journal*, *16*(10), 3603–3610. doi:10.1109/JSEN.2016.2531122

Hinton, G. E., Osindero, S., & Teh, Y. W. (2006). A fast learning algorithm for deep belief nets. *Neural Computation*, *18*(7), 1527–1554. doi:10.1162/neco.2006.18.7.1527 PMID:16764513

Karmakar, G., Chowdhury, A., Das, R., Kamruzzaman, J., & Islam, S. (2021). Assessing trust level of a driverless car using deep learning. *IEEE Transactions on Intelligent Transportation Systems*, *22*(7), 4457–4466. doi:10.1109/TITS.2021.3059261

Kohl, C., Knigge, M., Baader, G., Böhm, M., & Krcmar, H. (2018). Anticipating acceptance of emerging technologies using twitter: The case of self-driving cars. *Journal of Business Economics*, *88*(5), 617–642. doi:10.100711573-018-0897-5

Krishna Chaitanya, G., & Maragatham, G. (2021). Object and obstacle detection for self-driving cars using GoogLeNet and deep learning. In *Artificial intelligence techniques for advanced computing applications* (pp. 315–322). Springer. doi:10.1007/978-981-15-5329-5_30

Kumar, A., Priyanka, S., Dhanashree, K., Praveen, V., Rekha, R. (2022). Efficient binary grasshopper optimization based neural network algorithm for bitcoin value prediction. *International Journal of Nonlinear Analysis and Applications, 13*, 53-60. . doi:10.22075/ijnaa.2022.6330

Lai, Y. K., Ho, C. Y., Huang, Y. H., Huang, C. W., Kuo, Y. X., & Chung, Y. C. (2018, October). Intelligent vehicle collision-avoidance system with deep learning. In *2018 IEEE Asia Pacific Conference on Circuits and Systems (APCCAS)* (pp. 123-126). IEEE. 10.1109/APCCAS.2018.8605622

Leng, J., Liu, Y., Du, D., Zhang, T., & Quan, P. (2019). Robust obstacle detection and recognition for driver assistance systems. *IEEE Transactions on Intelligent Transportation Systems, 21*(4), 1560–1571. doi:10.1109/TITS.2019.2909275

Liu, G., Bao, H., & Han, B. (2018). A stacked autoencoder-based deep neural network for achieving gearbox fault diagnosis. *Mathematical Problems in Engineering, 2018*, 2018. doi:10.1155/2018/5105709

Nguyen, V. D., Van Nguyen, H., Tran, D. T., Lee, S. J., & Jeon, J. W. (2016). Learning framework for robust obstacle detection, recognition, and tracking. *IEEE Transactions on Intelligent Transportation Systems, 18*(6), 1633–1646. doi:10.1109/TITS.2016.2614818

Prabhakar, G., Kailath, B., Natarajan, S., & Kumar, R. (2017, July). Obstacle detection and classification using deep learning for tracking in high-speed autonomous driving. In 2017 IEEE region 10 symposium (TENSYMP) (pp. 1-6). IEEE. doi:10.1109/TENCONSpring.2017.8069972

Sanil, N., Rakesh, V., Mallapur, R., & Ahmed, M. R. (2020, January). Deep learning techniques for obstacle detection and avoidance in driverless cars. In *2020 International Conference on Artificial Intelligence and Signal Processing (AISP)* (pp. 1-4). IEEE. 10.1109/AISP48273.2020.9073155

Sreedhar, B. K., & Shunmugam, N. (2020, July). Deep Learning for Hardware-Constrained Driverless Cars. In *2020 IEEE 44th Annual Computers, Software, and Applications Conference (COMPSAC)* (pp. 26-29). IEEE.

Yin, H., & Berger, C. (2017, October). When to use what data set for your self-driving car algorithm: An overview of publicly available driving datasets. In *2017 IEEE 20th International Conference on Intelligent Transportation Systems (ITSC)* (pp. 1-8). IEEE.

Zhang, Z., Xu, H., Chao, Z., Li, X., & Wang, C. (2014). A novel vehicle reversing speed control based on obstacle detection and sparse representation. *IEEE Transactions on Intelligent Transportation Systems*, *16*(3), 1321–1334. doi:10.1109/TITS.2014.2360337

Zhao, X., Wu, H., Xu, Z., & Min, H. (2020). Omni-Directional Obstacle Detection for Vehicles Based on Depth Camera. *IEEE Access: Practical Innovations, Open Solutions*, *8*, 93733–93748. doi:10.1109/ACCESS.2020.2993934

Chapter 6
Human-Machine Interactions:
A Synthesis of Threats and Opportunities

Priyadarsini Patnaik
Birla Global University, India

ABSTRACT

Research on human-computer interaction (HCI) has been widely developed since the 1960s as a result of the rapid growth of information systems. Designing ergonomic human-computer interfaces is the purpose of human-computer interaction (HCI). Nonetheless, the user had complete control of the computer, but they were mostly static. The management of increasingly complex and coupled systems has become indispensable to both the public and private sectors over the past few years. Therefore, humans and machines must be able to adapt to unforeseen circumstances by maintaining degrees of freedom. Though modern machine manages all the activities that humans have done in the past, processing information as well as dealing with dynamic situations but uncertain factors play a crucial role in these dynamic situations, which are difficult to control. The trend the industry is experiencing demands a human-machine cooperative (HMC) approach to handle the new challenges such as cyber security. The chapter explains how HMC might improve HCI to address problems emerging from this change.

INTRODUCTION

As the new Industrial era begins, a new perspective is presented focusing on the centrality of humans. With the Fourth Industrial Revolution, many changes have been made to the individual as well as workforce's foundations. Human-machine interactions have been profoundly altered by Industry 4.0, (Nardo, Forino, & Murino,

DOI: 10.4018/978-1-7998-9710-1.ch006

2020). These interactions are enabled with the help of Artificial Intelligence (AI), Internet of Things (IoT). In order to emphasize the importance of Human-machine interactions identification and visualization technologies serve as the main interface for human-machine interaction, (Lu, Jiang et al. 2021, Lodgaard, & Dransfeld, 2020) . The importance of the human factor will be deeply studied through the implementation of the personalised AI which can imitate the cognitive functions that we attribute to the human mind (Syam, Sharma, 2018) and it is expected that AI will be the most frequent technology adopted by the stakeholders, (Columbus, 2019).

Almost all research and industrial practices have been turned upside down by the Fourth Industrial Revolution, which has led both to technological innovation, (Kagermann et al., 2013). The convergence of Industry 4.0 with digitalization also led to Digital Transformation, (Deloitte 2015) and are integral to the adoption of digital technology. Over the past several years, digital transformation has also permeated many other areas. AI and IoT were once regarded as a frontier area of research, but today, artificial intelligence (AI) and Internet of Things (IoT) are the important area of research in virtually all fields (Metaxiotis et al. 2003), However, the new term 'Society 5.0', advancing technology to improve the well-being of society across the board and implementing many of the technologies into everyday life, is intended to influence society over the long term which requires wellbeing of society with the online rights of citizens in highly dynamic situations.

In technology, industry, and scholarly fields, artificial intelligence (AI) is booming. A burgeoning field of technology, artificial intelligence (AI) is far from a new concept. AI holds tremendous potential for dramatically changing the way marketing is done, but its applications have not yet been widely adopted. The application of AI in management arena is a vital area for researchers. So, understanding of artificial intelligence's long-term implications by examining its present and future applications, planning its implementation, and assessing its possible uses and where to make improvements is vital now. AI provides new ways to create and distribute value to companies, corporations and for end users. Hyper customization, scalable experiences, and deep insights comprise the innovations in this area and researchers predict that marketing strategies and consumer behaviours will change as a result.

Intelligent Automation evolves day after day and week after week, it allows to live in a more humane world. Although being in control is an important aspect of human-machine interaction, (Zanatto, et al., 2021) little is known about the effect of trust on this automation. Despite of India is emerging as a global IT superpower, asserting its digital sovereignty but there are many security challenges in terms of data privacy and financial transactions. In 2019, market size for cyber security stood at $161.07 billion and is expected to grow by 14.5% between 2020 and 2025, according to industry reports, (Businesswire.com) The growth of cyber security market signifies the more challenges in cyber securities. So there is a need to invest more in advanced security mechanism, cyber welfare, cyber security talent gap to bridge the cyber security challenges and to safeguard data privacy and online rights of citizens.

Objective of the Study

According to Statista, the global market for Iot end users solution is expected to grow a staggering 1.6 trillion by 2025. It shows that the Internet of things IoT is the next big thing in technology. As the IoT market continues to grow, and more and more devices enter the world, this creates a whole new set of challenges for IoT security. For these cyber security issues, there is a lack of trust in users to adopt these technologies. So, this study tries to address this research gap with possible solutions to overcome the cyber security issues along with trust. The study focuses on a specific human connection feature that has a helpful impact on the human-autonomy relationship. In this article, the author explored one possible approach for rebuilding trust, which is a crucial component of a successful relationship.

LITERATURE REVIEW

Human–Machine Interaction (HMI)

A human–machine interaction (HMI) model is created for the human operator (HO) executing many simultaneous tasks and is distinguished by a strategy function, (Chandler A. et.al.2007). Making machines behave like humans is one of the major objectives of human–machine interaction (Xiao et al., 2020). So, automation is becoming increasingly important in human-machine systems, (Riley, 1989, October).

Modern interactions with technology are progressively shifting away from conventional human usage of computers as tools and moving towards the formation of human connections with autonomous units that perform activities on behalf of human (Ewart, et al., 2018 Hancock, 2017). One of most crucial component of HMI systems is humans. Hence, the suitable design of a novel type of technologies with highly automated systems should be prioritised. The core concepts of HMI technologies include fulfilling requirements of the application for user services and improving user experiences, (O. A. Araromi et.al.2020). In HMI, the physiological and developmental parameters can be continually monitored in real time using flexible and stretchable sensing devices, (S. Yao et.al.2020). The precision and user acceptance of the interaction process are substantially determined by convenient and sophisticated sensing devices (M. Cianchetti et.al.2018) With the development of increasingly independent, autonomous systems that appear to be of a separate class than ordinary automation, Peter Hancock (Hancock, 2017) suggested that society is in the midst of a profound transformation and experts are working at tremendous speed to design these new technologies by offering direction to the creators of these future models . In the context of this, the psychological and human factors consequences may be overlooked. The author of this study offered a human-cantered approach to ensure that future human autonomy interactions with user's requirements and preferences.

The human-machine relationship is ever-changing, reflecting the adaptability of autonomy and the human perspective, within human-machine interaction. In place of this, collaborative and interactive paradigms should model dynamic interactions between people using human-machine modelling as their foundation. So, autonomous systems should be able to identify incidental information gathered from others through the use of human skills. As a result of this consideration, and influenced by human-human interaction models, it is argued that future autonomous design should be influenced by social science. Hence, research on human-human relationships and human autonomy should be broadened to include:

1. A new approach to human-autonomy based on more than simply "interactions," and
2. A reframe of human-autonomy relationships as a "relationship." (Madhavan & Wiegmann, 2004, 2007a).

Three observations can be made in terms of how a person interacts with the computer system when attempting to satisfy his or her intention.

1. There are four stages to human interaction with computers, each with a different goal, a different method, and a different need while interacting with that machine (Norman, 1984).
2. It is known that each method for building the interface has its own benefits as well as its own shortcomings. There are also differences in the trade-offs between each stage of the user interface (Norman, 1983a).
3. A variety of messages and interactions can be exchanged between a human and a machine Confusion and misunderstanding can occur if the levels do not match. Knowing the user's intentions is often necessary to determine which level is appropriate (Norman, 1981 a, 1983b)

In an ideal intelligent interactive environment, machines can detect external stimuli similar to humans and simulate recognition, processing and understanding of external stimuli, and has the ability to calculate emotions. (Hall et.al, 2014) show that an expressive machine can communicate comparative human feelings, which is viewed as more affable, humanized and refined, hence having higher pleasant interaction. Likewise, many studies have investigated and planned humanoid or creature molded PCs (robots), (Mirnig et.al., 2015, Bennett et al., 2014) enthusiastic voice or non-verbal communication (Knight & Simmons, 2016) making human–machine collaboration more like human–human collaboration.

Presently, emotional computing has certain examination establishment in machine feeling and expression framework. A few scientists portray the powerful course of mimicking enthusiastic articulation through unconstrained passionate exchange and outer enthusiastic incitement, (Kumar, 2016). Emotional transfer model advances the improvement of emotional registering to cutting edge man-made consciousness, yet additionally changes the hypothesis of full of feeling figuring into relevant numerical models, (Clavel, C., et.al.2009). Hence it can be accepted that different characters produce various people's intellectual encounters. Be that as it may, the situation presently is absence of a solitary feeling which can't understand robot with character, (Creed & Beale 2008; Mower, Mataric, Narayanan, 2009). Since, the idea of human–machine feeling association ought to be founded on the exact transmission of enthusiastic data which identifies with successful feeling response, this study considered emotion as a crucial part of human machine interaction.

Emotion and HCI

HCI (Human - computer interaction) is creating a new research field recently which underscores that the planned PC framework and users of it exist together amicably. The research on the mix of feeling and HCI has drawn in wide consideration of numerous specialists in emotional figuring (Zhang et al., 2016). To work on the helpfulness of HCI model, Springett et al. investigated their persuasions in games, online business, and in education, (Springett, 2015) Moreover, feelings are applicable in multimodal HCI to perceive different parts of emotional phenomena. Displaying feelings through HCI assists to create emotional intelligence insight in cutting edge human–computer interaction, (Kumar, 2016).

Emotion Simulation

The pursuit of flawless human-like reasoning is one objective of artificial intelligence, (Xiao et al., 2020) In other words, the computer would function similarly to humans but are not confined to reasoning, memories, and emotions. Prior studies on enhanced artificial intelligence are substantial and primarily concentrate on emotional simulations. (Stryker, & Nancy. 1988). was the first to propose the integration of emotion into computers. Zhang et al. proposed a model on emotional reactivity to sensory stimuli, (Zhang & Li 2014) Wang et al. have outlined the studies in emotion modelling and offered the theoretical concepts involved, (Wang Zhiliang, 2009). Furthermore, Wang et al. triggered the cognitive-affective paradigm developed in HCI. According to that theory, it may better simulate the effective and progressive emotion process and humanistic intelligence level. (Liu et al., 2012). Dazhi Jiang incorporated the memory element into a robotic emotional model influenced by personality, and offered appropriate responses to various scenarios (Dazhi Jiang, et al., 2020)

Cyber Security

People have various perspectives on cyber security. Researchers are looking to solve the basic problem of understanding a wide range of cyber security encounters in order to develop effective and integrated cyber security services. Researchers investigate the vocabulary of network security throughout three major phases of life i.e., young people, middle-aged people, and elderly people. Utilizing techniques on checklists of cyber security characteristics produced by each age group, researchers were able to create a comprehensive picture of cyber security. Because of the pervasiveness and continual advancement of technology, cyber security events are likely to affect more individuals than it has ever been. Users are sometimes referred to it as the weakest link in surveillance systems (Ryan West et.al.2009), thus it's critical to comprehend the factors that may influence their threat protection actions and judgments. Differential hazards in cyber security, (James Pierce et.al.2018) undermine this goal; that is, the idea that information security can vary across people and be "socially dependent and differently applied" based on the attributes like gender, age, and experience. Responses to technical and societal development are influenced by the age, not because various life phases bring with them varied social, economic, and ecological settings and difficulties (Pinquart et al., 2004). Certain aspects of cyber security will be much more important to people at different periods of their lives. As a result, efforts to comprehend what cyber security entails for digital natives. This is important not just for designing higher security interfaces, but also for better understanding how diverse situations affect cyber threats.

Previous research has found significant age-related variations in cyber security threats and activities. Younger folks, for instance, are much more likely than that of other age categories to be victims of cyber-bullying (Kierkegaard, 2008) and are more vulnerable to hacking (Sheng et al., 2010), and are much more likely to disclose passwords, (Monica Whitty et.al.2015) as well as other personally identifiable information on social networking sites (Dowell et al., 2009). Old-aged individuals are less likely to use PIN or bio - metric authentication for their gadgets (Harbach et al., 2016), in which attackers provide benefits for breaching security (Daniela Oliveira et.al.2017). In comparison to younger demographics, older individuals have indeed been reported to participate in less privacy-related activities on social media sites, publish less information on the web, and exhibit greater concerns about the other user privacy (Kezer et al., 2016). a) Children, b) Working Adults, and c) Seniors. The larger the object, the more common the characteristic. Individuals' inclination to disclose security events is influenced by their perception of the personal significance of cyber security information and risks, (Briggs et al., 2017). As a result, the different ways in which different groups perceive, describe, and debate cyber security should become a significant design consideration for human – computer interaction, security alerts, training materials, and instructional materials. Only a small amount of research has looked at how variations in cyber security-related practices are mirrored in, or even driven by, changes in the context of the statement "cyber security" to different ages.

Cyber Security and HCI

IT security, or cyber security, is concerned with securing computers and their data from criminal behaviour, (Moallem, A. (Ed.) 2018). HCI plays a huge role in cyber-security despite of user error is the primary cause of security breaches. In the field of technology, users are described as advocates, and they serve as the human interface between users and technology. Through the appropriate use of computer technology, IT experts should ensure the users can efficiently, secure and effectively accomplish their goals. By taking proper security precautionary measures, users can avoid common attacks such as phishing, social engineering, malware, and insecure browsing (Rowe et al., 2010)

Cyber Security and Trust

Protection isn't guaranteed by any technology and there is a risk that people may encounter in these automated systems thus exposing themselves to cyber threats, even when technical controls are in place, (Butavicius et al., 2020) Automated tools, including cyber security tools, are more effective when trust is established, (Brown, et al., 2016) Hence, systems cannot be adequately protected by relying on technological security. It is vital to understand the safe and secure behaviour in order to improve security so that the underlying factors can be better predicted to address vulnerabilities. The science of cyber security used to be a reactive, narrowly oriented, and business-oriented science of cyber security risk assessment. To make the cyber security risk assessment process models more accurate, humans and risk factors must be incorporated into one model, and metrics that go beyond classical vulnerabilities must be used (confidentiality, integrity, accessibility, or CIA), (Oltramari et al., 2015)

Trust in HCI

Human-computer trust is defined as the *"extent to which a user is confident in, and willing to act on the basis of, the recommendations, actions, and decisions of an artificially intelligent decision aid"* adapted from McAllister (1995, p.25). HCT enables the system to take action without interference in instances when it is relevant. HCT is sometimes described as the user's willingness to follow the system's recommendations while they have enough evidence to make a decision about the system's trustworthiness. However, there have already been cases where such systems have made erroneous or even life-threatening judgments, resulting in unforeseen and tragic outcomes. Whenever a user becomes more susceptible and unsure of how the system will behave, researchers and developers alike must determine how much faith an end user places in the advice or suggestions offered by such intelligent and smart systems.

Research Gap

As technology progresses and much more advanced types of automation become accessible, network administrators must face extremely tough judgments on how much automation to employ about when to use it. Previously, automation decisions were taken by allocating tasks to humans or machines based on which was better suited to each. This criterion is now frequently difficult to implement. Moreover, there is a lack of trust experienced in users' acceptance, processor's decisions, or the operator's proclivity because of over-reliance on automated processes. Due to the dependence on technology, cybercrime has resulted in threats and vulnerabilities so that individuals, businesses, and other organizations are also experiencing trust issues and their cyber-trust has broken. Even though trust affects so many aspects of user interactions with technology, in prior research study, technology-mediated interactions have been evaluated and measured in the context of user technology interaction (Söllner et al., 2012; Benbasat & Wang 2005; Sousa et al., 2014; Gulati et al., 2017, 2018) but the question of how trust affects human-machine interaction remains unanswered. So, the present study tried to evaluate trust with regard to HCI in order to fill this gap. Also, a human computer trust scale,, is empirically evaluated in the present study.

Proposed Model and Hypothesis Development

Figure 1.

Proposed model and hypothesis development

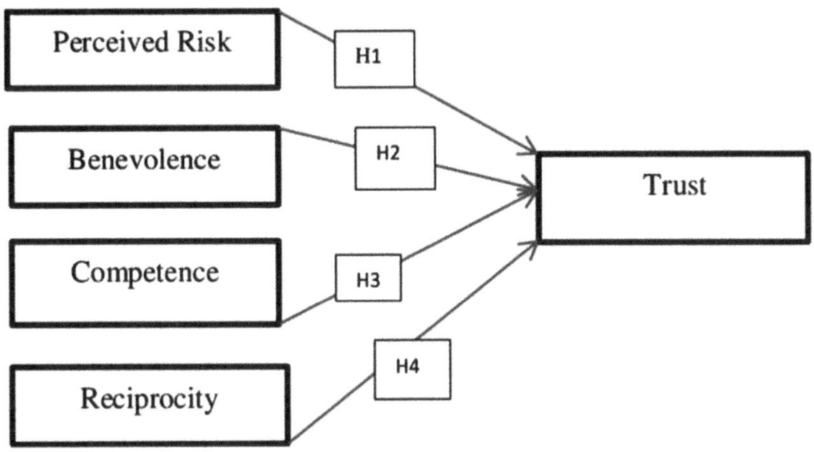

The major goal of this study was to see if the features of a "human computer trust model (HCTM) developed by Sousa et al., (2014)" capture the essence of trust and can be used to quantify it. The major attributes were borrowed from Sousa, et al., (2014) and Gulati et al., (2018) which would predict trust in user technology interactions. Finally, this study considered constructs such as perceived risk, Benevolence, Competence and Reciprocity as four independent variables and users trust as dependent variable in this model.

Perceived Risk

The term perceived risk refers to a subjective assessment by the user of the probability of specific incidents occurring while interacting with a technical artifact, and the level of concern they have regarding the consequences of those actions. It is particularly important to consider perceived risk when interacting with AI systems. The end user and programmer find it difficult to predict how a system reaches conclusions in such situations, or whether the advice it gives can be trusted. Hence, users' perception of risk associated with technological interactions is statistically significantly related to their trust in the system. The first hypothesis arises from this as

H1. Users' perceived risk associated with their trust in that technology.

Benevolence

Beneficence is traditionally seen as the trustee acting in the best interests of the trustor and always prepared to help those who need it (Mayer, Davis, and Schoorman, 1995). Technology's ability to assist the end user in accomplishing specific interaction goals allows it to provide adequate, effective, and adaptive assistance from a technological perspective (Gulati, Sousa, and Lamas 2017, 2018; Benbasat and Wang 2005). The trust in a technology can translate into greater trust and continued use of the system if a user perceives that it will provide the help needed (Mcknight et al., 2011). Technology is associated with trust when it is benevolent. A technological artefact that is benevolent is associated with a high level of trust. Consequently, the study can hypothesise:

H2. Technology with higher levels of benevolence have higher trust levels..

Competence

The competency of a system refers to its ability to perform all functionalities and features effectively (Mcknight et al. 2011). It is more likely that a user will trust and act on software recommendations and advice that is perceived as competent by the user (that is to say, it has all the functionality needed to achieve the desired result).Users are therefore likely to trust an appropriate advice and recommendation provided by a technology if they perceive it to be competent (Gulati, Sousa, and Lamas 2017, 2018). Hence, a system's user will be more likely to trust it if they have a high level of competence with it. This leads to the third hypothesis:

H3. As a system's competence increases, so does the trust of its user..

Reciprocity

Reciprocity is characterized by give-and-take relationships. In this sense, the computer as social actor paradigm can be understood, which teaches people to view computers as social actors even ascribed to them a personality and gender(Nass, Steuer, and Tauber 1994). They demonstrated that a computer is more likely to be used by people when they feel it has helped them, Gulati, Sousa, and Lamas (2017) mention that if people perceive that computers are helpful to them, they will adopt the system and use it continuously over time in order to develop a symbiotic relationship..' Interactions with a system are generally expected to be reciprocal.. In addition to responding to their needs, they expect the system to do so with knowledge. Consequently, the chances of a user trusting the system are much higher when it does so and the fourth hypothesis can be postulated as:

H4. Trust and reciprocity are positively related in human technology interactions.

RESEARCH METHODOLOGY

The research was conducted through primary and secondary data collection. Content analysis has performed by subject experts and questionnaires were circulated through various online platforms in PAN India. Online questionnaire had a screening question of the users of different applications to determine if there is a human-machine interaction. Further the study conducted EFA and regression analysis to confirm the factors and the relationship among them.

Table 1. Demographic profile of the respondent

Category	Variable	Frequency	Percentage
Gender	Male	137	68.5
	Female	63	31.5
Age	20 -30	47	23.5
	31-40	43	21.5
	41-50 50 and above	65 45	32.5 22.5
Application User User Experience (System interaction)	Enternmnt Application Desktop Application Gaming Application Other Application Beginner Competent Expert	79 38 46 37 48 69 83	39.5 19.5 23 18.5 24 34.5 41.5

Demographics Description

Over 200 surveys have been returned complete, out of the 273 distributed. Of these, 73 did not pass a data screening. The survey was having a number of 137 male respondents (68.5% of the total sample) and 63 (31.5%of the total sample) female respondents. Similarly, age group of 20-30 was 23.5% of the total sample, 31-40 was 21.5%, 41-50 was 32.5%, 50 and above was 22.5% of the total sample. There were 4 different kinds of application users were in this study. Such as Entertainment Application users (39.5% of the total sample), Desktop Application users (19.5% of the total sample), Gaming Application users (23% of the total sample) and other application users (18.5% of the total sample). The application users were divided into 3 segments. Such as beginner (24% of the total sample), competent (34.5% of the total sample) and expert (41.5% of the total sample)

HCT (Human Computer Trust) Measurement Instrument

This study identified the dimensions underpinning HCT and their indicators resulted in the next segment of research. There was a merger of similar concepts and elimination of those outside the scope of this study. The new instrument is built upon five constructs. Using the initial set of items assigned to these constructs, the overall HCT instrument was composed of these scales. In a field study with users, the construct validity and scale reliability of the instrument was assessed.

Table 2. The measurement of research variables

Variables	Items	Scale
Perceived Risk	7	Gulati, Sousa, and Lamas 2017, 2018)
Benevolence	5	
Competence	5	
Trust Reciprocity	2 3	

Table 3. Factor loadings

Factor	Item Loadings	Cronbach's α
Perceived Risk		.937
PR1	.717	
PR2	.746	
PR3	.878	
PR4	.951	
PR5	*.838*	
PR6	.868	
PR7	.870	
Benevolence		.914
B1	.741	
B2	.967	
B3	.938	
B4	.663	
B5	.920	
Competence		.935
C1 C2 C3 C4 C5 **Trust**	.917 .811 .796 .926 .860	
TR1 TR2 Reciprocity R1 R2 R3	.848 .749 .547 .681 .745	.813 .818

Table 4. KMO and Bartlett's Test

Kaiser-Meyer-Olkin Measure of Sampling Adequacy.		.718
Bartlett's Test of Sphericity	Approx. Chi-Square	5676.234
	df	231
	Sig.	.000

Table 5. Model summary

Model	R	R Square	Adjusted R Square	Std. Error of the Estimate
1	.683[a]	.467	.456	.6687
• **Predictors:** (Constant), Reciprocity, Competence, Benevolence, Perceived risk				

Table 6. ANOVA[a]

Model		Sum of Squares	df	Mean Square	F	Sig.
1	Regression	76.270	4	19.068	42.647	.000[b]
	Residual	87.185	195	.447		
	Total	163.455	199			
• **Dependent Variable:** Trust						
• **Predictors:** (Constant), Reciprocity, Competence, Benevolence, Perceived risk						

Table 7. Coefficients[a]

Model		Unstandardized Coefficients		Standardized Coefficients	t	Sig.
		B	Std. Error	Beta		
1	(Constant)	.382	.659		.580	.563
	Perceived risk	.117	.038	.171	3.082	.002
	Benevolence	.052	.044	.065	1.186	.237
	Competence	.759	.072	.589	10.485	.000
	Reciprocity	.038	.101	.020	.372	.710
• **Dependent Variable:** Trust						

From the above regression analysis, it shows the following results: R 2 =.467 which explains 46.7% of the total variation. F value was 42.647 which were significant at significance level. The above regression results indicate that Perceived risk, Benevolence, Competence, Reciprocity are positively related to trust of users..

From the regression analysis (Table No.-6) and the above model summery (Table No.-7,8), Regression Equation of trust of users can be denoted as:

$$\left(\begin{array}{l} Trust\left(TR\right) = .412 + .117\, Perceived\, risk + .052\, Benevolence \\ +.759\, Competence, +.038\, Reciprocity \end{array}\right)$$

FINDINGS

The goal of this study was to see whether characteristics of the HCT model hold valid in human- computer interactions. Principal Components Analysis was used to analyse the measuring instrument in this scenario (PCA). Based on the EFA and regression analysis, findings were presented and the results strongly support all the postulated hypotheses i.e. H1, H2, H3 and H4, which confirmed that perceived risk, benevolence, competence, and reciprocity are positively associated with trust that a user has in HCI.

Study Limitations

There are certain limitations to this study that can be used to guide future research. This research focused on intelligent systems that are intended to assist with decision-making and trust building which may utilise a variety of technologies, but only those that give advice or make judgments. It's also believed that the users are either experts or novices in the task domain. Future research can consider various level of expertise, and their impact on trust level. Trust is a valid concept that changes over time. It can be lost or recreated based on the circumstances and/or the end user's evaluation of their actions and intentions (Mcknight et al. 2011). So, future study may consider how HCTS performs in different situations with respect to trust.

CONCLUSION

Digital trust must be developed and sustained through cyber security. Cyber security should be viewed as a new interaction enabler instead of a barrier, and trust should be embraced as a positive driver of growth alongside the current perspective of "cyber security as an obstacle". Unpredictable cyber risks are becoming a more significant threat to safety and security. Hence, human machine co-operation is extremely critical to trust and rely human machine interaction to handle the new challenges such as cyber security. In the context of this study, human computer trust model was proposed and HCT instrument was designed for this study with a good level of internal consistency and construct validity. Thus, HCT-relevant constructs are selected for research. This instrument can be useful in exploring HCT structure and dynamics within the context of implementing computers in decision-making processes. It has been demonstrated to be both reliable and valid, thus making it useful in exploring HCT structure and dynamics within computer-aided decision-making processes. Based on the findings of this study, trust in Human Computer Interaction model proved useful. The scale performed well in terms of internal reliability and construct validity. However, trust on HCI was associated with the perceived risk, benevolence, competence, reciprocity. It is proposed that future research focuses on other predictors of trust, investigating the benefits of assuring users that technical controls are fallible.

REFERENCES

Araromi, O. A., Graule, M. A., Dorsey, K. L., Castellanos, S., Foster, J. R., Hsu, W. H., & Wood, R. J. (2020). Ultra-sensitive and resilient compliant strain gauges for soft machines. *Nature*, *587*(7833), 219–224. doi:10.103841586-020-2892-6 PMID:33177670

Benbasat, I., & Wang, W. (2005). Trust in and Adoption of Online Recommendation Agents. *Journal of the Association for Information Systems*, *6*(3), 72–101. doi:10.17705/1jais.00065

Bennett, C. C., & Šabanović, S. (2014). Deriving minimal features for human-like facial expressions in robotic faces. *International Journal of Social Robotics*, *6*(3), 367–381. doi:10.100712369-014-0237-z

Briggs, P., Jeske, D., & Coventry, L. (2017). The Design of Messages to Improve Cybersecurity Incident Reporting. In *International Conference on Human Aspects of Information Security, Privacy, and Trust*. Springer. 10.1007/978-3-319-58460-7_1

Brown, P., Christensen, K., & Schuster, D. (2016, September). An Investigation of Trust in a Cyber Security Tool. *Proceedings of the Human Factors and Ergonomics Society Annual Meeting*, *60*(1), 1454–1458. doi:10.1177/1541931213601333

Butavicius, M., Parsons, K., Lillie, M., McCormac, A., Pattinson, M., & Calic, D. (2020). When believing in technology leads to poor cyber security: Development of a trust in technical controls scale. *Computers & Security*, *98*, 102020. doi:10.1016/j.cose.2020.102020

Cianchetti, M., Laschi, C., Menciassi, A., & Dario, P. (2018). Biomedical applications of soft robotics. *Nature Reviews. Materials, 3*(6), 143–153. doi:10.103841578-018-0022-y

Clavel, C., Plessier, J., Martin, J. C., Ach, L., & Morel, B. (2009, September). Combining facial and postural expressions of emotions in a virtual character. In *International Workshop on Intelligent Virtual Agents* (pp. 287-300). Springer. 10.1007/978-3-642-04380-2_31

Columbus, L. (2019). State Of AI And Machine Learning In 2019. *Forbes Magazine, 8.*

Creed, C., & Beale, R. (2008). Psychological responses to simulated displays of mismatched emotional expressions. *Interacting with Computers, 20*(2), 225–239. doi:10.1016/j.intcom.2007.11.004

de Visser, E. J., Pak, R., & Shaw, T. H. (2018). From "automation" to "autonomy": The importance of trust repair in human-machine interaction. *Ergonomics, 61*(10), 1409–1427. Advance online publication. doi:10.1080/00140139.2018.1457725 PMID:29578376

Deloitte. (2015). *Industry 4.0—Challenges and solutions for the digital transformation and use of exponential technologies*. Study of Deloitte Consulting. https://www2.deloitte.com/content/dam/Deloitte/ch/Doc uments/manufacturing/ch-en-manufacturing-industry-4-0-24102014.pdf

Elizabeth, B. (2009). Clustering of Internet risk behaviors in a middle school student population. *The Journal of School Health, 79*(11), 547–553. doi:10.1111/j.1746-1561.2009.00447.x PMID:19840232

Gulati, S., Sousa, S., & Lamas, D. (2017). Modelling Trust: An Empirical Assessment. In *IFIP Conference on HumanComputer Interaction* (pp. 40–61). Springer.

Gulati, S., Sousa, S., & Lamas, D. (2018). Modelling Trust in Human-Like Technologies. In *Proceedings of the 9th Indian Conference on Human Computer Interaction* (pp. 1–10). ACM.

Hall, J., Tritton, T., Rowe, A., Pipe, A., Melhuish, C., & Leonards, U. (2014). Perception of own and robot engagement in human–robot interactions and their dependence on robotics knowledge. *Robotics and Autonomous Systems, 62*(3), 392–399. doi:10.1016/j.robot.2013.09.012

Hancock, P. A. (2017). Imposing limits on autonomous systems. *Ergonomics, 60*(2), 284–291. doi:10.1080/00140139.2016.1190035 PMID:27409152

Harbach, M., De Luca, A., Malkin, N., & Egelman, S. 2016. Keep on Lockin' in the Free World: A Multi-National Comparison of Smartphone Locking. In *Proc. of the 2016 CHI Conference on Human Factors in Computing Systems (CHI '16)*. ACM. 10.1145/2858036.2858273

Hu, X., Wu, J., Wu, M., & Hu, J. (2021). Recent developments of infrared photodetectors with low-dimensional inorganic nanostructures. *Nano Research*, 1–13. https://www.businesswire.com/news/home/20210126005806/en/Global-Cybersecurity-Market-2020-to-2025---Growth-Trends-and-Forecasts---ResearchAndMarkets.com

Jiang, D., Wu, K., Chen, D., Tu, G., Zhou, T., Garg, A., & Gao, L. (2020). A probability and integrated learning based classification algorithm for high-level human emotion recognition problems. *Measurement, 150*, 107049. doi:10.1016/j.measurement.2019.107049

Kagermann, H., Wahlster, W., & Helbig, J. (2013). Acatech–National Academy of Science and Engineering. *Recommendations for implementing the strategic initiative INDUSTRIE, 4.*

Kezer, M., Sevi, B., Cemalcilar, Z., & Baruh, L. (2016). Age differences in privacy attitudes, literacy and privacy management on Facebook. *Cyberpsychology (Brno), 10*(1), 1. doi:10.5817/CP2016-1-2

Kierkegaard, S. (2008). Cybering, online grooming and ageplay. *Computer Law & Security Review, 24*(1), 41–55. doi:10.1016/j.clsr.2007.11.004

Knight, H., & Simmons, R. (2016, May). Laban head-motions convey robot state: A call for robot body language. In *2016 IEEE international conference on robotics and automation (ICRA)* (pp. 2881-2888). IEEE.

Kshirsagar, S. (2002). A multilayer personality model. In *Proceedings of the 2nd international symposium on Smart graphics* (pp. 107-115). ACM. 10.1145/569005.569021

Kumar, J., & kumar, J. (2016). Affective modelling of users in HCI using EEG. *Procedia Computer Science, 84*, 107–114. doi:10.1016/j.procs.2016.04.073

Liu, X., Xie, L., Wang, Z., & Fu, D. (2012). Cognitive-affective regulation process for micro-expressions in active field state space. In *2012 IEEE 2nd International Conference on Cloud Computing and Intelligence Systems* (Vol. 2, pp. 831-835). IEEE. 10.1109/CCIS.2012.6664292

Lodgaard, E., & Dransfeld, S. (2020). Organizational aspects for successful integration of human-machine interaction in the industry 4.0 era. *Procedia CIRP*, *88*, 218–222. doi:10.1016/j.procir.2020.05.039

Lu, L., Jiang, C., Hu, G., Liu, J., & Yang, B. (2021). Flexible Noncontact Sensing for Human–Machine Interaction. *Advanced Materials*, *33*(16), 2100218. doi:10.1002/adma.202100218 PMID:33683745

Madhavan, P., & Wiegmann, D. A. (2004, September). A new look at the dynamics of human-automation trust: Is trust in humans comparable to trust in machines? *Proceedings of the Human Factors and Ergonomics Society Annual Meeting*, *48*(3), 581–585. doi:10.1177/154193120404800365

Madhavan, P., & Wiegmann, D. A. (2007). Similarities and differences between human–human and human–automation trust: An integrative review. *Theoretical Issues in Ergonomics Science*, *8*(4), 277–301. doi:10.1080/14639220500337708

Mayer, R. C., Davis, J. H., & Schoorman, F. D. (1995). An Integrative Model of Organizational Trust. *Academy of Management Review*, *20*(3), 709–734. doi:10.2307/258792

McAllister, D. J. (1995). Affect- and cognition-based trust as foundations for interpersonal cooperation in organizations. *Journal of The Academy of Management*, *38*(1), 24–59.

Mcknight, D. H., Carter, M., Thatcher, J. B., & Clay, P. F. (2011). Trust in a Specific Technology: An Investigation of Its Components and Measures. *ACM Transactions on Management Information Systems*, *2*(2), 12. doi:10.1145/1985347.1985353

Metaxiotis, K., Kagiannas, A., Askounis, D., & Psarras, J. (2003). Artificial intelligence in short term electric load forecasting: A state-of-the-art survey for the researcher. *Energy Conversion and Management*, *44*(9), 1525–1534. doi:10.1016/S0196-8904(02)00148-6

Moallem, A. (Ed.). (2018). *Human-Computer Interaction and cybersecurity handbook*. CRC Press. doi:10.1201/b22142

Mower, E., Mataric, M. J., & Narayanan, S. (2009). Human perception of audio-visual synthetic character emotion expression in the presence of ambiguous and conflicting information. *IEEE Transactions on Multimedia*, *11*(5), 843–855. doi:10.1109/TMM.2009.2021722

Nardo, M., Forino, D., & Murino, T. (2020). The evolution of man–machine interaction: The role of human in Industry 4.0 paradigm. *Production & Manufacturing Research*, *8*(1), 20–34. doi:10.1080/21693277.2020.1737592

Nass, C., Steuer, J., & Tauber, E. R. (1994, April). Computers are social actors. In *Proceedings of the SIGCHI conference on Human factors in computing systems* (pp. 72-78). ACM.

Nicole, M., Ewald, S., Weiss, A., Kühnlenz, B., Dirk, W., & Manfred, T. (2015). Can You Read My Face? *International Journal of Social Robotics*, *7*(1), 63–76. doi:10.100712369-014-0261-z

Norman, D. A. (1981b) A psychologist views human processing: Human errors and other phenomena suggest processing mechanisms. *Proceedings of the International Joint Conference on Artificial Intelligence*.

Norman, D. A. (1983a). Design principles for human-computer interfaces. *Proceedings of the CHI 1983 Conference on Human Factors in Computing Systems*.

Norman, D. A. (1984). Four stages of user activities. In *INTERACT '84, First Conference on Human-Computer Interaction*. North Holland.

Norman, D. A. (1981a). Categorization of action slips. *Psychological Review*, *88*(1), 1–15. doi:10.1037/0033-295X.88.1.1

Oliveira, D., Rocha, H., Yang, H., Ellis, D., Dommaraju, S., Muradoglu, M., Weir, D., Soliman, A., Lin, T., & Ebner, N. 2017. Dissecting spear phishing emails for older vs young adults: On the interplay of weapons of influence and life domains in predicting susceptibility to phishing. In *Proc. of the 2017 CHI Conference on Human Factors in Computing Systems*. ACM. 10.1145/3025453.3025831

Oltramari, A., Henshel, D. S., Cains, M., & Hoffman, B. (2015). Towards a Human Factors Ontology for Cyber Security. *Stids*, *2015*, 26–33.

Phillips, Repperger, Kinsler, Bharwani, & Kender. (2007). *A quantitative model of the human–machine interaction and multi-task performance: A strategy function and the unity model paradigm*. doi:10.1016/j.compbiomed.2006.12.007

Pierce, Fox, Merrill, & Wong. (2018). Differential Vulnerabilities and a Diversity of Tactics: What Toolkits Teach Us About Cybersecurity. *Proc. ACM Hum.-Comput. Interact., 2*. doi:10.1145/3274408

Pinquart, M., & Silbereisen, R. K. (2004). Human development in times of social change: Theoretical considerations and research needs. *International Journal of Behavioral Development*, *28*(4), 289–298. doi:10.1080/01650250344000406

Riley, V. (1989, October). A general model of mixed-initiative human-machine systems. In *Proceedings of the Human Factors Society Annual Meeting* (Vol. 33, No. 2, pp. 124-128). SAGE Publications. 10.1177/154193128903300227

Rowe, D. C., Lunt, B. M., & Ekstrom, J. J. (2011, October). The role of cyber-security in information technology education. In *Proceedings of the 2011 conference on Information technology education* (pp. 113-122). 10.1145/2047594.2047628

Sheng, S., Holbrook, M., Kumaraguru, P., Cranor, L. F., & Downs, J. (2010). Who Falls for Phish?: A Demographic Analysis of Phishing Susceptibility and Effectiveness of Interventions. In *Proc. of the SIGCHI Conference on Human Factors in Computing Systems (CHI '10)*. ACM. 10.1145/1753326.1753383

Söllner, M., Hoffmann, A., Hoffmann, H., Wacker, A., & Leimeister, J. M. (2012). Understanding the Formation of Trust in IT Artifacts. *Proceedings of the International Conference on Information Systems (ICIS 2012)*, 2, 1038–1055.

Sousa, S., Lamas, D., & Dias, P. (2014). A Model for HumanComputer Trust. *International Conference on Learning and Collaboration Technologies*, 128–137.

Springett, M., Law, E. L. C., & Coulson, M. (2015). Integrating the strengths of cognitive emotion models with traditional HCI analysis tools. *Universal Access in the Information Society*, *14*(2), 203–214. doi:10.100710209-013-0335-y

Stryker & Nancy. (1988). The society of mind. *Systems Research, 5*(3), 266–266.

Syam, N., & Sharma, A. (2018). Waiting for a sales renaissance in the fourth industrial revolution: Machine learning and artificial intelligence in sales research and practice. *Industrial Marketing Management*, *69*, 135–146. doi:10.1016/j.indmarman.2017.12.019

Wang, Z. (2009). *Artificial Emotion*. Machine Press.

West, R., Mayhorn, C., Hardee, J., & Mendel, J. (2009). The weakest link: A psychological perspective on why users make poor security decisions. In *Social and Human elements of information security: Emerging Trends and countermeasures* (pp. 43–60). IGI Global. doi:10.4018/978-1-60566-036-3.ch004

Whitty, M., Doodson, J., Creese, S., & Hodges, D. (2015). Individual differences in cyber security behaviors: An examination of who is sharing passwords. *Cyberpsychology, Behavior, and Social Networking*, *18*(1), 3–7. doi:10.1089/cyber.2014.0179 PMID:25517697

Xiao, G., Ma, Y., Liu, C., & Jiang, D. (2020). A machine emotion transfer model for intelligent human-machine interaction based on group division. *Mechanical Systems and Signal Processing*, *142*, 106736. doi:10.1016/j.ymssp.2020.106736

Yao, S., Ren, P., Song, R., Liu, Y., Huang, Q., Dong, J., O'Connor, B. T., & Zhu, Y. (2020). Nanomaterial-Enabled Flexible and Stretchable Sensing Systems: Processing, Integration, and Applications. *Advanced Materials*, *32*(15), 1902343. doi:10.1002/adma.201902343

Zanatto, D., Chattington, M., & Noyes, J. (2021, July). Sense of Agency in Human-Machine Interaction. In *International Conference on Applied Human Factors and Ergonomics* (pp. 353-360). Springer.

Zhang, J., Chen, M., Zhao, S., Hu, S., Shi, Z., & Cao, Y. (2016). ReliefF-based EEG sensor selection methods for emotion recognition. *Sensors (Basel)*, *16*(10), 1558. doi:10.339016101558 PMID:27669247

Zhang, Y., & Li, L. (2014). A Personality Model based on NEO PI-R for emotion simulation. *IEICE Transactions on Information and Systems*, *97*(8), 2000–2007. doi:10.1587/transinf.E97.D.2000

Chapter 7
Implementation of E-Health Services Using Machine Learning for Smart Cities

Mukesh Joshi
Graphic Era Hill University, India

Naveen Tewari
https://orcid.org/0000-0002-6104-5005
Graphic Era Hill University, India

Sandeep Budhani
Graphic Era University, India

ABSTRACT

Algorithms based on machine learning can be used for analyzing large amounts of varied data with exact results. One can use machine learning in offering healthcare services in the areas of interpretation, prognostics, and diagnostics in healthcare services for smart cities. In today's world we are using information systems very widely in offering various health services, and also digitization of the patient's information generates large amounts of information/data. While performing these activities, the data is stored in many systems and in various formats. Thus, after making a detailed analysis, the authors came to know that about 80% of health-related data is still in unstructured format. While combining the large amounts of mottled data, it needs high computing power and fine methods that enable fast execution and automated designing of machine-learning algorithms, which enables the analysis of the complex health data with exact results. Offering of e-health service through machine learning for smart cities makes the services more proactive.

DOI: 10.4018/978-1-7998-9710-1.ch007

INTRODUCTION

About eHealth Services

eHealth is also termed under many heads like eHealth, eHealth care or using of digital technologies for improving and facilitating health care services. In this we proposed the usage of various technologies like internet, smart phones etc. Basically, eHealth is a concept which is used to deliver the quality health care services to the patients by various health care providers.

The Need for eHealth

eHealth is the need of today's world as it provides enhanced documentation of health record including effective tracking of patients. It helps us to retrieve the patient's record during pandemic time which in turn facilitate the health providers to provide improved health services in timed approach. However, high cost of health care services is also promotes us to use eHealth services as per the Indian context. Also, the field of telemedicine has also been emerged with eHealth services, which (telemedicine) facilitates us in spreading health services up-to remote level.

eHealth Technologies

eHealth as termed by Agniel et al. (2018), allows us to make the use of digital technologies extensively. For Example, usage of Internet, promotes our population to commune with health care providers by e-mail facility also; it makes us to easily access personal health records, to make research on health-related issues, and at last it also allows us to make the exchange of medical information on person-to person basis. Various audio/video solutions like polycom can also be used for transferring the information related to patient's health on real time basis. Kiosks can also be used to offer basic health facilities on real time basis. Kiosks are the unattached gadgets (generally PCs), can be utilized in giving intuitive eHealth data to the client. They can likewise be use in gathering informative contents through clients one can promote DVD drives, USB drives, and various other devices can also be promoted to store the data cautiously. Different, ongoing mobile phones are planned with their own computing and network capacities likewise they are viable with applications (or applications) downloading, which permits clients to promptly get to different wellbeing related data. The vast majority of the advances drew in with e-wellbeing are open to all clients, including to those, who are enduring with different debilitations like various impairment or other health challenges.

BENEFITS AND BARRIERS

There are different advantages and obstructions for suppliers (who offered the administrations) and buyers (who use e-wellbeing). It likewise has some ideal effects that like its use by doctors for modernized medication requesting frameworks which inturn lessens the danger of hostile medicines impacts through choice sensitive networks. Also, robotized and mechanized frameworks can build the orders for recommended expectancies, for example, yearly health checkups, mammography, and prostate assessments. In more serious circumstances, e-wellbeing has likewise been utilized by different crisis clinical work force and furthermore turned into the specialists on call for conference during catastrophic hazards and during war like circumstances in military combat zone. In rustic and remote spaces of our country this innovation has likewise been utilized by various essential medical services suppliers for giving interviews to patients through direct linkage to metropolitan based subject matter experts. E-wellbeing can likewise be utilized in a spreading clinical training for essential and proceeding with instruction for oppressed regions. Worldwide shared drives have likewise been profited from propels in e-wellbeing by making the accessibility of medical care data promptly accessible to medical care experts and buyers.

Also, various benefits for purchasers are their which progresses the usage of advance components in e-prosperity, for example, the ability to orchestrate arrangements over the Internet for direct commitment the eHealth organizations to entrance steps. Distinctive crisis center and other extraordinary thought foundations have their pages which intrun gives the nuances of their prepared experts and suggestion various organizations for the patients. This thought moreover supportive for individuals facing some genuine ineptitudes. Escalated this one can abide locally by permitting provider patient correspondence through various modes like text, sound, or video conferencing to really look at privately arranged headway.

Obstructions to the usage of e-prosperity by clinical benefits suggested by, Ahmad et al. (2018) providers join a shortfall of financial impulses and a shortfall of reimbursement to help its use inside and across affiliations. Moreover, the hidden combination of new e-prosperity headways habitually moves back set up processes (due to the assumption to learn and adjust expected to execute new devices and contraptions) before the more-streamlined structure is set up. Various deterrents to e-prosperity development fuse costs, for instance, those related with hardware and programming purchases and support and upgrades, and the shortfall of rules concerning the plan and content of e-prosperity information, particularly private patient prosperity related information, which has real and monetary repercussions for providers as to obligation and carelessness security.

Maybe the greatest impediment to inevitable client usage of e-prosperity is the supposed automated segment. The high level hole can be portrayed as uniqueness in permission to mechanized developments, particularly the Internet. However people on one part of the hole move towards those headways and have the data expected to use them, people on the contrary side of the division normally don't. Notwithstanding the way that induction to electronic correspondences is reliably extending in both made and less-made countries, the additions are not consistent, and abnormalities in availability and capacity level continue on. Issues like cost, schooling level, social appropriateness, and consistence with rules for those with insufficiencies (like, Americans with Disabilities Act of 1990) are basic hindrances in the use of e-prosperity. Various issues join those related with the affirmation of assurance and protection and informed consent, similarly as those related to the straightforwardness of e-prosperity headways.

MACHINE LEARNING

Computer based intelligence is the examination of PC estimations that can chip away at normally through experience and by the usage of data. It is seen as a piece of mechanized thinking. Computer based intelligence is the area that enables computers to become skilled at without being unambiguously adjusted. ML is conceivably the most exciting innovation that one would have anytime run over. As it is understandable from the name, it gives the computers that make it more like individuals: The capability to learn. Artificial intelligence is viably being used nowadays, perhaps in a ton a more noteworthy number of spots than one would expect. Artificial intelligence estimations use obvious data as commitment to expect new yield regards.

Machine Learning Classification

Conventional AI is normally organized by how a computation finds out some technique to end up being more precise in its assumptions. There are four fundamental systems: controlled learning, independent learning, semi-managed learning and backing learning. The kind of computation given by Sarver et al. (2003) about data analysts to choose and employ depends upon what kind of information they need to predict.

Supervised Learning

In this kind of AI, data analysts supply estimations with stamped getting prepared data and portray the variables they necessitate the calculation to review for the relations. Both the facts and the yield of the estimation are shown.

Unsupervised Learning

This kind of AI allowed the estimations that instruct on unlabeled data. The estimation investigates instructive records attaining for any huge affiliation. The data that estimations train on similarly as the assumptions or suggestions they acquiesce are destined.

Semi-supervised Learning

This method for managing AI, it integrates a mix of the two obtainable before types. Data scientists may pact with a calculation by and huge stamped getting ready data, but the model is legitimate to research the data in segregation and cultivate its own admiration of the educational file.

Reinforcement Learning

Data specialists usually use support sorting out some way to assist a machine with completing multi-step processes for which there are evidently described values. Data analysts program an estimation to do a assignment and give it positive or negative signs as it works out how to wrap up a task. However, by and large, the judgment chooses its own, what steps to convey the way.

Today, AI is utilized in a wide scope of uses. Maybe one of the most notable instances of AI in real life is the suggestion motor that controls facebook's news channel.

Facebook utilizes AI to customize how every part's channel is conveyed. Assuming a part often stops to peruse a specific gathering's posts, the proposal motor will begin to show a greater amount of that gathering's action prior in the feed.

In the background, the motor is endeavouring to build up known examples in the part's web-based conduct. Should the part change examples and neglect to peruse posts from that gathering before long, the news source will change as needs be.

Methods of Selecting ML Tools

The method involved with picking the right AI model to tackle an issue can be tedious if not drew nearer deliberately.

- **Step 1:** Adjust the issue to potential information inputs that ought to be considered for the arrangement. This progression needs support from information researchers and specialists who have a profound comprehension of the issue.
- **Step 2:** Gather information, design it and mark the information if fundamental. This progression is regularly driven by information researchers, with assistance from information wranglers.
- **Step 3:** Select the algorithm(s) which is use to utilize and test to perceive how well they are performing. This progression is basically done by the information researchers.
- **Step 4:** Keep on fining tune yields until they arrive at an adequate degree of exactness. This progression is normally completed by information researchers with criticism from specialists who have a profound comprehension of the issue.

SMART CITIES

Splendid metropolitan networks set data and mechanized development to endeavour to make better decisions and work on the individual fulfilment. More careful, continuous data empowers associations to watch circumstance as they come to pass, perceive how solicitation plans are changing, and respond with speedier and less expensive courses of action.

Three layers coordinate to make a splendid city mumble. First is the development base, which consolidates a base measure of cutting edge cells and sensors, related by quick correspondence associations. The ensuing layer includes unequivocal applications. Making a translation of unrefined data into alerts, comprehension, and movement requires the right instruments, and this is where advancement providers and application engineers come in. The third layer is usage by metropolitan regions, associations, and individuals overall. Various applications succeed given that they are by and large embraced and sort out some way to change lead. They encourage people to use travel during off-hours, to shift directions, to use less energy and water and to do as such at different events of day, and to reduce strains on the clinical consideration structure through preventive dealing with oneself.

Machine Learning Tools for Smart Cities

The accessibility of savvy machines through Artificial Intelligence (AI) given by Sarver et al. (2003) and Machine Learning (ML) (Google's, 2017) has been pivotal in raising the idea of brilliant urban communities. You would now be able to fuse sharp figuring programs converged with human insight to make a digital actual space that incorporates traffic sensors, camcorders, climate sensors, brilliant meters, and so forth Information securing is completed consistently to outline noteworthy bits of knowledge focused on towards shrewd city arranging.

AI applications might possibly work on the exactness of treatment conventions and wellbeing results through algorithmic cycles.

For instance, profound learning, a sort of mind-boggling AI that copies how the human cerebrum capacities, is progressively being utilized in radiology and clinical imaging. Utilizing neural organizations that can gain from information with no management, profound learning applications can distinguish, perceive and dissect dangerous sores from pictures.

Quicker handling rates and cloud frameworks permit AI applications to distinguish inconsistencies in pictures past what the natural eye can see, supporting diagnosing and treating sickness.

Future progressions in AI in medical care will keep on changing the business. AI applications a work in progress incorporate an indicative apparatus for diabetic retinopathy and prescient investigation to decide bosom disease repeat dependent on clinical records and pictures.

Identifying Diseases and Diagnosis

This is the essential ML based feature in clinical benefits, it is the distinctive verification and investigation of disorders and hardships which are by and large measured to be firm to-dissect. This can consolidate everything from injurious developments which are not easy to find throughout the hidden phases, to other heritable contaminations. IBM Watson Genomics is an ideal representation of how planning intellectual enlisting with genome-based development sequencing can enhance in building a fast end. Berg, the bio-pharma giant is using AI to cultivate healing drugs in areas like oncology. Pvital's PReDicT (Predicting Response to Depression Treatment) by Lou, et al. (2013), plans to cultivate a monetarily achievable strategy for diagnosing and furnishing management in scheduled medical circumstances.

Drug Discovery and Manufacturing

This is the primary medical employments of AI, which states in starting phase drug revelation method. This moreover joins R&D headways, for instance, forefront sequencing and exactness drug which can assist in finding discretionary methods for healing of multifactorial diseases. During the current time, the AI strategies incorporate independent acknowledging which can perceive strategy in facts lacking in giving any estimates. Errand Hanover made by MS (i.e. Microsoft) is using ML-based advancements for quite a while together with making AI-enabled development for dangerous development handling and altering medicine combination for AML (Acute Myeloid Leukemia) by Meenakshi & Maragatham (2019).

Medical Imaging Diagnosis

Man-made intelligence and significant education are in cooperation accountable for the headway advancement called Computer Vision. It noticed affirmation in the InnerEye drive made by MS which manages picture characteristic devices for picture assessment. As AI ends up being more reachable and as they fill in their illustrative limit, desire to observe supplementary sources of information from vacillated clinical descriptions become a piece of the AI-enabled decisive cycle.

Personalized Medicine

Tweaked meds can't solely be more reasonable by mixing individual prosperity with insightful assessment and yet is prepared are for extra investigation and better disease assessment. As of now, specialists are confined to perusing a specific game plan of investigations or determine the jeopardy to the serene ward on his demonstrative past and available genetic information. The AI in drug is building exceptional strides, and IBM Watson Oncology is at the cutting edge of this advancement by using patient clinical narration to help with delivering diverse dealing decisions. After a short time, we will see more contraptions and biosensors with complex prosperity assessment limits punch the market, allowing additional data to open up for such best in class ML-based clinical consideration progresses.

Machine Learning-based Behavioural Modification

Social transformation is a critical part of anticipatory remedy, and in view of the fact that the time the augmentation of AI in clinical benefits, endless new organizations are jumping to the fields of dangerous development countering and unmistakable confirmation, patient handling, etc Somatix suggested by Huenerfauth et al. (2009) is a B2B and B2C-based data examination association which has conveyed a ML-based application to see movements which we build in our normal schedules, allowing us to appreciate our unaware lead and carry out significant enhancements.

Smart Health Records

Staying aware of ground-breaking prosperity records is an intensive cycle and remembering that development has had its impact in working with the data section process, really even as of now, a bigger piece of the cycles put away a lot of work to wrap up. The central occupation of AI in scientific thought is to alleviate cycles to keep instance, attempt, and capital. The report request measures of using vector-based machines and ML-enabled OCR techniques are steadily meeting, for instance, Cloud Vision API by Google and AI enabled MATLAB's for writing. MIT is today at the forefront of cultivating the best-in-class time of sharp, insightful prosperity data, which in turn will combine ML-based expenses as of the start to assist with examination, experimental treatment thoughts, etc.

Clinical Trial and Research

AI has few likely apps associated with the ground of experimental preliminaries and examination. As anyone in the pharma business would notify about, experimental preliminaries cost a great deal of time and cash and can require a very long time to finish by and large. Applying ML-based prescient investigation to distinguish possible clinical preliminary competitors can assist scientists with drawing a pool from a wide assortment of important informative elements, for example, past specialist visits, online media, and so forth AI has likewise found use in guaranteeing constant checking and information access of the preliminary members, viewing the best example size as tried, and utilizing the force of electronic records to decrease information based blunders.

Crowd Sourced Data Collection

Openly supporting is amazingly well known in the clinical field currently, promoting researchers and experts to get to a gigantic proportion of data moved by individuals reliant upon their own permission. This lives prosperity information has unprecedented results in the manner drug will be seen behind the line. Apple's Research Kit grants customers to get to instinctive applications which pertain ML-based facial affirmation to endeavor to treat Asperger's and Parkinson's given by Singh et al. (2019) sickness. IBM actually united along with Medtronic to unwind, store up, and make existing diabetes and insulin data consistently subject to the freely upheld information. With the movements being made in IoT, the clinical consideration industry is at this point finding new habits by which to employ this data and tackle outrageous to-dissect cases and assist in the in general enhancement of end and medication.

Better Radiotherapy

One of the most pursued uses of AI in medical care is in the area of Radiology. Clinical picture investigation has numerous discrete factors which can emerge at a specific snapshot of time. There are numerous injuries, malignant growth foci, and so on which can't be essentially displayed utilizing complex conditions. Since ML-based calculations gain from the large number of various examples accessible close by, it becomes more straightforward to analyze and find the factors. One of the most well-known employments of AI in clinical picture examination is the grouping of articles, for example, sores into classes like typical or strange, injury or non-sore, and so forth Google's Deep Mind Health is effectively helping specialists in UCLH create calculations which can identify the distinction among solid and carcinogenic tissue and further develop radiation therapy for the equivalent.

Outbreak Prediction

Re-enacted knowledge-based advances and AI are today similarly being place to use in checking and expecting scourges all through the planet. Today, scientists approach a great deal of data accumulated from satellites, progressing electronic media revives, site information, etc Fake neural associations help to assemble this information and expect everything from gastrointestinal affliction eruptions to outrageous steady powerful contaminations. Anticipating these eruptions is especially valuable in immature countries as they need basic clinical establishment and informative systems. A fundamental delineation of this is the ProMED-mail, an Internet-based uncovering stage which screens propelling diseases and emerging ones and gives scene reports constantly.

HOW IMPACT OF MACHINE LEARNING ENCHANCES HEALTH CARE SYSTEM OF REMOTE AREAS

Recordkeeping

Simulated intelligence in prosperity informatics can streamline recordkeeping, including electronic prosperity records (EHRs). Using AI to additionally foster

EHR the board can deal with getting thought, decrease clinical consideration and administrative costs, and improve exercises.

One model joins standard language taking care of, which enables specialists to catch and record clinical notes, abstaining from manual cycles.

Man-made intelligence computations can in like manner make EHR the board structures more direct to use for specialists by giving clinical decision help, modernizing picture assessment and planning telehealth propels.

Data Integrity

Openings in clinical benefits information can achieve AI computations making incorrect gauges, which can unfavorably influence choice creation in clinical settings.

Since clinical benefits data is at first expected for EHRs by Kaur, et al. (2019), the data ought to be prepared before AI computations can effectively use it.

Prosperity informatics specialists are at risk for staying aware of data decency. Activities that prosperity informatics specialists perform fuse social event, exploring, gathering and filtering the data.

The mix of AI, wellbeing informatics and prescient examination offers freedoms to further develop medical services processes, change clinical choice help apparatuses and assist with working on persistent results. The guarantee of machine learning changing medical care lies in its capacity to use wellbeing informatics to anticipate wellbeing results through prescient examination, prompting more exact finding and therapy and further developing doctor bits of knowledge for customized and partner therapies.

AI can likewise offer extra benefit from prescient investigation by deciphering information for chiefs to reveal process holes and further develop in general medical care business activities.

Clinical Management in the Intensive Care Unit

The ICU regularly focuses on the most seriously sick patients in the emergency clinic, taking into consideration the arrangement of extremist lifesaving therapy. When conceded, the state of patients is intensely checked to work with the rebuilding of their physiology to ordinariness. Some clinically-important boundaries, like pulse (HR) and circulatory strain (BP), are checked as habitually as consistently. Nonetheless, because of the limit of human asset distribution, quite a bit of this data isn't utilized by clinical staff. Moreover, basic consideration medication is one of the quickest developing clinical fields as far as tolerant numbers, and the intricacy of basic ailment requires understanding of various collaborating boundaries. Subsequently, the ICU has turned into a capable region for the improvement of novel informatics for the continuous handling of floods of physiological information, fully intent on giving clinical choice help to patient administration.

Forecast of patient results is a promptly clear utilization of well being informatics and has a long the past, which has framed a reason for later developments. As per the reports in 1981 by Alsaadi et.al (2019) distributed the intense physiology score, as a part of the intense physiology, age, and ongoing wellbeing assessment (APACHE) framework. With the new expansion in the amount of accessible information, and the advances in AI procedures that permit these information to be handled, there is an extraordinary chance for making precise patient-explicit forecast models, which in this way permit an appraisal of individual patients, appropriate for influencing the executives choices. Late work by Johnson et al. applied a half and half framework consolidating a hereditary calculation with molecule swarm advancement to make an ICU seriousness score that utilized a tightfistedly chosen include set. In any case, these models utilize total synopses of patient composition over the primary hour or day in the ICU (like the greatest HR). Compelling the examination to considering just early-confirmation information in this manner is, as the creators depict, because of the conventional setting of ICU seriousness scores for use at affirmation.

A new audit of all transience expectation concentrates around ninety four examinations that contained 118 prescient instruments, from these 118 apparatuses, 34 were interesting given by Jin et al. (2020). This showed that there is an absence of wide take-up of more complex mortality forecast models. Be that as it may, there exist promising methodologies attempted lately. Author Jin et al. (2020) present a model which intended to anticipate youngster dismalness utilizing persistent fundamental sign checking information from the initial 3 hours of the children life. The model initially sifted the persistent imperative sign information into long-and transient fluctuation signals. For every class of patient, a likelihood dispersion for each fundamental sign was assessed, and the probabilities of dismalness from every one of these appropriations were amassed into a solitary score utilizing regularized calculated relapse. The subsequent score, named the PhysiScore, had incredible separation for different neonatal morbidities when contrasted with all the more generally used seriousness scores. To discover whether the elements of the pulse could additionally illuminate a prescient model, similar creators then, at that point, apply an expansion of Latent Dirichlet Allocation (LDA) shown by Pandl, et al. (2020) for time-series information. The model makes subjects for each instance series which are pooled across patients. These themes allude to ideas, for example, sound or lung, and control the circulation of autoregressive models which depict the information. The outcome was approved with 30 newborn children and showed that the time allotment a youngsters adapted was considered to spend in the solid theme was related with seriousness of that children condition.

Near models have been used to screen grown-up ICU patients. Lehman by Jin et al.(2020), apply a trading vector autoregressive (SVAR) model to the components of HR and BP time-series for patients in the ICU. This model equivalently summarizes the time-series using autoregressive modes and usages the degree of time in each mode as a marker of prosperity or reality. The makers showed that BP components alone can beat existing ICU scores. Also, they show that the development of BP components to the by and large used APACHE IV score moreover chipped away at its AUROC from 0.82 to 0.85.

One especially fruitful wellbeing informatics framework is the HeRO screen, which has been approved by means of a randomized controlled preliminary. This framework gives early forecast of neonatal sepsis by consolidating different proportions of HR inconstancy together with the standard deviation of R-R Intervals, test lopsidedness, and test entropy. These measurements caught a typical example in HR saw in youngsters with sepsis: fast decelerations joined with low inconstancy. A correlation of the rule and study populace showed that for each 48 low birth weight newborn children, the HeRO screen saved one life.

Regardless the different time-series signals acquired in the ICU, there furthermore exists a tremendous gathering of supportive information held inside the clinical notes created during a patients reside in the ICU. These remarks ordinarily enclose information with respect to past clinical history, suggestive information, existing patient status, management plans and discussion of end-of-life care decisions. These notes address an underused, yet outstandingly significant, wellspring of information, and are by and by much of the time recorded electronically, making them more sensible for motorized arrangement. Various researches shows the utility of reconciliation of information enclosed in these notes into ICU-based systems The system removes a lot of 50 focuses from the case notes for patients, using a LDA model got from the super 500 words from the notes. The limit of the model to expect long stretch mortality extended AUROC from 0.711 to 0.788 by Jin et al. (2020) when such data were associated with the ICUs earnestness score. Further investigation by comparable makers showed that relative show could be refined by an estimation executed at the bedside. The AUROC of the last choice model for anticipating mortality (with no information from the ICU remarks) was 0.76, while the extension of took in subjects from the remarks chipped away at this to 0.84.

Clinical Management Outside the ICU: Wearable Sensors

Most of patients in clinics are ordinarily mobile, and in this scenery patient condition is frequently checked simply by manual perception by the medical staff. It is set up that the utilization of wearable sensors for dealing with the consideration of such populaces can work on tolerant results. A new report also observed that remote checking is related with further developed endurance to emergency clinic release, continuing in-emergency clinic heart failure. Nonetheless, regardless of such examinations, proof proposes that the present utilization of frameworks dependent on wearable sensors in clinic is a long way from ultimate. This has been ascribed to frameworks being utilized improperly, trying to oversee patients for whom such checking gives no advantage. This last option impact has been ascribed to the insufficiency of current rules for separating patients to decide if extra checking will be of advantage.

The norm of clinical administration for patients on intense clinic wards is ordinarily to have crucial signs estimated physically every 4-6 hours. Proof that wearable sensors are is being utilized in clinical exercise to build the recurrence of estimation of the crucial signs, determined to recognize the falling apart understanding. Robotized examination of the patterns of indispensable signs gained from wearable sensors has been utilized under the watchful eye of wandering post-usable patients fully intent on deciding whether the time-series information from a patient seem like those of ordinary or in danger patients. This last option concentrate on utilized Bayesian Gaussian cycles (GPs) to perform bunching of whole univariate time-series; a comparable methodology was portrayed by various researches in which numerous time-series of essential signs were demonstrated at the same time utilizing perform various tasks Gaussian cycles, for the observing of mobile patients. Important bodily function information from wearable sensors were intertwined with the physical perceptions through the clinical staff for intensely sick, walking patients in utilizing GPs and one-class support vector machines. Such techniques have been reached out by the utilization of wellbeing informatics dependent on outrageous worth hypothesis, a part of measurements used to show external information, and which have been exhibited in a patient checking study in a US clinic. The proof base for the utilization of wellbeing informatics frameworks dependent on AI was addressed in, where techniques for assessing the adequacy of a framework utilized for distinguishing the falling apart quiet in an enormous Emergency Department surroundings were portrayed.

Gaussian cycles are non-parametric Bayesian strategies for performing time-series investigation, which are turning out to be progressively embraced by probabilistic wellbeing informatics frameworks dependent on AI. They offer a principled way of adapting to loud or deficient information, as commonly experienced in routine clinical administration. We note in fleeting that further uses of this methodology outside the domain of sensor-based examination incorporate the appraisal of hotness openness disorder, jungle fever diseases, and the administration of patient charging records.

Other late assessments, for instance, concentrating on using existing sensors (like ECG terminals) for the evaluation of respiratory rate (RR) from accessories of in like manner strolling patients. This last space of usage has moreover been tended to by the use of heartbeat oximeters, again tattered by strolling patients. The penetration of such structures into real clinical organization is limited, and this is a point would in general make in which computations were made to chip away at both the strength of RR appraisal from beat oximeters, and to widen their battery life so much that they could be used in the office without normal substitution of batteries. The usage of heartbeat oximeters to measure other fundamental signs, for instance, changes in BP from the thump travel time (the stretch between a QRS complex from the ECG and the relating thump of a finger-worn oximeter), has been all over portrayed in the composition, but another report cases to have dealt with the obvious shortfall of life that has as of late thwarted the use of such techniques in clinical practice.

Clinical Management Outside the ICU

Past the customary checking of indispensable symbols, there is a prosperous asset of information gathered from the standard consideration of patients in numerous clinics. This huge information issue commonly implies that main explicit subsets of information are utilized for clinical administration, though wellbeing informatics frameworks dependent on AI offer the latent for illuminating clinical consideration by intertwining data from the EHR. This segment considers those EHR-based examinations that relate to applications slight the ICU (where the last option were depicted in segment 1).

The EHR is a rich wellspring of information for some roads of examination outside the ICU, giving information to both conventional the study of disease transmission studies and preparing information for the advancement of prescient calculations. Be that as it may, patient records are regularly to some extent absent, inaccurate, deliberately deceptive, and disconnected. Especially pertinent for administered AI examination, patient marks might be puzzled by the way that care is being given. Some authors also states, few of the related tests of separating EHR information that reflects genuine patient physiology.

The coordinating science and the bedside (i2b2) drive, the UK bio-bank, and the Kaiser Permanent Research Program on Genes, Environment, and Health (RPGEH) shown in Pandl et al. (2020) additionally all comparatively intend to join a lot of clinical and genomic information; Jensen et al. give extra models. The EHR-based phenotyping endeavors depicted above have to a great extent been engaged after recognizing patients who can be portrayed by flow sickness phrasing. In any case, with developing admittance to EHR information, it is currently becoming conceivable to distinguish novel patient subgroups. While numerous complex infections have customarily been considered as solid substances, almost certainly, better grained phenotypic sub-types exist. Various researches. have outlined the issue as far as tensors, utilizing nonnegative tensor factorisation to all the more consequently distinguish patient aggregates.

Various researches, for instance, checked out the circumstance and occurrence of ICD9 demonstrative codes in the EHRs of more than 13,000 chemical imbalance range problem patients. Utilizing progressive bunching, they found three particular patient directions, comparing to a gathering of patients portrayed by seizures, a gathering by multisystem messes, and a gathering by mental issues. They are based upon such methodologies by fostering a probabilistic model to group patient directions of different clinical markers identified with scleroderma. Also,. utilized Gaussian cycle relapse over serum uric corrosive estimations, joined with a two-layered neural organization, to make an inadequate list of capabilities of time-series directions. When proposed into two aspects, the learned capabilities showed great separation between known illnesses and furthermore proposed conceivable extra hidden infection subtypes. The objective of such exploration is principally theory producing, fully intent on deciding whether patients inside phenotypic subgroups are joined by a typical genomic pathway, or then again in the event that they may react to specific kinds of treatment. Such investigation isn't expected to be quickly translatable into the facility, however educates our clinical arrangement regarding these issues and focuses towards promising future spaces of exploration.

While the investigations introduced above empower the investigation of physiological components that might be engaged with human infection, there are numerous instances of as of late created frameworks that are intended to be consolidated into an insightful EHR framework. Huang et al. utilized highlights got from the EHR to make an evaluating device for melancholy with sensitivities and specificities comparable to those accomplished by essential consideration doctors. Various authors also. made a danger separation model for becoming tainted with clostridium difficile by consolidating the time sensitive connection between clinical factors in the EHR; the subsequent framework had the option to foresee patient danger substantially more effectively than when considering highlights in total. Authors also utilized transient Bayesian organizations to anticipate fuel occasions for ongoing obstructive pneumonic infection patients dependent on EHR information; their framework had the option to foresee intensification inside 24 hours with an AUROC of 0.90 in an approval study.

For a bigger scope, an EHR information model can be mined and joined with organized genomic information to create early prognostic danger pointers for normal ailments. The objective of such examinations is for the subsequent instruments to be embraced into experimental practice as new transmission procedures. One illustration of the fruitful interpretation of a genomics-based informatics device has been the improvement of quality articulation based calculations for bosom disease anticipation. Researchers also utilized administered and unaided AI procedures on quality articulation information from bosom disease cells to give enormous improvement over active experimental prognostic markers. These strategies have now been formed into a business device (MammaPrint, Agendia, Netherlands) that is utilized for bosom malignant growth guess in the facility. The representation of such complex information was tended to by Ding et al., who fostered a perception stage utilizing unaided learning strategies. Various researches also give an audit of such informatics strategies and information types that have been utilized for foreseeing disease guesses, fully intent on working on the clinical administration and handling of such patients.

Clinical microbial science is one more ground that is set to be changed all the way through genome sequencing and steady informatics devices installed inside the EHR. While genome sequencing, joined with AI calculations, has been utilized for HIV drug obstruction forecast for a long time, comparative approaches are presently being applied for bacterial diseases. For example, various models have outlined the plausibility of utilizing entire genome groupings from Staphylococcus, pneumonia, and Escherichia coli to decide bacterial medication obstruction. It has also been outlined the initial moves towards formation of an AI based anti-microbial obstruction forecast device determined to give suitable medicines.

CONCLUSION

We close by stressing that the field of wellbeing informatics frameworks dependent on AI, drawing on unique data types from the ICU, the more extensive clinic, and from (conceivably extremely intricate) EHR information, is in its outset. While most of medical clinics in the created world have carried out EHR frameworks or some likeness thereof, the coordinated utilization of the enormous amounts of information that emerge from such frameworks isn't utilized at range. This article has overviewed ongoing improvements in this ground, in which the clinical administration of patients has been impacted by wellbeing informatics frameworks dependent on AI, or in which frameworks for performing such administration are being developed. We see in shutting that the boundaries to-section for such action are generous and are impacted by the accessibility of multidisciplinary groups drawing on both clinical aptitude and professionals from the data sciences. The ramifications for research in this field are that clinical administration of patients will turn out to be more reliant upon AI frameworks with the steadily expanding amounts of information that are being gathered in routine consideration, from the ICU through to release, in most medical care frameworks.

REFERENCES

Agniel, D., Kohane, I. S., & Weber, G. M. (2018). Biases in electronic health record data due to processes within the healthcare system: Retrospective observational study. *BMJ (Clinical Research Ed.)*, 361. doi:10.1136/bmj.k1479 PMID:29712648

Ahmad, M. A., Eckert, C., & Teredesai, A. (2018, August). Interpretable machine learning in healthcare. In *Proceedings of the 2018 ACM international conference on bioinformatics, computational biology, and health informatics* (pp. 559-560). ACM.

Alsaadi, M., Lisitsa, A., Khalaf, M., & Qasaimeh, M. (2019, August). Investigating the capability of agile processes to support medical devices regulations: The case of xp, scrum, and FDD with EU MDR regulations. In *International Conference on Intelligent Computing* (pp. 581-592). Springer.

Huenerfauth, M., van Rossum, G., & Muller, R. P. (2009). *Introduction to python*. Harvard University.

Jin, C., Chen, W., Cao, Y., Xu, Z., Tan, Z., Zhang, X., ... Feng, J. (2020). Development and evaluation of an artificial intelligence system for COVID-19 diagnosis. *Nature Communications, 11*(1), 1–14.

Lou, Y., Caruana, R., Gehrke, J., & Hooker, G. (2013, August). Accurate intelligible models with pairwise interactions. In *Proceedings of the 19th ACM SIGKDD international conference on Knowledge discovery and data mining* (pp. 623-631). 10.1145/2487575.2487579

Meenakshi, K., & Maragatham, G. (2019, March). A review on security attacks and protective strategies of machine learning. In *International Conference on Emerging Current Trends in Computing and Expert Technology* (pp. 1076-1087). Springer.

Nott, G. (2017, June 23). *Google's research chief questions value of "Explainable AI."* Computerworld. https://www2.computerworld.com.au/article/621059/google-research-chief-questions-value-explainable-ai/

Pandl, K. D., Thiebes, S., Schmidt-Kraepelin, M., & Sunyaev, A. (2020). On the convergence of artificial intelligence and distributed ledger technology: A scoping review and future research agenda. *IEEE Access : Practical Innovations, Open Solutions, 8*, 57075–57095.

Singh, R. K., Dalal, S., Chauhan, V. K., & Kumar, D. (2019, March). Optimization of FAR in intrusion detection system by using random forest algorithm. *Proceedings of 2nd International Conference on Advanced Computing and Software Engineering (ICACSE).*

Tamayo-Sarver, J. H., Hinze, S. W., Cydulka, R. K., & Baker, D. W. (2003). Racial and ethnic disparities in emergency department analgesic prescription. *American Journal of Public Health*, *93*(12), 2067–2073. doi:10.2105/AJPH.93.12.2067 PMID:14652336

Chapter 8
LoRa:
A Novel Technology for Reduced Power Consumption of Street Lights Using IoT in Smart Cities

Ghanshyam Parmar
Natubhai V. Patel College of Pure and Applied Sciences, CVM University, India

Monika Patel
Natubhai V. Patel College of Pure and Applied Sciences, CVM University, India

ABSTRACT

The main concern of this chapter is to getting information regarding the controlling of streetlights using the IoT technology and the help of LoRa technology. LoRa is a novel wireless protocol designed for long range with low power consumption for communication. This chapter provides, firstly, why energy consumption should be reduced. The chapter explains IoT and how the IoT industries are coming up with new technologies. The chapter contains different technologies with basic information and pros can cons, including Wi-Fi HaLow, Bluetooth and (BLE), LPWANs, Cellular (3G/4G/5G), Zigbee and other Mesh protocols, and RFID. Further, the chapter describes the features of LoRa and LoRaWAN with its architecture. Why the LoRa technology is the best for controlling streetlights with its working is explained. The main focus of this chapter is the devices used in smart streetlights and how to control the streetlight. The chapter describes the working procedure of streetlight management through IoT and LoRa technique.

DOI: 10.4018/978-1-7998-9710-1.ch008

INTRODUCTION

Need of Reducing Power Consumption of Street Lights

Artificial light helps us to see in the dark. Artificial lights are required at every place such as at our homes, place where we are working (FLASHNET, 2021), and also the streets we used for wander. The invention of this artificial light becomes the core part of our lives, also in current scenario street lighting is the basic requirements of urban as well as rural parts of the world for safer life for drivers and pedestrians. Below are the two reasons which creates the need for reducing power consumption of street lights

- **Costs For Municipalities:** For safety of citizens of rural and urban there is an authority from local and municipal level, which provides street lighting facilities. This facility incurred major cost for the government at local level. The lamps used in street lighting are generally outdated which are also inefficient which consumes high energy. Another issue is the obsoleted street lighting systems requires more maintenance. To find the exact location of fault in lamps which is burnt or not in working condition is harder to find. This obsoleted system consumes approximately 50% of a typical city's entire energy bill.
- **Environmental Concerns:** The higher consumption of energy is not only concern with urban or local level, but it is concern with world level. When taking the consideration as globally hundreds of millions of streetlights are used which is almost 19% of electricity usage and which is already emits the level of CO_2. The UN had stated that in 2050 more than 68% with respect to the world population will live in urban from rural, which creates issues regarding the use of limited resources. It is difficult to sustain municipalities with this old and outdated streetlights system to secure the safety of current as well as future residents.

OVERVIEW AND WORKING OF IoT

A system which consists of interconnected computing devices which has sensors and ability to process the data as well as exchange that data to other such type of devices that attached with humans, animals or any other objects over internet or other communication networks are called Internet of Things (IoT) (Witeck, What is internet of things (IoT)?, 2021). These computing devices has their own unique identifiers (UIDs) by which they can communicate with each other.

The applications of IoT are a device implant in a person for heart monitor, a biochip transponder used in an animal within a farm, for measuring and giving alert to the driver using built-in sensors in an automobile or any other man-made device or any other natural object is used by assigning Internet Protocol (IP) address for communicating and transferring data over a dedicated network.

Manufacturing organizations are currently using IoT (Industry 4.0) for creating a network of the physical objects for monitoring in real-time and operate them for higher efficiency and make decisions for increase productivity, also satisfy customer's need, provide customer service effectively and increase the value of the business.

Working of IoT Technology

An embedded system which a web-enabled IoT smart device consists of sensors, communication hardware and a processor. It will send and receive data as per the dedicated purpose from the network connected. The collected data from the environment will be share through IoT gateway to the cloud-based system for processing or will be processed locally and analyzed that for further actions.

This is the normal working procedure but sometimes these smart devices exchange information with such other devices and also act as per the information they get. In IoT systems the work done is almost automatic so no human intervention required, but human can interact when they required for set-up the devices and giving instructions of getting data or information from such devices.

TECHNOLOGIES USED FOR COMMUNICATION WITH IoT

Wi-Fi HaLow (Low Power, Long Range Wi-Fi® for IoT)

Wi-Fi HaLow™, is from IEEE 802.11ah which expanding spectrum of Wi-Fi by reducing one gigahertz (GHz) so, it will increase the range and consumes low power then Wi-Fi. The requirements for IoT systems are basically (Figueroa, 2021) longer range and lower consumption of battery power which is fulfilled by Wi-Fi HaLow used in creating smart industries, smart buildings, smart agriculture, and smart city.

The necessary thing in sensor networks applications and in wearables is the connectivity with low power consumption which is enabled by Wi-Fi HaLow. The range also matters because the longer the range less devices are used which will reduce the cost for number of devices to create a network. Another important ability is to penetrate the walls and other barriers for smooth connection with other such smart devices in the environment.

Figure 1. Features and benefits of Wi-Fi HaLow

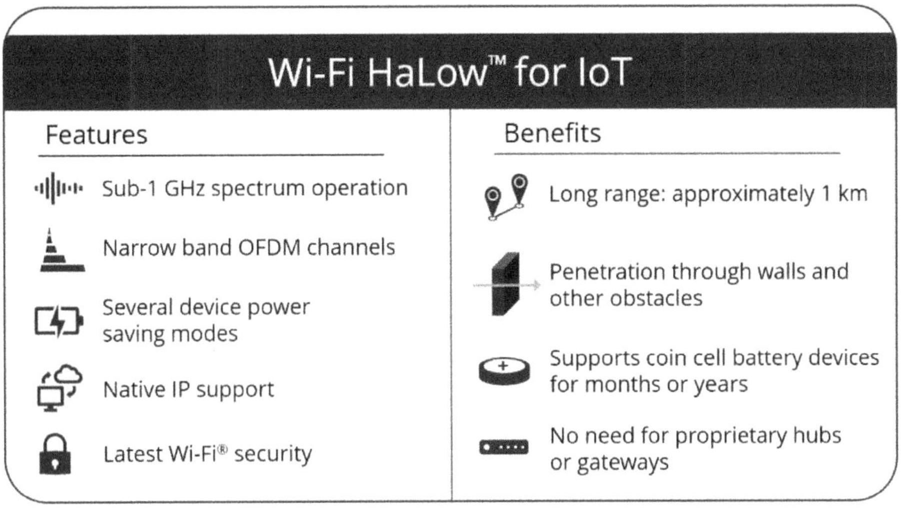

The Figure 1 (Figueroa, 2021) shows the features and benefits of Wi-Fi HaLow like approximately 1 km long range connectivity using sub-1 GHz spectrum, narrow band OFDM channel for penetrating walls and other obstacles, a power saving mode for devices to consume less power in devices. Supports coin cell battery devices which runs for months or years. Easy connectivity to cloud using native IP, and exclusive hubs and gateway is not required for connectivity with latest Wi-Fi security.

Bluetooth Low Energy (BLE)

A wireless communication technology which is used for short distance and consumes less power is smart devices for communication is Bluetooth Low Energy (BLE). Nowadays people are using smart watch, smart phone, fitness tracker and wireless headphones are using BLE ((KCT), 2021) for communication and sharing information between your devices.

The Bluetooth Special Interest Group (SIG) has created a new Bluetooth standard called BLE with the low energy consumption feature. SIG allowed manufactures to use BLE in the existing devices at low power communication interface. Also, they had created a new device named beacon with a small coin sized cell battery which consumes low power and lasting for months or even years.

BLE has a wide scope of potential outcomes and is executed in a wide arrangement of fields like wellbeing, wellness, security, home computerization, home diversion, shrewd industry and IoT (Internet of Things). It's additionally not far off in the cell phones and PCs that we utilize each day.

The BLE was firstly used in smartphones by Apple's iPhone 4s in the year of 2011. Today, most of all Android and iOS smart devices are using BLE technology for interact and communicate with other such devices. Today, the majority of Android and iOS devices on the market incorporate BLE for communication and interaction with other devices.

Working of BLE Devices

To see how BLE functions, we need to characterize the terms like central as client and peripheral as server where BLE devices are working as a central or peripheral like client and server.

Central (Client)

A computer of smartphone works as a central which is the device that receiving and scanning requests or commands and then accepts responses.

Peripheral (Server)

The peripheral may be any sensor like temperature, hart rate monitor, headphones, fitness trackers which is mainly a BLE device (Apps, 2021) that transmit data packets in set of intervals and also receives commands and requests, and returns responses.

Figure 2. Working of BLE device

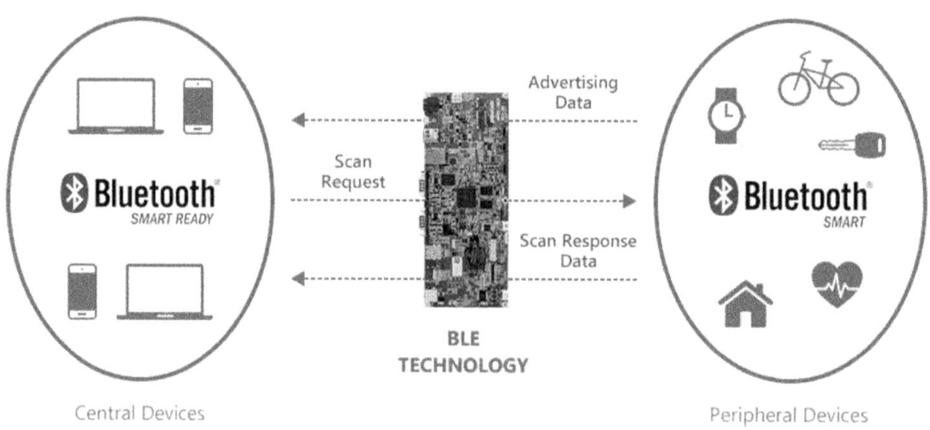

Figure 2 shows (Apps, 2021) the working of BLE device where once the central device is discovered by BLE and establishing the connection the services offered by the BLE can be used.

Low-Power Wide Area Networks (LPWANs)

LPWA is the general term used for such technologies which consumes low power for communicating wirelessly is a network excluding cellular, satellite or Wi-Fi technologies. The range of LPWANs (Paessler, 2021) greater than the other low power technologies like NFC or Bluetooth.

Respect to sending data over long distances with low power consumption, the restriction will be the amount of data must be small at a time in LPWAN. However, the cellular networks with LTE – 5G technology offers amount of data as gigabit per second where LPWAN offers much less as only in kilobits. But on other hand the communication range is far more as up to 500 miles or more.

LPWA networks are not suggested for accessing audio, video, voice or text messaging for consumer and for commercial use, because it is completely developed for the communication in IoT and machine-to-machine (M2M).

Managing devices for business or home like lightbulbs, refrigerators of Nest thermometers for manage room temperature will easily managed by Wi-Fi connection. But using Wi-Fi connectivity some devices are not working properly.

In the country of America, farms are connected with an irrigation channel that stretches over miles. There are thousands of water pumps and gates are attached with this chain of watercourses and channels. This irrigation channel is scattered in ones (private) land which has crops and its length is in miles. For checking that pumps and gates are running properly in a day and there are no issues in irrigation is just unrealistic. If any pump is not working at the end of irrigation channel and get the location and repair that will takes hours and another one fails just behind your current location may be known on next day.

Much of the irrigation channel runs through private land, between crop fields, and across several miles. As such, following the path even once per day to ensure that the pumps are running properly is impractical, not to mention of limited value. If a pump near the far end fails, it might be hours until you get to it. If one fails behind you, you might not know until the next day.

Actually, in such situation where cellular network is not working properly because the area where crops fields are, there is no cellular network is available. But, a satellite connection works with very high cost and high-power consumption is required. In such situation the LWPA network works efficiently with the high range and low power consumption.

LWPA network transmits limited data like just send status of a water pump, where a single data sends an overall status. If pump is working properly is responds one data as normal operation, and series of data which represents error message, and another one data represents the amount of water propelled by pump per day. Farmers can monitor every pump from the console of the system. The different errors like failing a pump, less water propelled by any pump of overheating condition in a pump are monitored and issues are resolved accordingly by the technician by getting notification in advanced.

Some other areas like monitoring and managing of traffic lights, weather stations located on hills, monitoring of lights in greenhouse scattered in acres, train crossing signals and much more where desistance is very long in miles and requires low power consumption is easily managed by LPWAN.

Different Technologies That Support LWPA Networks

1. **LoRa Alliance:** A non-profitable association which had defined and implemented the Low-Power Long Range Wide Area Network LoRaWAN (Paessler, 2021) protocol is called LoRa Alliance. Some big names like Alibaba, Cisco, IBM, Charter Communications, and SoftBank are serve as members in LoRa Alliance technology.

Currently SEMTECH is the only one company that makes chips which uses the open standard LoRaWAN protocol. The topology used by the technology is star-of-star. Devices connect to gateways, and the gateways connect in a more traditional manner over regularly connected IP networks.

2. **Sigfox:** A French company Sigfox offers exclusively LPWA network in France which has the largest LPWA network. An unlicensed frequency of 868 or 902 MHz band is used in LPWA networks, which is an ultra-narrowband radio transmission with long range distance coverage. But, Sigfox (Paessler, 2021) supports only 100 or 600 bits per second as data transfer rate which is actually low.

If any application requires bidirectional communication, then Sigfox fails because it has limited communication response where it gives only a maximum of 12-bytes payload in uplink message. Sigfox, theoretically delivers communication up to 30+ miles and uses star network topology, where any broadcast within the range can be received by any base station.

3. **MIOTY:** MIOTY is designed for IIoT (Industrial IoT) (Kapur, 2021) applications which is also a low-power, wide-area network (LPWAN) protocol. An unlicensed frequency of 868 MHz band in Europe and 915 MHz band in US is used in LPWA networks. MIOTY supports 512 bits/S per second as data transfer rate up to 20 kms.

The technology used by the MIOTY is the Telegram Splitting which is developed and patented by Fraunhofer IIS. In this technology a message is divided into multiple sub-packets which are transmitted at different times and frequencies. By sending packets using Telegram Splitting technology MIOTY gets less rate of packet error, even in a crowded spectrum and no other networks can interference it.

MIOTY sending subpackets at low airtime so, the chance of interference with another message becomes very low. However, if only 50% of sub-packets received at the gateway, the Telegram Splitting will reassemble full message. But, if the location of sensor is in underground or hard-to-reach this will create enormous problems.

4. **RPMA:** Random Phase Multiple Access (RPMA) (Paessler, 2021) is a technology offered by Ingenu with the range of 3 to 6.5 miles (non-line-of-sight). This covers the Dallas / Fort Worth area using only 17 network towers by the company. RPMA can connect devices which are underground as well as it penetrates concreate walls also.

The spectrum used by RPMA is 2.4 GHz which make interference with other technologies like Wi-Fi, Bluetooth and older wireless phones because they are using the same 2.4 GHz in many countries which is a radio type with single signal in all over the world.

5. **LTE-M:** The standard LTE-M (Paessler, 2021) is published by the 3rd Generation Partnership Project (3GPP) Release 13 specification is a 4G cellular-based technology. Devices which are using cellular technology consumes more power, so for using cellular technology in LPWA networks, the 4G wireless chips specially designed with a Power Saving Mode. In the Power Saving Mode most of the time the chip will turned off, it will turn on at programmed time interval by this the power saving occurs.

The wireless chips use half-duplex method for communication so it uses less power when they are turn on, however it's transfer rate is much slower as only 100 kbits/s compare to 4G connection. LTE-M standard of LPWAN works only where the user gets the LTE network. The drawback of this technology is that the devices which are only supporting 2G network will not work.

The LTE-M standard is not working for IBM devices; however, the major four U. S. carrier support this standard are Verizon, AT&T, Sprint, T-Mobile and several chip manufacturers like Qualcomm.

6. **NB-IoT:** The standard Narrowband Internet of Things (NB-IoT) (Paessler, 2021) is also published by the 3rd Generation Partnership Project (3GPP) Release

13 specification which is sometime called CAT M2 also. The communication technology of NB-IoT is DSSS modulation with maximum of 250 kbit/s uplink with multi-tone and 20 kbit/s with single tone. The main advantage of using the NB-IoT devices is its battery life which is approximately 10 years but where the network is avail not at remote area.

NB-IoT works on a new physical layer and signals and supports the devises which has 2G, 3G and 4G carriers. NB-IoT networks are currently deployed in China Mobile, China Telecom, Deutsche Telekom, and others.

7. **EC-GSM-IoT:** EC-GSM-IoT stands for Extended Coverage-GSM-IoT (Paessler, 2021) which uses GSM protocol used by the many cellular networks outside of the United States. This protocol also uses existing LTE-M network and consumes low-power. This standard supports all the devices which uses 2G, 3G and 4G networks.
8. **Weightless:** Weightless uses unlicensed spectrum (Paessler, 2021) which is an open standard technology. So, user can use hardware and network operator as per their choice. There are three types of weightless. Weightless-W is an unlicensed band and used in between the TV stations. Weightless-N uses unlicensed narrowband protocol, whereas Weightless-P is also a narrowband protocol which uses the 12.5 kHz and gives bidirectional communications.
9. **DASH7:** DASH7 is developed by Haystack Technologies, (Paessler, 2021) which is basically used for the devices that are not static but in moving nature with the frequency of 433 MHz. Based on the interference and location the range for connection from base to the device is from 200 meters to 2 kms. Respect to the power consumption the company says that it lasts for several years and the devices can produce energy from the solar cells also. DASH7 uses a public key encryption method with AES 128 cryptography for communication with device clocking to hide them from scanners.

Zigbee and Mesh Protocols

Zigbee is an open global standard of a wireless technology that supports wireless IoT networks which offers low-cost and low-power consumption. The IEEE 802.15.4 standard is used by the Zigbee technology (Konezny, 2021) operates on unlicensed bands with 2.4 GHz, 900 MHz and 868 MHz of physical radio frequency. IEEE gave approval to the Zigbee stack for using 802.15.4 specification in 2003.

The 802.15.4 specification offers low-cost as well as low-power consumption battery-operated devices by using packet-based radio protocol. The battery life of

Zigbee device is for several years because the specification supports different types of network topologies.

The Zigbee 3.0 Protocol

The members of Zigbee Alliance are from different companies (Amazon, Google, Apple, Comcast, IKEA, Legrand, etc.) have created and approved the Zigbee protocol. The Zigbee protocol is a reliable wireless network architectures which delivers secure and easy-to-use wireless data communication. The Zigbee Alliance has more then 300 leading technology firms, semiconductor manufacturers, service companies and OEMs companies.

The applications or devices of commercial and industrial are generally working with noisy Radio Frequency environments, where Zigbee 3.0 protocol is properly designed for this type of environments. Zigbee 3.0 certification supports exchange of information between devices which are from different manufacturer.

Features of Zigbee Protocol

- Supports multiple network topologies such as point-to-multipoint and point-to-point, and mesh networks
- Uses short burst transmission which increases battery life
- It processes high volume of data messages with low latency.
- It uses Direct Sequence Spread Spectrum (DSSS)
- One can create 65,000 nodes per network
- It gives secure data connection by 128-bit AES encryption
- Avoid resource contention, repeats and acknowledgements

Radio Frequency Identification (RFID) Systems

There are three main components in RFID systems: RFID tags, reader, application system. The Figure 3 (Xiaolin Jia, 2012) shows the working of RFID system with the use of its components.

Figure 3. Working of RFID system

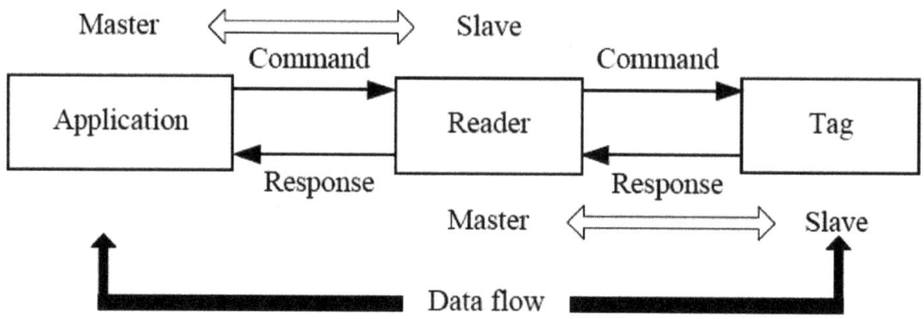

RFID Tags

It is the one type of transponder which can transmit as well as respond based on the objects with it attached. There are basically two types of RFID tags named active or passive. The tags which are partly of fully operated by power and which can communicate by itself to other devices or other tags which can read the communication information is called active tags. On the other end passive tags are not power operated and they can't communicate by self, but they can operate by the tag readers, just their information can be read by the readers. The components of tags are mainly microchip, and a coiled antenna with the purpose of storing data.

Reader

It consists of control unit and RFI – Radio Frequency Interface module which transmit and receive communication sequence. The main function of reader is to send signals to activate the RFID tags and start sending the data between the tags and the application system as per the communication sequence.

Application System

Based on the requirements the application system works as a database or a data processing system. The function of application system is to initiates the activities of RFID tags and readers. The RFID tags contains the information and unique product ID which is quickly detected electronically as well as tracked and controlled. application software initiates all readers and tags activities. RFID provides a quick, flexible, and reliable way for electronically detecting, tracking and controlling a variety of items. In RFID systems reader sends energy to the tags using radio transmission and then tag will releases its unique identification code to reader and then reader transmit that ID to information processing system. The data sent to the application system will be processed and proper information will transmit back to the reader which is a device that shows the proper output.

RFID technologies are successfully applicable in supply chain management system, production process control system, and objects tracking system as well as in manufacturing, healthcare, marine terminal operation, environment monitor, agriculture, transporting, retailing, payment, warehousing, logistics and supply. Table 1 (Patel, 2021) shows the comparison of LPWAN Technologies.

Table 1. Comparison of LPWAN technologies

Standards	Technologies			
	Weightles -W	**Sigfox**	**LoRaWAN**	**IEEE 802.11ah**
Frequency	TV whitespace (400-800 MHz)	868 / 902 MHz ISM	433 / 868 / 780 / 915 MHz ISM	License exempt bands below 1 GHz
Channel	5 MHz	Ultra-Narrow Band	868Mhz or 915MHz	1 / 2 / 4 / 8 / 16 MHz
Range	Urban: 5km	Urban: 3-10km Rural: 30-50km	Urban: 2-5km Rural: 15km	Up to 1km
Transmit Power	17 dBm	10uW to 100mW	EU: <+14dBm US: <+27dBm	1mW to 1W
Packet Size	>10 byte	12 bytes	Used Defined	7991 bytes to 65535 bytes
Governing Body	Weightless SIG	Sigfox	LoRa Alliance	IEE 802.11 working group
Topology	Star	Star	Star of Star	Star and Tree

WHAT IS LoRa?

LoRa is a LPWAN technology operated on (Patel, 2021) physical layer which is invented by Nicolas Sornin and Olivier Seller from France and patented by SEMTECH Corporation.

LoRa modulates the signals by patented Chirp Spread Spectrum (CSS) technique on Sub-Ghz ISM (Industrial Scientific and Medical) which is a narrow band signal operates on wider bandwidth which gives bidirectional communication. The signals are to be difficult to detected by the intruders because it spreads around the noise floor by which the actual signal cannot be detected by intruders.

LoRaWAN is a combination of LoRa devices and networks which is enabled with smart IoT applications. This is the best solution of the issues of our planet like pollution control, disaster management, energy consumption management, reduction in natural resources, efficiency of infrastructure. Using the LoRWAN and smart IoT applications will make all the things smart like smart city, smart homes and buildings, supply-chain management, communities, agriculture, smart street lights, smart parking and many more will create smarter planet.

LoRa Alliance

The manufactures of LoRa devices, (Patel, 2021) developers of LoRaWAN applications and group of companies work together to develop low power and low – cost standard with secure and long-range communication is the LoRa Alliance which has increased the applications of IoT in different way. and The LoRa Alliance is a group of companies, developers and manufacturers all over the world that come together to develop a standard for low power, low cost, secure and long-range solution of ever increase domain of internet of things. The LoRa Alliance was established in March 2015 and Figure 4 shows the current sponsored members of Lora Alliance.

Figure 4. Current sponsored members of Lora Alliance

The LoRa alliance offers memberships like Sponsors, Contributors, Adopters, Institutions. As per the memberships each have different levels of privileges in LoRa alliance group. The non-profit organizations help to create common standards for LPWAN for deploying in all over the world for sharing and standardizing the knowledge between the other members. The Figure 5 shows the more than 500 members of LoRa alliance.

Figure 5. Members of LoRa alliance

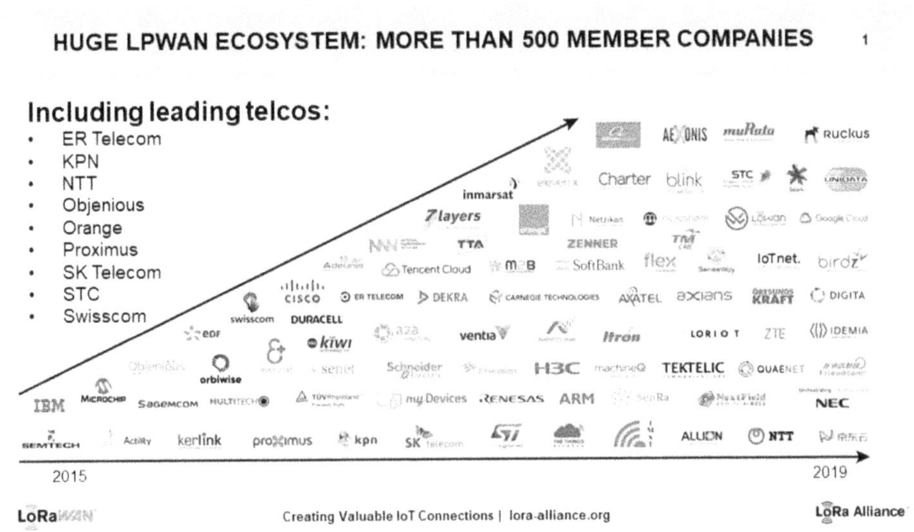

LoRaWAN

LoRaWan is a protocol for establish (Patel, 2021) communication which uses LoRa modulation technique. A French start up CYCLEO company had invented the LoRa modulation technique and in 2012 it was acquired by SEMTECH. LoRaWAN protocol works on the Media Access Control Layer (MAC Layer) which is the second layer of OSI model.

Advantages of LoRaWAN protocol over the other wireless communication protocols:

- Bi-directional communication
- AES encryption supports End to end security
- Over-the-Air Activation (OTAA) is the more secure for Things in network.
- Low power and low cost with long range communication
- Adaptive data rate (ADR) optimizes data rate, airtime and energy consumption.

LoRaWAN consists gateways, end devices, network servers and application server by star topology. Compared to using mesh network topology, star network topology reduces the energy consumption of end devices, because in mesh topology all devices are constantly communicates with each other which drains the power of end devices.

Whereas LoRaWAN uses overlapping star network topology in which gateway is used so, the communication occurs with needy devices which reducing the energy consumption.

In LoRaWAN network Adaptive Data Rate (ADR) facilitates low power consumption of end devices. For this facility gateways are used in which packets are arrived from different sources to the device at the same time by regulating the radio frequency and data rate.

The Figure 6 (EETimes, 2021) shows the different companies within different areas in India where LoRaWAN is used.

Figure 6. Different areas in India where LoRaWAN is used

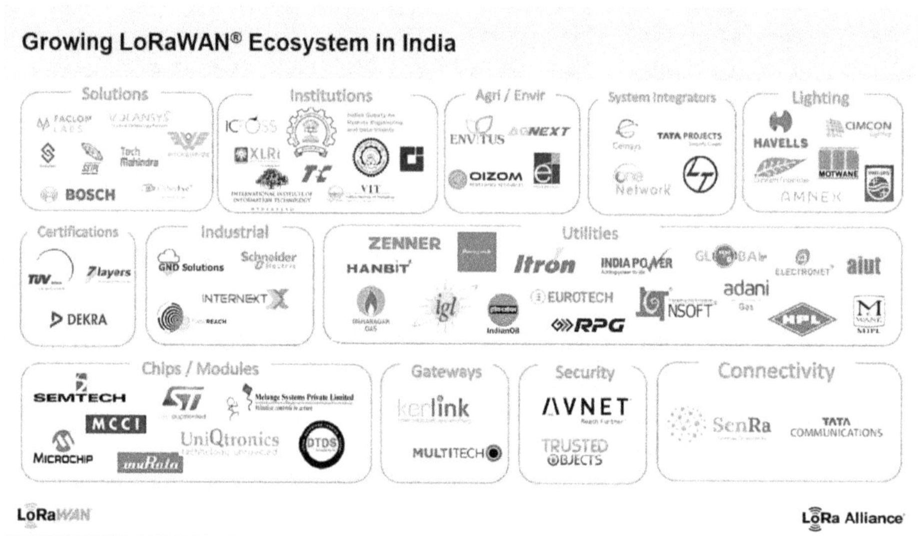

Technology

A star-of-stars topology is used in LoRaWAN (Alliance, 2021) network architecture with gateways which send information from central network server to and from end devices.

For connecting gateways and network servers standard IP connections are used which creates a transparent bridge which converts Internet Protocol packets to and from Radio Frequency packets. The characteristics of LoRaÒ physical layer supports

Long-Range in wireless communication by creating single-hop link between the single end-device and one or many gateways with bi-directional communication.

The specification of LoRa physical layer and LoRaWAN protocol does not use specific model for deployment like private, public, shared or enterprise so the industry has the freedom to use the technology to use with its extents. This specification is deployed by LoRa Alliance which are the collaborating members of the open association.

Classes

There are basically three different classes supported by LoRaWAN (Alliance, 2021) for end devices as per the needs and range for communication in different application.

LoRaWAN has three different classes of end-point devices to address the different needs reflected in the wide range of applications:

1. Class A – Lowest power, bi-directional end-devices: The default class which must be supported by all LoRaWAN end-devices, class A communication is always initiated by the end-device and is fully asynchronous. Each uplink transmission can be (Alliance, 2021) sent at any time and is followed by two short downlink windows, giving the opportunity for bi-directional communication, or network control commands if needed. This is an ALOHA type of protocol.

The end-device is able to enter low-power sleep mode for as long as defined by its own application: there is no network requirement for periodic wake-ups. This makes class A the lowest power operating mode, while still allowing uplink communication at any time.

Because downlink communication must always follow an uplink transmission with a schedule defined by the end-device application, downlink communication must be buffered at the network server until the next uplink event.

2. Class B – Bi-directional end-devices with deterministic downlink latency: In addition to the class A initiated receive windows, class B devices are synchronized to the network using periodic beacons, and open downlink 'ping slots' at scheduled times. This provides the network the ability to send downlink communications with a deterministic latency, but at the expense of some additional power consumption in the end-device. The latency is programmable up to 128 seconds to suit different applications, and the additional power consumption is low enough to still be valid for battery powered applications.
3. Class C – Lowest latency, bi-directional end-devices: In addition to the class A structure of uplink followed by two downlink windows, class C further reduces

latency on the downlink by keeping the receiver of the end-device open at all times that the device is not transmitting (half duplex). Based on this, the network server can initiate a downlink transmission at any time on the assumption that the end-device receiver is open, so no latency. The compromise is the power drain of the receiver (up to ~50mW) and so class C is suitable for applications where continuous power is available.

For battery powered devices, temporary mode switching between classes A & C is possible, and is useful for intermittent tasks such as firmware over-the-air updates

WORKING OF LoRaWAN ARCHITECTURE

For communication between two end devices Aloha method is used by the LoRaWAN networks (LoRa, 2021) where, any one or more end sensors senesces any change in their environment then end devices send data to network server for communication through gateways. After sending the data end devices waits for one to two seconds for the response from the server before going to sleep. End devices generally in sleep mode, so they consume less than one microampere of energy. This power saving method extends the battery life of end devices as 10 to 15 years.
The Figure 7 shows the LoRaWAN network architecture.

Figure 7. LoRaWAN network architecture

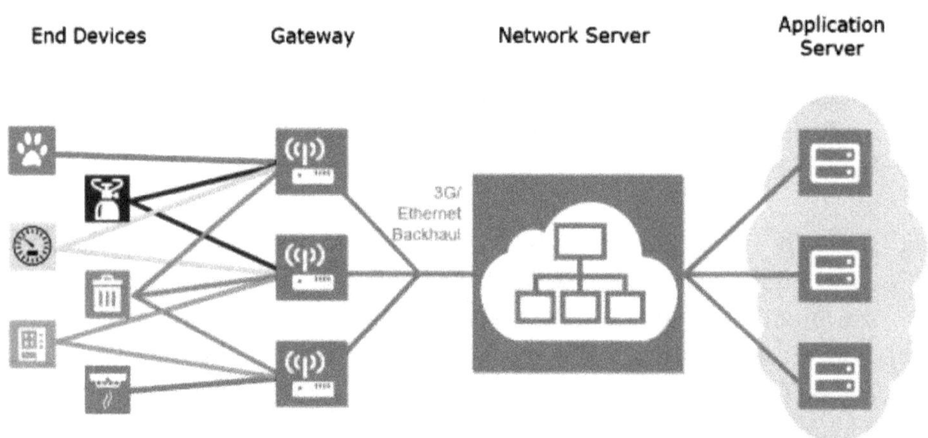

Procedure

- End devices in LoRaWAN network broadcast the packets asynchronously.
- Within the network one or more gateways will pick up these packets.
- The multi data rate radio – frequency and multi – channel devices are attached with gateways which can detect and scan packets using its any active channel.
- The packets are demodulated. The gateways are simple devices with no intelligence, so it passes packets to the core network.
- LoRaWAN has the central component called core network server which executes all the necessary actions with its intelligence on data get from the gateways.
- The operations of network server are: message consolidation, routing packets, managing network, and communicate with other servers (application servers).

WHY USE LoRaWAN INSTEAD OF Wi-Fi, BLUETOOTH, ZIGBEE, CELLULAR OR OTHERS?

There are plenty reasons for not using 2.4GHz, WiFi, Zigbee, and Bluetooth protocols. The first disadvantage of 2.4GHz is (Kilts, 2021) its poor range and its poor strength to propagate through walls, floors and other objects. As all knows that there will be a less range between the two floors in WiFi connection as well as in Zigbee network. On the other side LoRa devices can communicate in long range in miles with open-air environment, without losing the connectivity with any obstructions like buildings.

The second one is that the spectrum is too noisy which reduces the link quality means low bandwidth whereas LoRa uses just 915MHz so, no interference will create with other WiFi networks.

The security and key management used by the WiFi and other protocols is just unreliable is the third disadvantage. It means if one will change the password of router, then subsequent WiFi devices must be updated. But there is no such type of issues in LoRa devices because they are battery operated and they have not such type of interface to change password. As well as in LoRaWAN there is a different mechanism for securing battery operated devices which has its own unique identification or a key pair is used for security.

The main fourth disadvantage is the consumption of battery, the WiFi, Bluetooth, Zigbee and cellular devices consumes more energy compare to LoRa devices. The LoRaWAN devices does not required to regularly communicate with gateway as well as base station such is required by other protocols. The LoRaWAN devices are just get into deep sleep mode when they are not active which benefits to long lasting battery life.

Advantages of Using LoRaWAN for IoT Applications

There are many advantages for using LoRaWAN sensors (Kilts, 2021) are used in commercial and industrial business IoT applications. The main benefit using these devices are cost and energy savings and the other benefits are listed below:

- Covers Wide Range: LoRaWAN technology covers wide and long range with operators over 100 countries. Based on the specification it connects the IoT devices into long range up to 30 miles even in rural areas where hardy cellular network is available as well as in dense urban locations and indoor spaces.
- Long-lasting battery life: Compare to other networks LoRaWAN devices consumes less power because of they are not constantly connected to network servers. So, LoRaWAN supports long lasting battery life until users are required to recharge or replace it.
- Unlicensed frequency bands: The unlicensed frequency bands is used by the LoRaWAN technology so, cost of paying service charges for license band is eliminated. systems use unlicensed frequencies to operate, no licensing fees are necessary as you would have in, say, a licensed cellular band.
- Low bandwidth: IoT technologies requires low data rates to exchange data and that's the LoRaWAN networks support the best.
- The architecture of LoRaWAN systems is relatively simple so its deployment is fast so it is user-friendly.
- Cost-effective: The LoRaWAN technology uses unlicensed low bands for connectivity and uses lower energy from battery so, overall cost will reduce.

SMART STREETLIGHT WITH LoRaWAN

In a typical city the major energy consumption at night is street lights, (Witeck, Smart Streetlight, 2021) the smart lighting helps to reduce the consumption which tends to lower the cost. The smart lighting has the intelligence which controls the street lights. The remote lighting feature of smart lighting controls and adjusting the intensity of light as well as turned on and off lights as per scheduled time which reduces the cost without sacrificing the public safety.

Another cost which affects cost of energy consumption is the cost of maintenance of street lights. Smart lights also send other issues and data through networking to other smart city applications.

The components and technology of smart streetlight are sensors, light-sensing photocell sensor, cameras and functionality of real-time monitoring. Another feature of adaptive lighting or intelligence street lighting is the advanced technology in the development of smart cities projects.

As per the local surroundings the municipality provides the proper amount of smart street lights for the safety and security of the citizens for their satisfaction. The smart street lighting infrastructure provides data which are also used in monitoring weather, traffic, and pollution by IoE (Internet of Everything) applications.

How Smart Streetlights Work

Usually, the smart streetlights infrastructure contains basically sensors and cameras but, as per the technology and features provided by the manufacturer there will be some differences in different infrastructure. The smart lighting provides the dynamic lighting means as per the requirement it will on and off as well as dimming when the object is nearer to them and for this feature one smart street light communicate with others called neighboring fixtures. The best feature of smart street lighting is that if a vehicle is detected then surrounding lights are lighten up until that vehicle passed away.

These types of features require more infrastructure and technology like seismic sensors, image sensors, sound sensors, water and weather detection sensors, speakers, and wireless transmitters for data transmission with neighboring smart street lights. The vendors of smart street lights offer other software which can help municipalities to make city as smart city by managing and monitoring the data available from smart street lights.

The U.S. Federal Highway Administration has published guidelines on how government agencies can implement smart streetlights.

Features of Smart Streetlights

The feature of smart street lighting is (Witeck, Smart Streetlight, 2021) based on the specific technology used by the planners of the city, the examples of common functionality are mentioned below:

- The movement detection feature controls the dynamic lighting.
- Data collected by smart street lights are used in environmental and weather monitoring.
- Digital signage is used for advertisement, parking availability and accident alerts updates.

- Inappropriate parking of vehicle done by drivers in illegal or open spaces will be informed to the officials is the feature of smart street lighting.
- Smart street lighting is also be used to extended cellular and wireless communications.
- The real-time data transferred from the smart street light is used to manage the traffic and track the congestion and speed of the vehicle as well detecting the accident or crime done at specific location.

Benefits of Smart Streetlights

Implementing smart streetlight systems offers (Witeck, Smart Streetlight, 2021) the following advantages:

- The feature of dynamic street light dimming control will reduce the energy consumption will tends to lower the energy cost.
- Increased the satisfaction of pedestrian by safety and security.
- The monitoring software helps to find the faulty street light which lowers the repair and maintenance.
- By setting the time duration for lighting the street lights will reduces the carbon emissions and light pollution.
- Increased lamp life and shorter response times to outages;
- Improved architectural planning based on real traffic patterns and insights; and
- Increased revenue opportunities, such as leasing poles for digital signage or other services.

Disadvantages of Smart Streetlights

- There are few challenges for upgrading lighting networks as a long-term vision. Because the initial investment is more but as long-term it will save money.
- There are multiple vendors are available with different technology platforms, so it is difficult to choose appropriate one.
- Plus, there are multiple applications and technology platforms, so choosing the right ones can be challenging. The lack of common standards across networks poses challenges as well.
- The consumer has less knowledge and benefits of the smart street lightings.

DEVICES USED IN LoRaWAN-BASED SMART STREET LIGHTING SYSTEM

LoRa Light Control Unit (LoRa LCU)

The Low Power RF technology (Guha, 2021) with 1 GHz wireless radio band is used in the Light Control Unit (LCU).

The street lights with LCU supports long-range in LoRa network for two-way communication to the cloud-based street light management software installed in the cloud server. Figure 8 shows LoRa Light Control Unit.

Figure 8. LoRa Light Control Unit

Each streetlight is enabled with LCU which provides a unique ID to every street lights. This control unit is installed on every street light and provides a unique ID to every light. The LCU will caried out several functions to the street lights such as:

- On/off control
- Dim level setting
- Current consumption tracking
- Ambient light sensing (optional)
- Lamp surface level temperature sensing (optional)

LoRaWAN Gateway

The LORaWAN gateway is a device which is used to create error free and seamless LoRa network in the region of the streetlights. The LoRaWAN gateway allows the LCU enabled street lights to communicate over the network. Figure 9 shows the LoRaWAN gateway used in smart street lighting system.

Figure 9. LoRaWAN gateway

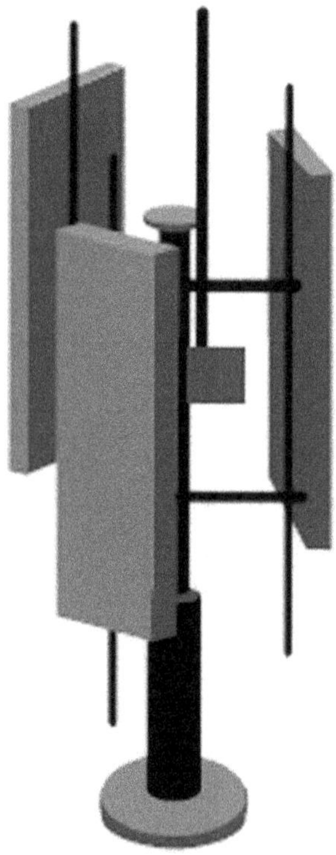

The arrangement of LoRaWAN gateways as they cover the maximum reach and as per the environmental and surrounding conditions. The LoRaWAN supports about 10,000 LCUs to communicate within its coverage. The LoRaWAN has options that allow it to communicate over the internet.

Remote Feeder Control (RFC) With GSM / LoRaWAN Gateway

For on/off control and tracking of energy consumption for multiple street lights is done with the use of Remote Feeder Control (RFC) which is connected through a power supply cable.

The Remote Feeder Control panel can be installed with any type of street lights and can be connected to a single phase or a three-phase energy meter for tracking energy consumption of lights. Figure 10 shows the Remote Feeder Control (RFC) with GSM / LoRaWAN Gateway.

Figure 10. Remote Feeder Control (RFC) with GSM / LoRaWAN Gateway

Remote Feeder Control panels can be controlled by a local manager by sending SMS messages to Remote Feeder Controls from authorized phones. Remote feeder control is an ultra-low-cost automation system for street lights that can be retro fitted to most streetlight systems without necessarily upgrading street lights to LED.

WORKING OF SEMTECH LoRaWAN-BASED SMART STREET LIGHTING SYSTEM

The features of LoRa technology-based street lighting are sensing, (SEMTECH, 2021) analytics, reducing energy costs, and cloud connectivity which gives safety and security to the citizens.

First, a street lamp is embedded with LoRa-enabled sensor. The functions of lamp are controlled by this sensor. The sensor of each lamps connects them to LoRa-based gateway. The messages and data from all nearby lamps are handled by each gateway. Other solutions provided by the LoRa such as parking sensors can connect to the same gateway for easy transmission of data in smart city-based applications. When gateway get the message from the sensors it will transmit that message to the network server for analysis. All the functions like send maintenance alerts, replacing burnt-out bulb or other problems of LoRa are controlled and monitored through the server remotely.

The Figure 11 (SEMTECH, 2021) shows the working of SEMTECH LoRaWAN street light system.

Figure 11. SEMTECH LoRaWAN street light system

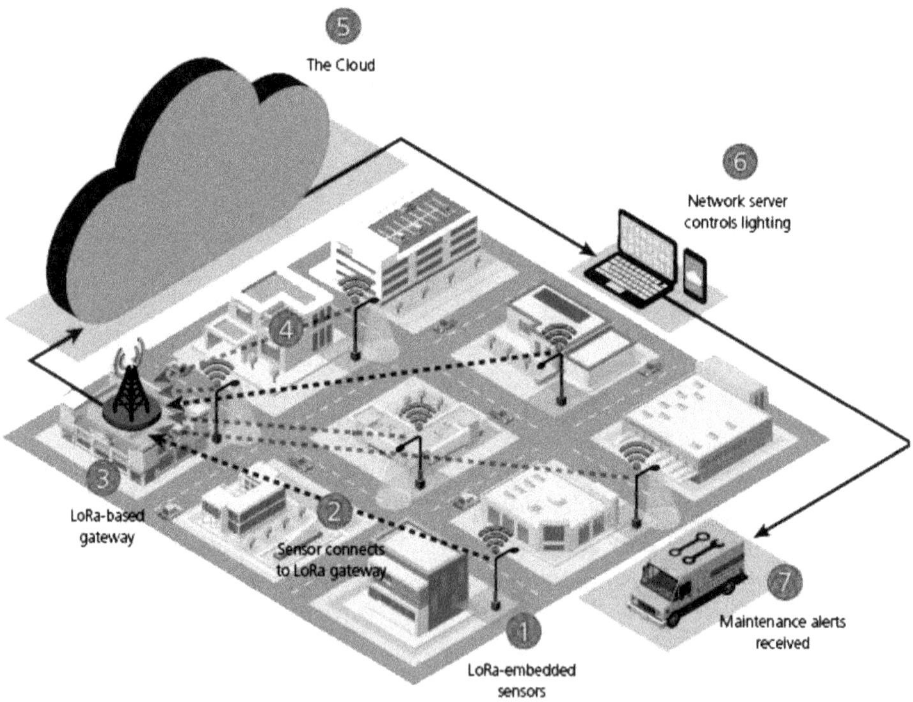

Table 2. Working of SEMTECH LoRaWAN street light system

1	Sensors embedded in each street light have the ability to control light functions.
2	LoRa Technology in the sensor connects the street light to a LoRa-based gateway.
3	The LoRa gateway aggregates data from all nearby street lights.
4	Sensors for other smart city applications connect to the same gateway.
5	The gateway sends information to the Cloud where the data is analyzed by an application server.
6	Application server controls lighting.
7	Server sends maintenance alerts for burnt out bulbs and other issues.

REFERENCES

Alliance, L. (2021). *What is LoRaWAN® Specification*. Retrieved from www.lora-alliance.org: https://lora-alliance.org/about-lorawan/

Apps, B. M. (2021). *Bluetooth Low Energy (BLE) Technology*. Retrieved from www.blemobileapps.com: https://www.blemobileapps.com/ble-mobile/

EETimes. (2021). *India's LPWAN Networks Expand to Meet Smart Cities IoT Demand*. Retrieved from www.eetimes.com: https://www.eetimes.com/indias-lpwan-networks-expand-to-meet-smart-cities-iot-demand/#

Figueroa, E. (2021). *Long range, low power Wi-Fi® for IoT*. Retrieved from www.wi-fi.org: https://www.wi-fi.org/discover-wi-fi/wi-fi-certified-halow

FLASHNET. (2021). *Everything you need to know about smart street lighting*. Retrieved from www.intelilight.eu: https://intelilight.eu/smart-street-lighting/

Guha, J. (2021). *Street Lighting*. Retrieved from https://www.neevenergy.com/street-lighting/: http://neevenergy.com/wp-content/uploads/2019/04/Intellivolt-Solution-Version-2.pdf

Jia, X. (2012). RFID technology and its applications in Internet of Things (IoT). In *2nd International Conference on Consumer Electronics Communications and Networks (CECNet)* (pp. 1282-1285). Yichang, China: IEEE.

Kapur, R. (2021). *What is MIOTY Technology?* Retrieved from www.everythingrf.com: https://www.everythingrf.com/community/what-is-mioty-technology

KCT. (2021). *BLE (Bluetooth Low Energy)*. Retrieved from www.blog.kct.ac.in: https://blog.kct.ac.in/ble-bluetooth-low-energy/

Kilts, S. (2021). *LoRa Sensors*. Retrieved from www.radiobridge.com: https://radiobridge.com/wireless-technologies/lora

Konezny, R. (2021). *Zigbee Wireless Mesh Networking*. Retrieved from www.digi.com: https://www.digi.com/solutions/by-technology/zigbee-wireless-standard

LoRa, S. (2021). *Understanding the LoRaWAN Architecture*. Retrieved from www.tech-journal.semtech.com: https://tech-journal.semtech.com/understanding-the-lorawan-architecture

Paessler. (2021). *What is LPWA?* Retrieved from www.paessler.com: https://www.paessler.com/it-explained/lpwa

Patel, D. (2021). *Low Power Wide Area Networks (LPWAN): Technology Review And Experimental Study on Mobility Effect*. South Dakota State University. Retrieved from www.openprairie.sdstate.edu: https://openprairie.sdstate.edu/etd/2667/

SEMTECH. (2021). *LoRa White Papers, Smart Lighting*. Retrieved from www.semtech.com: https://www.semtech.com/lora/resources/lora-white-papers

Witeck, C. (2021). *Smart Streetlight*. Retrieved from https://internetofthingsagenda.techtarget.com/definition/smart-streetlight: https://internetofthingsagenda.techtarget.com/definition/smart-streetlight

Witeck, C. (2021). *What is internet of things (IoT)?* Retrieved from www.internetofthingsagenda.techtarget.com: https://internetofthingsagenda.techtarget.com/definition/Internet-of-Things-IoT

Chapter 9
Machine Learning Applications to Smart Cities:
A Comparative Study

Nikita Malik
Maharaja Surajmal Institute, Guru Gobind Singh Indraprastha University, India

Menal Dahiya
Maharaja Surajmal Institute, Guru Gobind Singh Indraprastha University, India

Nipun Walia
Maharaja Surajmal Institute, Guru Gobind Singh Indraprastha University, India

ABSTRACT

A city becomes a smart city when it employs ICT (information and communication technology) to share data with the public, improve government services' quality, and develop operational efficiency. The key objective of a smart city is to optimize the operations of the city and encourage economic growth through data analysis and the use of smart technologies. This chapter focuses on exploring the IoT and soft computing technologies' role in the development of smart cities. These methods are used for designing optimal policies efficiently for complex problems around smart cities. Different case sectors of a smart city are explored, and various machine learning-based algorithms are discussed to help, improve, and resolve their issues. Additionally, comparisons are drawn between the techniques adopted in existing systems and the results they yield, along with identification of the limitations they present.

DOI: 10.4018/978-1-7998-9710-1.ch009

INTRODUCTION

With the rapid increase of Internet technology day by day, the lifestyle of urban society has upgraded automatically. Along the same lines of using technologies to assist mankind in resolving difficulties, smart cities are a futuristic approach for alleviating the complications that stem from growing population and speedy urbanization, in order to support the government and public.

The label "smart city" is used interchangeably with similar terms like "intelligent city" or "digital city" (Albino et al., 2015). These variants are categorized into three dimensions (Harrison et al., 2010; Nam & Pardo, 2011): Technology, People and Community. Offered definitions appear to have dual interpretations: (a) smart city is a living solution that integrates different life facilities such as transportation, power, and buildings in an efficient manner to improve the services for its citizens; (b) smart city exemplifies the importance of sustainability of resources and applications for future generations.

Smart cities can deliver requisites such as socio-economic development and quality of life, which are lacking in modern day cities. They employ ICT technologies to monitor critical infrastructure such water works, electricity, traffic etc. to create cities that are sustainable, efficient and all in all more livable. Making a city smart encapsulates learning and working on what, how, and why. To function, smart city technologies require the processing of enormous quantities of data, or "Big Data". Big Data and artificial intelligence (AI) are interlinked. AI can efficiently shift through large quantities of Big Data to generate data predictions and cost-effective solutions to fuel smart city technologies. Big Data has been described in terms of the three "Vs" as "high-volume, high-velocity and/or high-variety information assets" which means massive datasets, processed very quickly (through the use of algorithms) and the use of different data sources, including combining different datasets. Big Data and AI are interlinked. AI refers to various methods "for using a non-human system to learn from experience and imitate human intelligent behavior. AI can efficiently shift through large quantities of Big Data to generate data predictions and cost-effective solutions to fuel smart city technologies.

This chapter includes five major sectors that help in making a city smart and with the help of various IoT technologies and ML algorithms these sectors enhance the functionality and efficiency:

- Social distancing model with the help of object detection and looking into various algorithms such as the R-CNN (Region-based Convolutional Neural Networks).
- Air quality indexing and weather detection using several IoT technologies and deep learning algorithms such as the LTSM, STDL, CNN, SVM etc.

- Traffic control working with algorithms like K-mean, random forest, SVM, Decision tree along with the role of supercomputers.
- Power and energy management with the help of Neural Network clusters and some deep learning frameworks like Blue Wave-Ai.
- Security from both physical threats (through various models and advanced ICT) as well as cyber/autonomous attacks (by learning from past attacks with the help of several neural network structuring algorithms).

A detailed comparison of the different technologies implemented for each aspect of the smart city is explained along with the dataset used, results obtained, issues resolved and the limitations identified from the existing literature, offering scope for future work.

SOCIAL DISTANCING

Need

Everything has changed since the outbreak of Covid-19 in 2019. The norms we are used to, the machinery, on which our world was moving, came to a halt for an unknown period. With no knowledge about either the cause or the treatment of this virus, there began chaos with countries rapidly losing their most precious asset, i.e., their citizens. The only solution that came forward to fight this war was to contain its spread as much as possible, and the best option for that was social distancing. The concept of social distancing became essential in smart cities all over the globe. To limit the spread of Covid, some of the popularly adopted measures include distancing in social places, imposing lockdowns, restricting travel, creating quarantine spaces, and preventing gathering of people at crucial workplaces. Figure 1 graphically depicts the rise and fall in cases worldwide over time since the first case was diagnosed, showing how mitigation strategies such as social distancing over the year 2020 have helped flatten the curve. In the current critical situation, it is paramount that coordination is achieved among various agencies at work and measures are effectively implemented to limit the spread of the disease. Several initiatives were taken, including making upgraded wearables available, announcing speakers/ ad campaigns to notify the public about hotspots, sudden outbreaks, etc., which gave a promising result (Gupta et al., 2020).

Figure 1. Worldwide Covid-19 cases over time, with and without practicing social distancing in 2020

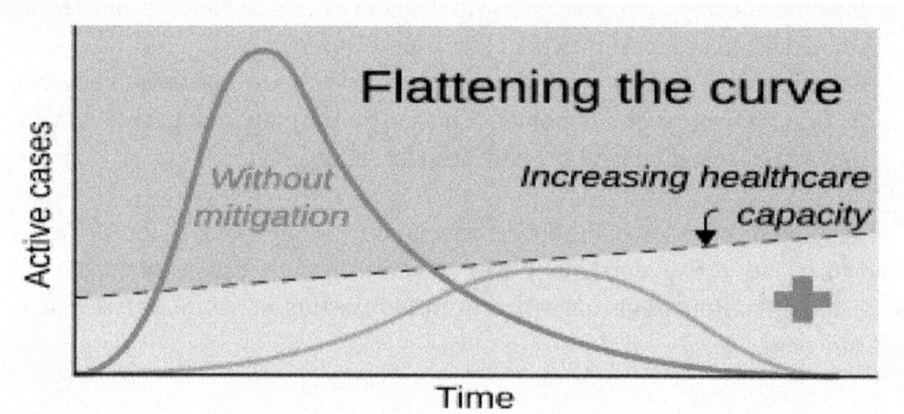

As achievable as this goal might sound, the reality turned out to be the contrary. Big cities like Delhi, Mumbai, New York, San Francisco, Brazil, and many more were the worst-hit cities despite issuing social distancing orders. This was a clear example that the cities were not properly prepared to handle such widely spread diseases. Having high-density populations, gathering at public spots, and using public transport made matters worse. To handle this situation and to ensure people and assets' safety, lower fatalities, and minimize the impact on the economy, it is needed that there is an understanding of situational intelligence and automated targeted response throughout the community. This made governments around the globe realize the need to shift towards being a smart city now more than ever (Surya & Yarlagadda, 2020).

With the help of IoT (Internet of Things) and data-driven technologies, social distancing through smart cities was brought into action, focusing on keeping a check on distancing protocols in public places to distribute data across various authorities and devices to contain the spread as much as possible. In this section, these methodologies and technologies employed by smart cities in reference to social distancing amid the Covid-19 pandemic are discussed. Additionally, a comparison of the various algorithms highlighting the role of IoT and machine learning in smart cities towards achieving social distancing is made, helping understand their effectiveness in the real world.

Role of AI and ML

When it comes to propagating such measures at a large scale, certain measures are required to maintain a smooth mechanism and a proper check to find faults in the plan of action. The most important but arduous part in maintaining social distancing is to keep a check on people on a large scale. It is practically impossible for any government, especially for India, to regularly check over a billion people. Along with supervision and proper guidelines to follow, warning the public at the right moment with speed and precision is difficult. Even deploying more workforce is not a viable solution as it will just increase the number of people with infectious contact and thus will work against the very same motive they were deployed for. Even the WHO (World Health Organization) recommends finding a more viable and technologically inclined solution to contain and reduce the contact as much as possible. This is where IoT and machine learning comes into action. The introduction of advanced technologies will change the course of action and help us fight the pandemic, prepare us for any such situation, and pave our way towards smart cities (Gupta et al., 2020).

With the collective efforts of several organizations, many techniques have been found and developed with algorithms and models such as Faster CNN (Convolutional Neural Network), YOLO (You Only Look Once), Object Detection, and several other deep learning methods. Nevertheless, just detecting is not enough to prevent the spread. Proper information distribution is required for both the agencies looking after it and the citizens and the medical and ground zero workers in real-time. If relying on a traditional message distribution system, even the fastest mode takes about a minute or two, and that is all it takes to get infected. Thus, faster information distribution is required, which means eliminating human intervention. Several IoT devices such as wearable watches or smart devices or even software dedicated to this can be used to relay the information, such as apps like Aarogya Setu used by the Indian government to track the movement using GPS (Global Positioning System) and Bluetooth, or the South Korean government using drones to look out for places of high human density (Pai, 2020).

Amidst groundbreaking research going on in several aspects of this area, the following is a brief discussion on the most relevant ones to develop an understanding of the AI/ ML-based approach to managing the situation:

Human Detection

This is the first phase where live feed is taken by CCTVs (Closed Circuit Televisions) and run through a YOLO-based Deep Neural Network (DNN), as illustrated in Figure 2. Once input, it assesses the feed as a 2-D picture and assesses the objects (like humans, cars, pets, etc.) and calculates the distance between them, providing a warning whenever a certain distance limit is breached.

Figure 2. A YOLO based Deep Neural Network
(Rezaei & Azarmi, 2020)

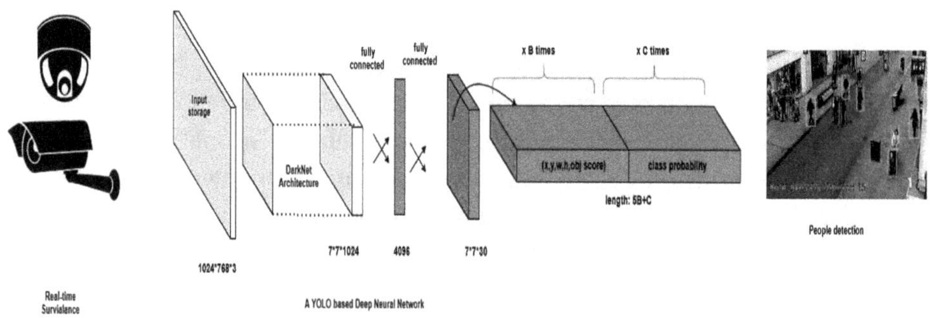

However, certain objects like animals or any other non-living object can create noise in the image and cause hindrance. Besides that, the algorithm requires a rich dataset to learn to differentiate between various ages, sizes, and gender. Moreover, the frame/image size also created an issue that was resolved through mosaic data augmentation that involved integrating four images in one. The evolution of this model went from YOLO v1 to v4, which saw many changes in the algorithms, making way for new possibilities in object detection.

Table 1. State-of-the-art techniques for designing a CNN-based model (Rezaei & Azarmi, 2020)

Input	Detection Core ®				Head ® Output	
Augmentation	**Activation**	**Backbone**	**Neck**	**Regularization**	**Dense**	**Sparse**
CutOut	ReLU	VGG	SPP	L1, L2	YOLO	R-CNN
MixUp	Leaky-ReLU	ResNet	PAN	DropOut	SSD	Faster-RCNN
CutMix	SELU	Darknet53	SFAM	Spatial DO	MatrixNet	CenterNet
Mosaic	Swish	Inception	NAS-FPN	DropBlock	FCOS	RepPoints

The whole progression can be studied as a process divided into input, processing/detection, and output. It is further divided into subcategories, a few of which are presented in Table 1 (Rezaei & Azarmi, 2020). Inputs can be extracted from various datasets for training such as MS Coco, ExDark, MaskedFaceNet, Google Open Image etc. (Mrinal, 2021), among the many different models used for activation like ReLU (Rectified Linear Unit), SELU (Scaled Exponential Linear Unit), Mish, Swish, etc, Mish performs the best with the fastest convergence rate and minimal noise (Rezaei & Azarmi, 2020).

Rezaei & Azarmi (2020) state that the YOLO-4 algorithm serves as the best option backbone architecture but being a cluster of several other techniques, it requires a deep study of its sub-techniques to find the most optimal one that suits our cause (Rezaei & Azarmi, 2020; Yang et al., 2021). To make the best use of the CNN approach, the receptive field can be expanded and the complexity can be enhanced using more layers. Else, models like Cross Stage-Partial (CSP), Dense blocks can be used in its place. Figure 3 presents a comparison among three backbone models based on their speed and number of parameters being able to intake, given 512*512 input resolution.

Figure 3. Comparison of three backbone models using an RTX 2070 GPU

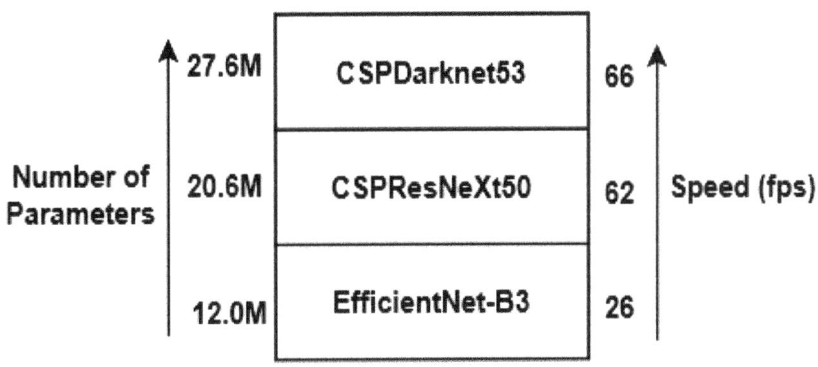

These days, a new set of layers is placed between the head and the backbone called the neck. They consist of multiple approaches to gather and deliver more precise image detection results. This also helps to solve the issue of detection of small objects, and therefore the algorithm can be used with any CCTV or drone. YOLO-3's spatial pyramid pooling layer (SPP) delivers one of the best results (Punn et al., 2020). Objects like people, table etc., are classified by the head of a DNN, along with calculation of the objects' size and the coordinates of the correspondent bounding boxes.

Usually, the head is of two types- one staged (dense) or two-staged (sparse). Two-stage detectors such as the Faster R-CNN (Region-based Convolutional Neural

Network) or Mask R-CNN generate RoI (Region of Interest) and send the regional data for object classification and box-based regression. This kind of model is highly accurate but delivers slower results than single staged detectors such as YOLO and SSD (Single Shot Multibox Detector) that take an input image and learn the class probabilities and bounding box coordinates, treating object detection as a simple regression problem. Such models achieve lower rates of accuracy but are much faster than two-stage object detectors. Providing them with an adequate dataset and adjusting the offsets can yield excellent results (Rezaei & Azarmi, 2020).

People Tracking

Along with accurate object detection, proper ID assignment and tracking comes as the next issue at hand. Although it raises privacy issues on account of being misused to track people from CCTVs or, say, monitor and check a child's growth using a baby monitor but by limiting its access to certain organizations and ensuring "health and protection over privacy," it has been passed.

Tracking techniques like Search Online and Real-Time detection (SORT) or Histogram of Oriented Gradients (HOG) can be some of the perfect techniques for the same (Punn et al., 2020; Rezaei & Azarmi, 2020). The Kalman filter with Hungarian optimization is also one of the preferred tracking tools with a result time of t+1 based on present measurement of time t and the human movement's mathematical modeling. The Hungarian algorithm is a combinatorial optimization algorithm that helps to assign a unique ID number to identify a given object in a set of image frames by examining whether a person in the current frame is the same detected person in the previous frames or not (Rezaei & Azarmi, 2020).

Inter-Distance Estimation

The idea behind this technology is for it to be used by every camera out there. However, when captured, conversion from 3-D to 2-D can cause a certain pixel distance (like parallel line pixel) intersecting with the horizontal lines, making them shorter and closer than they are. This effect is called the perspective effect. To calibrate the loss of the z-axis when converted into 2-D images, camera calibration with z=0 is needed to eliminate the perspective effect. The camera location, its height, angle of view, as well as optics specifications (i.e., the camera intrinsic parameters), must also be known. Basic calibration performed according to all these considerations should do the trick.

Comparison

After discussing a basic architecture, some of the models can now be discussed on their overall working based on aspects like recall rate, fps, and precision rate. These tests include results from several organizations (Rezaei & Azarmi, 2020; Yang et al., 2021). When tested with several datasets Faster-RCNN model showed good generalizability, however its speed came as an issue which was suspected due to the computational cost of regional proposal technique. It also failed to detect several occurrences of social distancing violation in places like malls. SSD failed on the same grounds as well. YOLO v3 and v4 performed relatively well as compared to their previous models with fps reaching near 23 but still gave issues when tested on compact places with multiple obstacles creating a noise in the input. The best results were given by DeepSOCIAL DNN models. Comparison of accuracy, recall-rate and speed of seven DNN models for the Oxford Town Centre dataset are presented in Table 2.

Table 2. Comparison of DNN models in terms of accuracy, recall-rate and speed (Rezaei & Azarmi, 2020)

Method	1				2	3	4
	DeepSOCIAL DNN Models				YOLOv3	SSD	Faster R-CNN
	α	β	γ	DS			
Accuracy (%)	99.8	99.5	99.6	99.8	84.6	69.1	96.9
Recall Rate	96.7	97.1	96.7	97.6	68.2	60.5	83.0
Speed (in fps)	23.8	24.1	23.8	24.1	23	10	3

Figure 4 shows the test results considering four large datasets- Pascal VOCO (Everingham, 2010), Microsoft COCO (Chen et al., 2015), Image Net ILSVRC (Russakovsky et al., 2015), and Google Open Images (GOI) Datasets V6+ (Kuznetsova, 2020), which include 16 million ground-truth bounding boxes in 600 categories. The datasets were annotated using the bounding-box labels on each image and the corresponding coordinates of each label. These tests became the pre-ground check for the algorithms developed. Most promising results were provided by the DeepSOCIAL team. They categorized four classes, to classify the crowd in each subsequent class, namely, man, woman, boy, and girl, and created a total of 3,762,615 samples. These were then used from training, including 257,253 samples from the COCO dataset and 3,505,362 samples from the GOI dataset, bringing an accuracy of 99.8% by the time of γ testing.

Figure 4. The number of annotated boxes per class in four common datasets. The horizontal axis is represented in logarithmic scale for better readability *(Rezaei & Azarmi, 2020)*

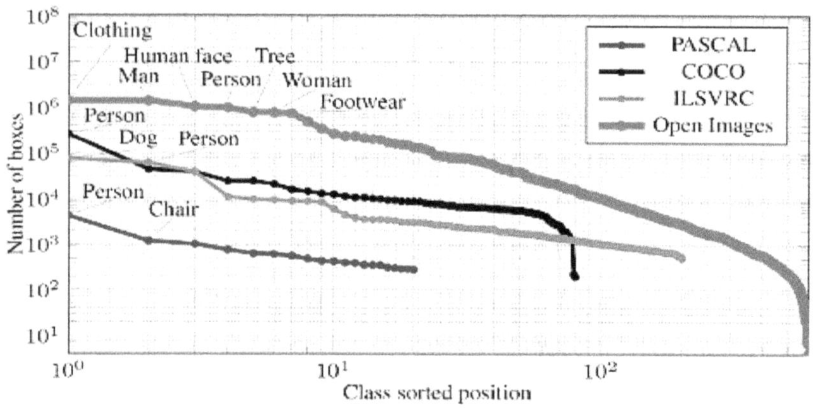

Even with these promising results, certain objections were raised by governments that allowed family members, couples, or people together to be exempted from this restriction. This is still a staggering issue that DeepSOCIAL has somewhat tackled.

Future Scope

Certain smartwatches are fully capable of being implemented with an alert system in them. An alert message with a sound can be delivered autonomously by the algorithms from the server assessing the surveillance footage. Even government apps like Aarogya Setu have made themselves open source and entirely available to add any required changes. Such solutions help maintain a social distance and be used for history-tracking to assess the infected person's contact and isolate them to prevent further spread.

While all these technologies help to cater and protect an individual, to look out for masses all together, certain measures can be taken such as using e-billboards and have them linked with a continuous feed monitoring of that particular section which might promote a sense of self awareness amongst the people. While a feed update, certain other information like a heat map can also be a viable outlook towards the problem. A critical density detection algorithm (Yang et al., 2021) along with an integration of IoT technology was tested out inside New York central used DeepSOCIAL and YOLO v4 to test their software and the results were pretty promising. DeepSOCIAL had its own algorithm to create a heat map as well which again gave promising results. Figures 5 and 6 represent the results of the algorithms.

Figure 5. Illustration of pedestrian detection using Faster R-CNN
(Yang et al., 2021)

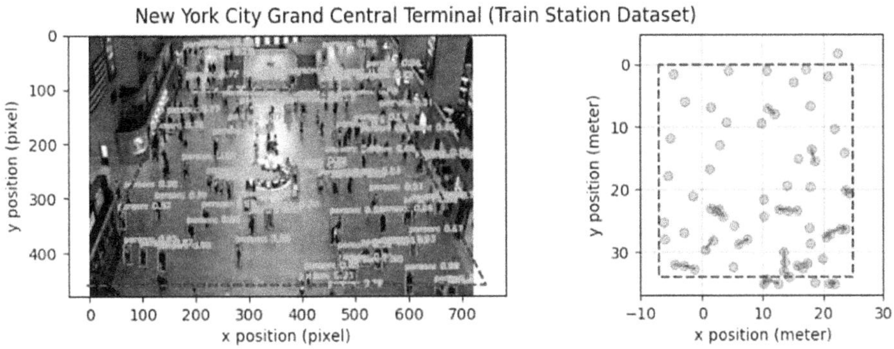

Figure 6. Accumulated tracking maps after 500 frames using DeepSOCIAL. Blue: low-risk, Red: high-risk
(Rezaei & Azarmi, 2020)

(a) BEV heat map (b) 2D Mosaic heat map

WEATHER AND AQI

Need

In the era where we have touched the line that divides our current cities from the future smart cities, certain factors are in serious need to be checked first- the weather being at the top of the list. The weather has always been a deciding and game-changing factor in all the events, from as small as farming to large projects such as the launch of missions like Mars Orbiter Mission (MOM). Predicting the weather is something researchers have been trying to do since the time Galileo first discovered the telescope and possibly even before that. However, the last decade has seen many successful results in predicting the weather, which became a deciding factor in major projects and helped reduce the damage caused by natural disasters by a prior prediction and early measures and evacuation if needed could be executed (Umer et al., 2021). Moreover, by assessing certain geographically strong locations, certain measures were taken to create renewable energy sources such as solar farms generating solar energy, wind parks having many windmills generating power with the help of high-speed winds, etc.

This is just one aspect of this field. As the world and humans are evolving, our environment is also changing and not for good. With the rise of the industrial age, different forms of pollution started degrading the environment. With a group effort of various organizations, this was brought to light, and measures to prevent it began. One such research was in predicting AQI (Air Quality Index). It is nationwide indexing and forecasting about the quality of air and the amount of matter such as SO2 (Sulfur dioxide), CO (carbon monoxide), ground-level ozone, and particle pollution (PM10 and PM2.5) in the air (US EPA, 2021).

All these events boosted their progress with the rise of new IoT technologies and the advancement of machine learning and neural networks. This section will discuss a few technologies and algorithms that have helped researchers worldwide to tackle this issue. This is divided into two parts, one where weather prediction is discussed- how it helped create various opportunities for generating renewable energy sources. The other explores the different algorithms used in predicting AQI and how it helps control air pollution.

Weather Prediction

Detection and Prediction

Weather detection is a mixture of various aspects like humidity, wind speed, and many more. These predictions are essential for a lot of factors such as farming and even power generation. Wind parks calculate their power output by studying the speed of wind flowing and even predicting future output capacity at a specific time of the year. Not only that, certain disasters can be prevented or controlled by the prediction of certain weather information. The Australian bush fire is an example of such a situation. Starting in 2019, the wildfire caused heavy destruction to the forest and wildlife, which lasted till 2020 as well. The study of direction and speed of winds helped a lot in controlling the situation. But as per the experts, this could have been prevented from happening in the first place if the study of humidity, wind speed, and temperature would have been more devoted (CDP, 2019).

Now to handle and study such situations, the first step is to collect raw data for processing and learning. The most cost-effective way to do that is by placing high-density sensing sensors in a wide range of areas collecting local weather information. The data then collected can be kept and sent to local server and local weather database. Processed data is then used to create datasets and remove as much noise as possible. For this purpose, many statistical insights can be used to find the relevant attributes, such as the covariance, box plot, central tendency, and many others (Umer et al., 2021). After several calculations, four classes came out: Rain, Fog, Thunderstorm and, No Event.

After this, an algorithm is needed to be selected. Since this is a problem of classification, certain algorithms that can be used are- K-nearest classifier, Random Forest, Classification and Regression tree (CART), Logistic regression or Gradient boosting classifier. Now that we have all the required segments of our cycle, we can start training our model for more precise and accurate results and predictions. Figure 7 shows the life cycle of the model (Umer et al., 2021).

Figure 7. Flow of prediction model
(Umer et al., 2021)

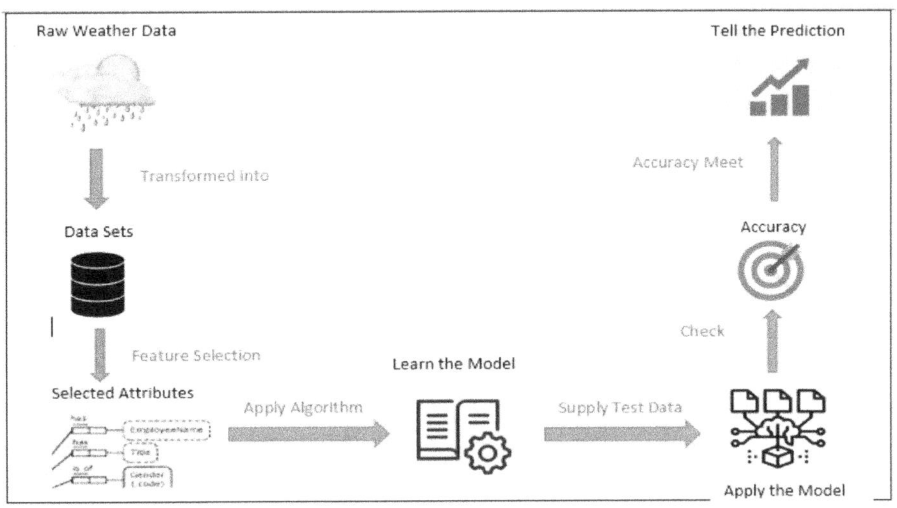

Renewable Energy Forecasting Approaches

With the increase in demand for resources for energy generation, classical or non-renewable resources are not viable solutions. Not only the fact that it takes millions of years of information but also that their use is degrading the environment. Thus, more and more demand for sustainable resources is seen in the last few decades. Furthermore, with the help of emerging IoT technologies, we can harness a significant amount of power, which includes predicting the right time and location as the prime factors that have been made more efficient with the help of several machine learning models implemented towards the cause.

Wind Power Forecasting

An estimation states that wind-generated electricity might contribute 19% of the total energy produced in the next ten years (Global W.E.C., 2015). To be able to achieve the goal, it is important to note the industrial cost and certain predictions as needed to maximize the output produced, which can be done by studying certain models. Wind power models can be categorized into several categories (Tascikaraoglu, 2018):

- Types of forecasting based on length of forecasting horizons ---long term/ short term
- Types of methods based on model structure----Model-based methods/ statistical methods
- Types of forecasting based on model output ---Point forecasting/probabilistic forecasting
- Types of models based on situations where data is used as input---Sapio-temporal models/ time-related models

As the term might suggest, the short term covers a period of a few days, while the long term covers a few months and, in some cases, years (Giebel et al., 2011). In the subsection discussed next, model-based methods, also known as Numerical Weather Prediction (NWP), prove to be highly efficient for a long time (such as the next day). But, as it approaches the short term (like in a few hours), its accuracy starts decreasing along with an increase in computational burden. That is where statistical methods prove to be effective as they are more accurate and can adjust themselves according to atmosphere changes. In situations of uncertainty in wind forecasts, the probabilistic model is given more preference than its counterpart. The incorporation of models in the last group describes the mingling of more than two models for achieving the accuracy in forecasting, especially for longer period.

Sapio-temporal is focused on the certainty that attributes of wind can be changed/ affected by their correlated sites and use this fact to provide more accurate results. In contrast, the time-based or temporal models looked only at time-based results. As proposed by Alexiadis et al., the use of ANN and SVM gives results with low error occurrence (Morales et al., 2010). Xie et al. Proposed the Trigonometric Direction Diurnal (TDD) model, a forecast approach for wind generation, which contemplate time series gathered from neighboring wind farms, stating its importance. Then, with Mirinda and Dunes' model, a multivariate time-series model was formed, taking both temporal and spatial factors into account.

Solar Power Forecasting

Around the world, about 580 GW of solar PV and 3.4 GW of thermal solar power were installed in 2019 (Bellini, 2020). An expected 87% total solar power capacity is to be achieved by 2030. The power harvested is affected by multiple factors like altitude, location, weather, season, sun position, atmospheric condition, etc. Properly predicting these factors plays an important role in adjusting the flow of power and control over or under-generation situations. The two approaches that could be adopted to maximize the solar irradiance are:

- Statistical methods
- Satellite-based methods,

The Least Absolute Shrinkage and Selection Operator (LASSO) regression method was proposed for short-term use, for linear regression. It was presented for sub-five-min solar irradiance forecasting using time series measured at a set of adjacent stations for different time lags (Yang et al., 2015).

For the long term, the use of ANN, linear model, SVM, GA (genetic algorithm) was proposed by Zagouras et al. Keeping in mind, the significance of unreliability about the forecasting for solar power operations, some approaches using spatio-temporal data have been discussed (Tascikaraoglu, 2018).

AQI Detection

With the emergence of IoT technologies, several detectors are made to detect the presence of SO2, NO, or any other particulate matter. Several local data stations can be created to collect the data passed from these sensors to local data houses. Even a mobile data collection van is being used to detect traces and possibly identify the main cause (Maran et al., 2021) (MTG, 2021). Figure 8 represents the data flow from sensors to the database. To process the data, though, several learning algorithms are being used across different locations. Some of them are:

Long-Short-Term-Memory

Mostly used method consists of input, output, and forget gate. This method uses pollution information, time-series data, and observation to predict data. "Mean absolute error" (MEA) and "Root mean squared error" (RMSE) is then used to calculate the performance further. A variation to the classic approach was tested in 2018 in which RNN and LSTM were used by an encoder-decoder (En-De) model, which allowed multiple layers of RNN to be stacked to increase the accuracy (Chang et al., 2020; Drewil & Al-Bahadili, 2021).

Spatiotemporal Deep Learning

Spatiotemporal (ST) data are spatial locations that correspond to multiple time series. This type of data is found in many dynamic systems, such as air quality readings. Spatiotemporal Deep Learning (STDL) mainly includes three modules namely spatial, deduction and a temporal module for prediction to deal with spatiotemporal features. One of the advantages of DTDL networks is extraction of the spatiotemporal features from the systems, which is why it is finer in prediction performance than temporal structures such as LSTM (Drewil & Al-Bahadili, 2021).

Convolutional Neural Network (CNN)

They are deep feed networks consisting of a series of convolutional layers used to extract the spatial features of neural networks. Samples are taken between two consecutive convolutional layers. These samples are used for maximal and intermediate aggregation. Researchers tested this in Seoul by taking a single image of the city and using CNN to solve the problem of Air Pollution Interpolation and its prediction. Then, they used spatiotemporal data (ST), which includes air pollution data, meteorological data, traffic volume, average driving speed, and air pollution indicators in outdoor areas. These spatiotemporal data were used in the proposed model, the Convolutional Long Short-Term Memory (Conv-LSTM), which has proven its superiority over the various models (Drewil & Al-Bahadili, 2021).

Figure 8. Flow diagram for air quality monitoring using IoT
(Maran et al., 2021)

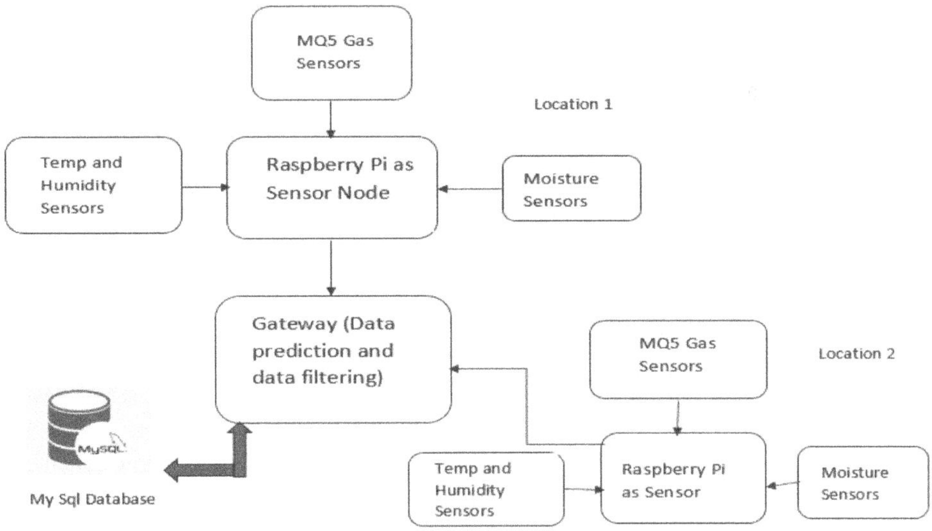

Future Scope

Even with the best of algorithms we have up to date, there are still a lot of uncertainties and factors that may affect or even change the prediction and findings all together. Thus, more and more data needs to be provided with variable aspects like humidity, irregular rain etc. and not just certain environmental changes that come with global warming and its effect on the environment. Moreover, not only are AQI predictions found and made, there is a need to find more and more ways to educate and inform people about it on a daily or a regular basis such that they can try to stay away or protected from high AQI zones.

ENERGY CONSUMPTION

Need

The previous section discussed how IoT and machine learning leads our way into becoming a smart city. However, an essential factor is still far from being discussed, i.e., energy consumption. From the day when AC/DC currents were introduced, the world changed completely. This change led to the industrial revolution, and ever since then, the power/energy demand has never been decreased to date. As we transition to smart cities, demand will also increase subsequently, which may cause an energy crisis soon. A proper and calculated use is required to be executed in order to keep the cycle moving smoothly.

To tackle this situation, many reviews are ongoing, and a constant study is being executed from the past few decades to monitor electricity usage, which in turn calculates future demand and even prevents wastage. Many results came out like a more improved distribution, management of waste, and best usage of resources available. Incorporating these suggestions, several sectors such as healthcare, industrial, security, education, and many more benefited. However, as we are progressing further towards becoming a smart city, the ever-increasing demands make it difficult to proceed with the traditional methods of mapping energy utilization. This is where advanced prediction and monitoring models assist us, and further help in tackling and possibly preventing any future energy crisis. This section will study such models and discuss their advantages and sectors where they need some improvement (Manimuthu & Dharshini, 2020; VE et al., 2021).

Energy Consumption in Industries

Industries are the most important sector in reference to a country's development and possibly the one with the highest energy demand and consumption. This section will discuss the existing literature and applications surrounding the energy requirements of industries, and predicting the same as accurately as possible.

Figure 9 shows the flow of proposed design by students of Sunchon National University (VE et al., 2021), based on their study of the industries of South Korea, which forms the basis for further discussion.

Figure 9. Flow of system to capture & analyze industrial energy data
(VE et al., 2021)

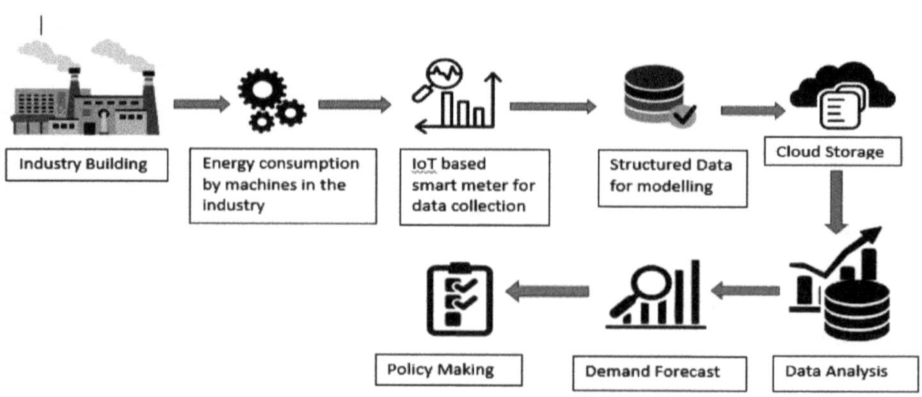

For the first step in the presented pipeline, the data needs to be collected from the industries. For that, smart sensors are being used to detect usage and consumption and this information is forwarded to local data houses for collection and preparation into proper datasets for processing. The data collected and sent through can vary from place to place due to several reasons. Thus, the sensors are calibrated according to the needs. Apart from geographical differences, factors such as time of day, weather, temperature, month, season, and whether it was a weekend a day before or not, also affect the variation in the collected data (Bacher et al., 2013). In their study, V E et al. (2021) configured the sensors to record data at every 15 mins in kWh, over a duration of twelve months. After that, the data is sent for analysis, which includes running it through ML algorithms to create prediction models. Popularly used algorithms include General Linear Regression (GLR), Support Vector Machine (SVM), Classification and Regression Tree (CART), CUBIST, K-Nearest Neighbours (KNN).

Besides using these features to increase the accuracy, they adopted feature engineering, which derives additional features from existing variables. This was followed by feature filtering, which involved generating heat maps every five weeks to work out the movement through algorithm and manual statistical tools.

The dataset holds several parameters or characteristics appropriate for obtaining the crucial characteristics that enhance energy consumption prediction for the industry. The Boruta package can be used to select all the appropriate factors (Kursa & Rudnicki, 2010). The Boruta algorithm is used to filter the variables, which further decreases the complexity of the model. It also helps to figure out factors that might be affecting the industries as a whole but might not be considered for certain sectors. For example, weather is considered as one of the most important aspects, but it might not matter that much when it comes to handloom industries.

Once the models are created, feature selection algorithms such as recursive feature elimination (RFE) can be used to find the optimal variable count required to reduce the root mean square error (RMSE), which evaluates the efficiency of an algorithm (Fan et al., 2014). V E et al. considered 18 predictors after the evaluation, which were then run through all regression models to reduce RMSE (VE et al., 2021). After prediction, the results are evaluated using the performance metrics of RMSE, mean absolute error (MAE), and coefficient of variation (CV). The data can then finally be prepared and submitted to the authorities, along with all the factors that may affect the energy production, as well as the solutions suggesting how it can be saved.

Following is a brief overview of the popularly used ML algorithms for creating prediction models to assess energy consumption in industries or households:

- GLR

General Linear Regression builds a generalization of the linear regression between the predictors and response and in (VE et al., 2021) it reveals that the dataset has an arbitrary model. Based on its distribution model, the connection between x (predictor factors) and y (reaction vector) is built using a connection feature. The design (X–Y) is depicted in equation 1.

$$g(E(y)) = x + O > yF \tag{1}$$

Here, g(•) is the link function, the predictor is x, the response variable is y, the offset variable is O, and the regression coefficient is β. The extra parameter in GLR boosts its flexibility (Apache Spark).

- Classification and Regression Tree

The CART model is used to make classification and regression trees according to the form of the target variable. A CART-based model is versatile enough to weigh misclassification expenses and reflect the previous distribution of likelihood in a classification task. Three impurity metrics can be used to detect CART model splits for a given target field. Furthermore, the target field is typically represented by Gini impurity, but the lowest squared deviation technique is used to select continuous goals without stating the option (Brownlee, 2020).

- K-Nearest Neighbors

KNN is a non-parametric learning algorithm that can be used to classify and/or predict data. In both circumstances, the K closest training examples in the feature space are used as input. Depending on whether KNN is used in classification or regression, the result differs. It is a simple learning algorithm still believed to be among the most efficient in practical applications (Harrison, 2018).

- Support Vector Machines

SVM aims to classify data in the highest hyperplane. To identify two groups that cannot be separated linearly, the data is converted and relocated to a high-dimensional feature space that can be separated linearly using kernel functions. Similarly, in this new high-dimensional feature space, the SVM performs non-linear data mapping with kernel functions before giving linear regression. On the one hand, the SVM model has higher generalization performance than conventional ANNs. Also, SVM could search for the optimal solution in the global region. Unlike other models, solutions from SVM generally do not get trapped into local regions, which results in the trouble of taking the local optimal solution as the best result in the global area. The disadvantage is that SVM cannot achieve the classification directly, which must be conducted through the dimension conversion. Besides, the training period of SVM is long. For the large-size problems, a large amount of computation time should be reserved (Gandhi, 2018).

- Cubist

The Quinlan M5 tree is the basis for the Cubist model. The Cubist technique generates a set of laws called "if-after-after." A linear multivariate model is associated with each of the created rules. Until the covariates meet the rule conditions, the appropriate model is utilized to evaluate the anticipated value. A linear mixture of the two models is used in the Cubistic method. The Cubist regression model is generally defined as follows- Many branches and leaves are produced throughout the growth of a tree. It is feasible to think of a series of "if-after-after" rules as branches, with the terminal leaves corresponding to a linear multivariable model (VE et al., 2021).

In their study (VE et al., 2021), all five ML-based prediction models were trained and tested in a 3:1 ratio of the dataset, using 10-fold-cross-validation. Table 3 shows the results of testing and training and helps in comparing the different models based on their evaluated performance along the three metrics of RMSE, MAE, and CV %. The model with the lowest RMSE, MAE, CV (%) is considered the best. According to VE et al. (2021), Cubist algorithm-based model performed the best. When compared to KNN, CART, and GLR, the performance of the SVM algorithm-based prediction model was significantly better. The SVM model demonstrated great stability in training and testing, thus ranking as the second-best option (VE et al., 2021).

Table 3. Performance evaluation of prediction models (VE et al., 2021)

Models	Training			Testing		
	RMSE	MAE	CV (%)	RMSE	MAE	CV (%)
GLR	4.6163	2.5387	16.8766	4.8586	2.5618	17.6764
CART	5.4005	2.9170	19.7434	5.6925	2.9569	20.7201
SVM	1.8930	1.5196	6.9205	1.9705	1.5438	7.1691
KNN	3.3078	1.4146	12.0930	3.6082	1.4826	13.1275
Cubist	0.1122	0.0322	0.4104	0.2410	0.0547	0.8770

Energy Consumption in HAN

HAN (Home Area Network) is the term used to discuss the local network that connects all the communication or, in our case, the power consumption of a house or apartment. Even though most of the techniques work the same as those discussed above for industries, we have fewer data to process per household or even take a group of houses like society. The data collection is way too small. Thus, there can be some compromise in terms of heavy algorithms to increase the processing speed.

Algorithms such as K-means clustering can be very beneficial and can be used to classify the data. The data used is collected from sensors and other measuring devices from appliances such as ACs, computers, etc. Mapping and collecting data can be done to check peaks in voltages and faults in circuits. All the data can be run through MATLAB. Thus, using GUI (Graphical User Interface) can be easily interpreted even by people with little or no knowledge on the matter.

Other than using algorithms as discussed, a library known as SVM can be used. This method can be cost-effective and runs without any large-scale security, which usually increases the computational time and thus increases output speed. The result on the run with just the AC's data, as performed by Manimuthu and Dharshini (2020), came with a 0.08 second error correction rate and a 3% error percentage.

Comparison

The related work and the ML-based approach which they adopt for their study on power needs, in the context of a smart city, are presented in Table 4.

Table 4. Comparative analysis of literature on energy consumption

Authors (Year)	ML-based Approach	Findings/ Results	Shortcomings Identified
V E et al. (2021)	Using Cubist algorithm to process the industrial data and provide insights on ways to reduce energy wastage and get maximum output	A more accurate and precise results were achieved	Results can possibly vary if the geographical location is changed as the results were completely based on industries in South Korea
V E et al., (2019)	Using SVM algorithm to process the household and industrial use. Suggestion of using hybrid models	Although, having commendable results with large chunks of data, SVM lacked in terms of speed	SVM can miss local optima and not reaching directly to classification can cause certain features being missed.
Manimuthu & Dharshini (2020)	Using smart sensing framework for early fault detection and using SVML to classify data	Data estimation (SVML) for both data flow and power flow in the simulation with a novel error value of only 0.05%, which is the least recorded value for smart home using SVML	Works efficiently in House Area Network (HAN) but still needs to be tested for industrial use.

Future Scope

Cubist shows promising results in the current day and time, and owing to its versatility, it is not bound to one type of dataset. Using Cubist can be a real game changer. Along with that, we include and consider several other factors such as solar energy generation, the weather, the sun's position, or in hydropower's case, the wave strength. These factors can also change the output of our energy needs' prediction for good. A hybrid algorithm approach is also proposed by (VE et al., 2021), using two or more algorithms, which may increase the processing speed and reduce the data load. Apart from that, new and advanced sensors can be built in the future to provide a clean and categorized dataset before reaching the database based on previous results, thus reducing the computational cost. In the future smart cities, billions of communicating IoT devices are projected to communicate wirelessly in the restricted spectrum of 5G and long-range technologies. This means that new agile technologies will have to overcome a large number of interferences without wasting power resources in IoT nodes (Lynggaard, 2020). Such issues are also in dire need to be looked at in order for existing technologies to improve.

TRAFFIC

Need

The invention of the wheel made traveling easy, and almost all trade and industries depend on travel, especially by road. The last two decades saw a rise in the production of automobiles. With them becoming affordable enough to be owned by the lower classes of society, automobiles saw an exponential rise on the road. The increase in number brought chaos on the roads, with authorities being unable to handle the rush manually. This introduced IoT in the game and the increased use of AI to find new ways to handle the ever-increasing traffic (Misbahuddin et al., 2015).

With cities moving to become smart cities, new and improved methods have been found to handle the traffic flow and make the roads function smoothly. Moreover, we are on our way to go completely electric, and with the introduction of driverless cars, a new set of rules and technologies are required to manage the traffic on roads. This section will discuss the workflow of managing traffic and how IoT and ML play their role in the same.

Traffic Management

Inefficient traffic management can be a significant hindrance in our movement towards becoming a smart city. This section will discuss some of the solutions to handle the traffic and ways to predict it beforehand to minimize the congestion on the roads.

Real-time Traffic Management

In 2019, countries like Rio de Janeiro, Bogotá experienced the worst traffic jams globally, with an average loss of 191 hours by a person stuck in traffic (Kopf, 2020). This amount of time wasted by the citizens creates an economic burden on the country and increases pollution, which again affects the citizen's health and the country overall. Real-time traffic management is the need of the hour. Humans alone are not capable of handling such traffic on their own. Thus, a real-time traffic management system is required to handle it, operating primarily on its own without human intervention.

An example of such a system can be controlling the traffic lights in order to regulate the traffic. One such system was implemented in the UK, called the SCOUT system. It is a fully adaptive system that regulates the traffic lights to reduce the stops and delays on the road (Robertson & Bretherton, 1991). After the success of this system, many similar and improved models came into effect (Crittin, 2004). Many of such models and systems led to systems based on RFID (Radio

Field Identification) technology which transform radio signals from vehicles into RF tags that can be tracked. Nowadays, this is done with GPS along with RFID (Misbahuddin et al., 2020).

So, with the idea to design a system that controls the traffic lights to handle traffic instead of a predefined time interval, two different approaches seemed to be the most cost-effective way of handling the traffic. The first approach introduced an ICRTMS (Intelligent Cross Road Traffic Management System), a system which proposed to have special sensors installed on roads (Salama et al., 2010). There were two types of those sensors, the first type installed at a proper place decided by the traffic department to have a full view of the road, and the other type categorized from s0 to s3 were laid out in strategic locations so that each uses photoelectric signals to measure the traffic weight on the road. This data is constantly updated in real-time to the main system installed in the control center, which provides the relevant instructions to the traffic lights and thus controls them. Salama et al. (2010) also suggested using RFID tags on emergency vehicles such as police cars, ambulances, fire brigades to send a specific signal to the furthest sensor, i.e., s3, so that necessary action can be taken timely. Figure 10 depicts the proposed working architecture of this model.

Figure 10. Architecture for ICRTMS
(Salama et al., 2010)

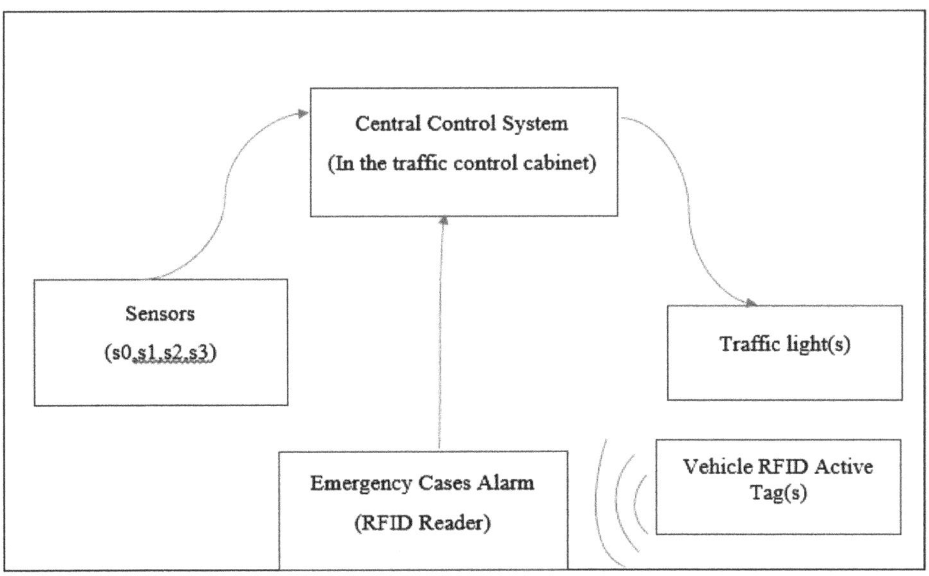

Architecture for ICRTMS

The other approach proposed a dynamic IoT-based solution, which involved tracking vehicles from RFID tags and controlling the traffic lights (Misbahuddin et al., 2020). Furthermore, they proposed a webcam to be installed at a certain point on the road that has access to the maximum distance and transmits regular data through HTTPS protocols from their software called GPIO (General Purpose Input/Output), which can be accessed through any device. This way, a real-time look of the traffic can be provided to the authorities and can help them implement necessary action if required. They also kept the possibility of software being hacked, thus transferring the data through the WebIOPi IoT framework that uses encrypted files to send the data. Besides the usual control, special access was also given to emergency vehicles (Misbahuddin et al., 2020).

Traffic Prediction Methods

With the rise in popularity of ITS (Intelligent Traffic System), there has been high usage of advanced hardware, development of more advanced sensors, and their fast production and implementation, introducing many impressive ideas which help regulate traffic. The ways to read and analyze the data collected by the hardware to regulate the traffic in real-time can help create and predict more and more future scenarios of possible reasons for blockage (Boukerche & Wang, 2020). The data is collected through several sensors, webcams, and other hardware devices and can be stored in a VC (Vehicular Cloud) (Salama et al., 2010). However, the problem with VC is that the redundant amounts of resources lying there create noise in all the future results. Thus, there is a need for a model that relates with the required data, according to the situation, and removes any redundant data available. But these models also need to be careful in terms of certain historical data as it can be used for particular references. Some of such models are (Sun et al., 2020):

- Regression model: The regression model aims to fit the dataset using a curve or a line by examining the connection between the dependent and independent variables.
- Example-based model: The example-based model handles the prediction job by evaluating the similarity between the input sequence and historical data samples and then making the final prediction using the found samples. The KNN model is one of the most preferred models of this type.
- Kernel-based model: The input data is mapped into a high-order vector space in a kernel-based model, making prediction problems simple to perform. SVM and RBF (Radial Basis Function) are the most suited to this type of model.

- NN model: An NN model is a sort of model created by mimicking the flow of information through neurons in the brain. In various types of NN models, the input data is transmitted via various network topologies. Using the activation function, the input data will be transformed into an activation signal in the network. Finally, the activation signal is used to make the final forecast. Models such as CNN, FNN, FFNN (Feedforward neural network) are the most used models of this type.
- Hybrid model: The final prediction in a hybrid model is made by integrating two or more prediction findings from distinct pre-prediction models.

Involvement of Big Data

With a large amount of data, such as traffic-related data, being used in today's day and age, its management becomes an expensive task. With the involvement of Big Data, this process tends to ease a bit. The nuggets example of Big Data is the Internet of Things. Among many features and layers, perception is one of the most useful in intelligent traffic systems, helping it analyze the data received from various sensors and other modes of sources. Through Big Data, integration of data becomes an easy task. Thus, it makes the available resources to be used at their optimum state and information easily shared with the public, which in turn helps formulate and execute the most efficient plan of action to handle the situation. Also, by providing other information such as alternate routes, weather, traffic conditions, etc., people will be able to obtain multiple travel plans according to their needs. Between the user and the traffic control system, data exchange and information transmission not only improves the user experience but also improves the efficiency and speed of data processing in public transportation, significantly saving the cost of operation (Chen et al., 2019). Figure 11 shows the different sectors or the areas of benefits of involving Big Data.

Figure 11. Big Data in the field of traffic control
(Chen et al., 2019)

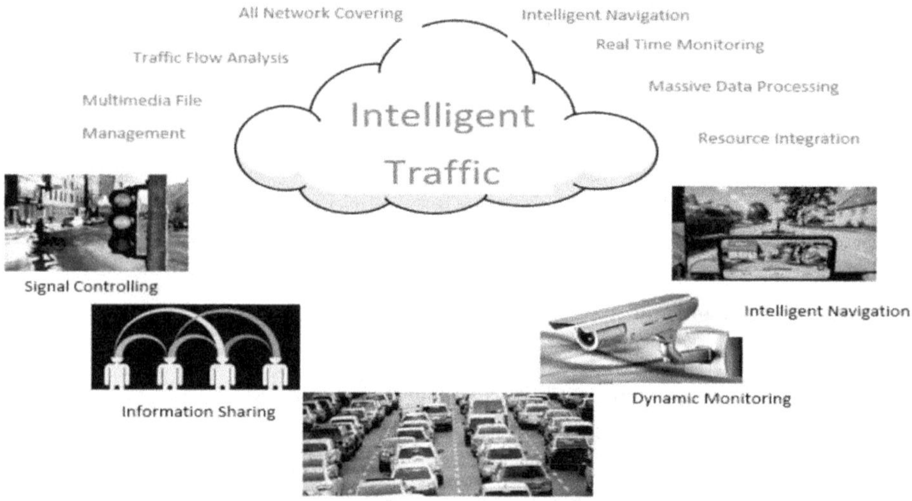

Conclusions

Both Misbahuddin et al. (2020) and Salama et al. (2010) have shown promising results in real-time tracking and management. However, a more informative and elaborate approach by Misbahuddin et al., (2020) gives it a slightly upper edge. Although both of them lack certain factors such as delay time, jitters, real-time update efficiency, possible road accidents analysis, etc., these can be resolved with the integration of Big Data and the use of GPS and satellite data. When it comes to prediction models, each has its flaws, as presented in Table 5.

Table 5. Comparison of different models (Boukerche & Wang, 2020)

Models ® Parameters -	**Regression**	**Example-Based**	**Kernel-Based**	**NN**			**Hybrid**
				FFNN	**RNN**	**CNN**	
Implementation difficulty	high	Medium	Medium	Less	Medium	Medium	Medium
Cost of implementation	Less	Less	Less	Medium	high	high	high
Dataset requirement	Less	Medium	Less	high	high	high	high
Deep-learning structure	û	û	û	ü	ü	ü	ü
Spatial-feature extraction	û	û	û	û	ü	û	ü
Temporal-feature extraction	û	ü	û	û	û	ü	ü
Prediction time cost	Less	Medium	Less	Medium	high	Medium	high
Maintain cost	Less	high	Less	high	high	high	high
Robustness	Less	Medium	Medium	Medium	high	high	high

Future Scope

With new and improving technologies, the traffic management system is getting better and better. With the involvement of Big Data in the system, the computation cost has decreased, and efficiency has increased drastically. Although improved, there is still much scope for improvement. As our society is moving towards becoming driverless, it opens many possibilities for new improvements and implementation. Linking the vehicle software with the traffic control systems will make it easy to track and communicate and reduce human error possibilities. Linking them with the upcoming 5G technology and providing customized information as per the user's journey can also improve the comfort of the journey and help save tons of resources (Pawłowicz et al., 2018).

SECURITY

Need

With the increase in the involvement of new technologies and everything going autonomous, the lives of many people have become easy. Still, few people find ways to use these same technologies to harm others and pose a threat to society. With the dawn of the digital world, where every detail of a person's life can be found on the internet, it has only added to the ease of taking over someone's identity and cause them harm in other ways. These attacks are executed not just on an individual's level, but also wide-scale hacks targeting firms or even nations are launched. An ideal society, in the presence of such security threats, can still find ways to prevent or stop such incidents. In this section, some of the ways of preventing security attacks and protecting from them are discussed. To resolve the issues at hand, ways to use new edge IoT devices and advanced ML approaches to help in detection and prevention of security incidents are presented.

Security in the Real World

With the evaluation of IoT technologies, new ways continue to come up for people to integrate them with their business and in their daily lives. Something as small as a Fitbit band to large-scale robots used in various industries, the technology has brought great change to the industries to perform jobs that would otherwise require much more production time and effort. But the integration of such technologies also created several security loopholes and risks which were then exploited by people for personal benefit or to harm others. Colonial pipeline hack in April 2021 in the USA can be a perfect example of this, wherein a group of hackers named Darkside hacked into the companies' mainframe and made them shut down the entire pipeline affecting the entire east coast and ultimately increasing the fuel prices (Turton & Mehrotra, 2021). Many such incidents have happened in the last decade which has made security to become an even more critical issue than ever. Some of the sectors having high possibility of attack are explored below, along with the ways to prevent them:

According to Li et al. (2019), a five-layer system consisting of perception layer, network layer, middleware, application and business layers was proposed to distribute the working of various industries. IIoT (Industrial Internet of Things) devices function in the bottom most layer and are typically found in open spaces. The layer above it is responsible for network connections, i.e., servers and APs (access points) for data transfers. Both the brain and fully-fledged ML applications run on the middleware.

The top two layers comprise of software and applications and are closer to the users. The general five layers in IIoT architecture are illustrated in figure 12.

Figure 12. Five-layer IoT architecture
(Magaia et al., 2020)

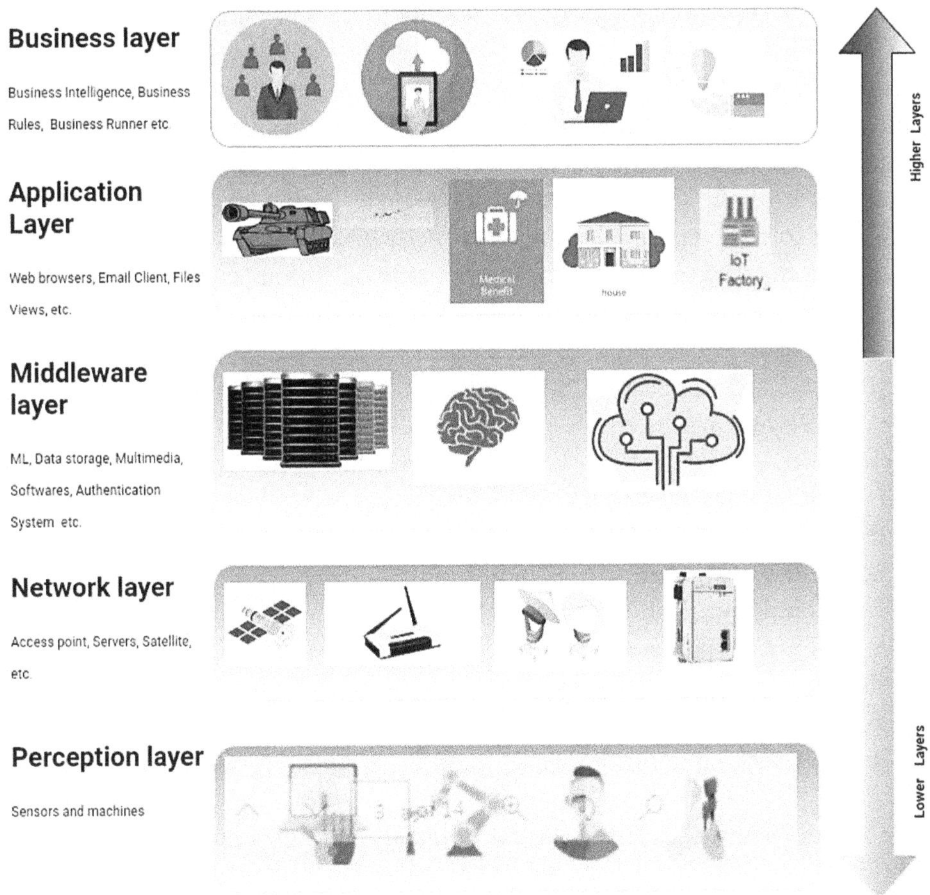

Using this architecture with deep learning integrated, although provides a lot of possibilities, but has a lot of loopholes as well. For instance, in the Colonial pipeline scenario, the most plausible cause of the hack found was one of the employees must have used the access password as a security credential somewhere else as well or might be using the same credential everywhere, which made the hacker work to only find the place with the least security to get those codes and enter into the pipelines' mainframe (Turton & Mehrotra, 2021). The lack of basic security knowledge is one of the prime causes of such cases. On an average, 40% of the time people either don't know about the firmware or know they have to update it but defer the process (Savjani, 2019).

As the architecture is divided into five parts, each of the parts has its own weak spots and vulnerabilities. In the perception layer, because IIoT devices operate in the outside environment, they are subject to a variety of physical attacks at the perception layer. An attacker's primary purpose is usually to steal or tamper with the sensor data. Man-in-the-middle attacks and DoS (denial of service) assaults are common in the network and middleware layers. Privilege escalation and structured query language (SQL) injections are two examples of threats that can affect business and applications in the same way that they might affect a traditional computer. Usually, the hackers target the software most due to the high accumulation of sensitive data. Also, these software and applications are usually where they inject malicious code to penetrate the system. Various attacks possible on various layers, few are listed below (Magaia et al., 2020):

1. Business and Application layer:
    a. Session hijacking
    b. Privilege escalation
    c. Return oriented attack
2. Middleware and Application layer-
    a. Man in Middle
    b. Integrity violation
    c. Eavesdropping
3. Perception layer
    a. Side-channel (power analysis, timing analysis)
    b. Depacking of the device

In terms of hardware security, it involves mainly two components- hardware trust and isolation of system. Basic concept behind these is the creation of cryptic keys. But certain issues are present here as well like high power consumption, clocks not being in sync in real-time, causing security risks. In terms of OS, isolation systems can run which allows the software to run without interference with the help of MMU (Memory Management Unit) but the issue here is there isn't a single centralized OS in IIoT. Running unwanted/harmful scripts in the OS can also be dangerous and is prevented by creating access specifiers and tokens. Most prior restriction of access use folders, processes, and imaginary objects, therefore creating a security system for IoT while utilizing the innate instinct about real objects is a compelling task (Magaia et al., 2020).

Preventive Solution Using ML and IoT Technologies

The different threats and the ways particular attacks can be specific to certain software or segments of IIoT are discussed previously, now solutions towards these and how ML supports the cause are discussed through Table 6.

Table 6. Comparison of preventive solutions using ML and IoT

Security Aspect	Publication	Application	Model	Technique	Function Approximation	Scalability	Reliability	Resource Constrained Device
Resource Management	Guo et al., 2019	Network	DRL	Service Function Chain (SFC)	Regression	Yes	No	Yes
	Luong et al., 2019	Network	DRL	SFC	Regression	Yes	No	Yes
	Sodhro et al., 2019	Device	ANN	Forward Central Dynamic and available	Regression	No	Yes	No
Authentication System	Chatterjee et al., 2018	Device	DNN	Radio Frequency Physical un-cloneable functions (RF-PUF)	Classification	No	Yes	No
	Ferdowsi & Saad, 2018	Device	DRL	Game Theory, Nash Equilibrium, LSTM	Classification	Yes	Yes	Yes
	Chauhan et al., 2018	Device	RNN	SVM, LSTM	Classification	Yes	No	Yes
	Liao et al., 2019	Device	DNN	SVM, LSTM	Regression	Yes	Yes	No
Surveillance	Hussain et al., 2019	Software	DNN	Bi-directional Long Short-term Memory	Classification	Yes	No	No
	Hussain et al., 2019	Software	CNN	LTSM	Classification	No	Yes	Yes
	Muhammad et al., 2019	Software	CNN	Hierarchical Weighted Fusion	Classification	No	Yes	Yes
	Muhammad et al., 2019	Software	DNN	Hierarchical Weighted Fusion	Classification	No	Yes	Yes
	Santana et al., 2019	Network	CNN	Siamese U-Nets and Semantic Segmentation	Regression	No	Yes	No
	Gao et al., 2020	Software	DNN	Two-path information propagation and stage-wise network refinement	Classification	No	Yes	No

Continued on following page

Table 6. Continued

Security Aspect	Publication	Application	Model	Technique	Function Approximation	Scalability	Reliability	Resource Constrained Device
Privacy Awareness	Liu et al., 2019	Software	CNN & DRL	Markov Decision Process	Classification	Yes	Yes	No
	He et al., 2018	Network	DNN	Post-Decision State Algorithm	Regression	No	Yes	No
	Zhang et al., 2019	Network	DNN	Proxy Re-Encryption	Regression	Yes	No	No
Attacks and Threads Detection	Ullah et al., 2019	Network	DNN & CNN	Term Frequency and Inverse Document Frequency and Logarithm of Term Frequency weighting techniques	Regression	No	Yes	No
	Azmoodeh et al., 2018	Device	DEL	Operational Code Sequence	Classification	Yes	No	Yes
	Abeshu & Chilamkurti, 2018	Device	DNN	Statistical Signal Processing	Classification	Yes	No	No
	Diro & Chilamkurti, 2018	Device	CNN	LSTM	Classification	No	Yes	Yes
	Aminanto et al., 2017	Network	DNN	Deep-Feature Extraction and Selection	Classification	Yes	No	No
	Li et al., 2019	Device	CNN	Dual Disaggregation and Aggregation DL models	Regression	No	No	Yes
	Tian et al., 2019	Software	DDL	Multiple Concurrent Models	Classification	Yes	No	No
	Garg et al., 2019	Software	CNN	Feature Selection and Grey Wolf Optimization.	Regression	Yes	Yes	No

Comparison

A comparative analysis of various aspects of security explored in the existing works is presented in Table 7.

Table 7. Comparison of Security aspects in existing works

Security Aspect	Publication	Advantages	Disadvantages
Resource Management	Guo et al., 2019	Even when using medium numbers of SFCs and MEC enablers for Blockchain, there is a small delay.	As MEC and SFC are used, it is not reliable
	Luong et al., 2019	Shows a decent response in terms of scalability and small processing.	Reliability can't be guaranteed
	Sodhro et al., 2019	Significant energy savings in data transfer communication by managing duty cycle with acceptable dependability	The method has a higher level of complexity to implement, and it has a substantial communication delay.
Authentication System	Chatterjee et al., 2018	Low error rate and very reliable	Depending on the application, there is a compromise between robustness and security.
	Ferdowsi & Saad, 2018	Dramatically increases the authentication rate as the target is to achieve a greater number of devices.	If there aren't enough targets, the success rate will be poor.
	Chauhan et al., 2018	High precision and it may be used on a device with little resources.	The RNN's training is too reliant on it.
	Liao et al., 2019	Accuracy is high and trains fast with data augmentation	It may require more resources.

Continued on following page

Table 7. Continued

Security Aspect	Publication	Advantages	Disadvantages
Surveillance	Hussain et al., 2019	High accuracy is achieved despite unsteady and low-quality image.	The cost of calculation is high, and it's not ideal for local processing.
	Hussain et al., 2019	Low computational power, allowing it to run on a local device	There is no way to notice unusual behavior, and there is no way to detect smoke or fire.
	Muhammad et al., 2019	It's suitable for both static and dynamic salience calculations.	There is no mechanism for learning sequences.
	Muhammad et al., 2019	Low computational power and honed for use in the industrial sector	Lacking in quantity and number of views, it is unsuitable for multi-view.
	Santana et al., 2019	Noise tolerance and multi-resolution capabilities	Only works in a petroleum and gas atmosphere.
	Gao et al., 2020	Capable of processing in a variety of angles and distances, as well as multiple object detection	Before the training, it is required to prepare and have certain information about the data, such as the size and any erroneous data.
Privacy Awareness	Liu et al., 2019	Error rate is lower even the quantity of devices increases	Time is a bad trade-off.
	He at al., 2018	Training speed for IoT devices is significant along with a fair balance of utility and privacy.	It may get significantly slow based on the neuron count.
	Zhang et al., 2019	For a limited number of devices, it is lightweight and effective.	In small numbers of devices, there is a significant overhead.

Continued on following page

Table 7. Continued

Security Aspect	Publication	Advantages	Disadvantages
Attacks and Threads Detection	Ullah et al., 2019	Increases data classification rate by combining the strengths of two DL techniques.	It can be computationally costly since it requires image processing, and it lacks scalability for the same reason.
	Azmoodeh et al., 2018	Code classification accuracy is high.	Limited testing and unsuitable for use in big networks.
	Abeshu & Chilamkurti, 2018	The environment that is both scalable and heterogeneous	There are too many simplistic comparisons and no performance evaluations provided.
	Diro & Chilamkurti, 2018	It is scalable and can be used in a heterogeneous setting.	Devices that are delay-sensitive are still having difficulties.
	Aminanto et al., 2017	Classifies attacks accurately and operates in a heterogeneous environment	It's just good for impersonation attacks and isn't ideal for devices with limited resources.
	Li et al., 2019	Attacks are quickly detected and classified.	Device interfaces are limited to physical interactions.
	Tian et al., 2019	For higher accuracy, it can be used in a variety of devices and with numerous distinct models.	Edge devices have a high computational cost.
	Garg et al., 2019	High-accuracy real-time reaction in a cloud environment	It doesn't have a malware detection system and doesn't understand the environment's diversity.

Future Scope

With the study on various techniques, we looked at, we can be certain that the security of the IIoT is protected by fullest at this moment but the possibility of new threats arising is always there. That's why researchers are working to learn and implement and retaliate on any such attacks but with the limit of being a human and being able to handle all the threats on their own. Thus, with the introduction of Deep Learning in the industry and making algorithms and technologies that learn and implement, the new age of security is been born, and with that creating it is the new Warfield. Figure 13 shows how a basic DL algorithm embedded in IoT works.

Figure 13. Representation of DL working architecture
(Magaia et al., 2020)

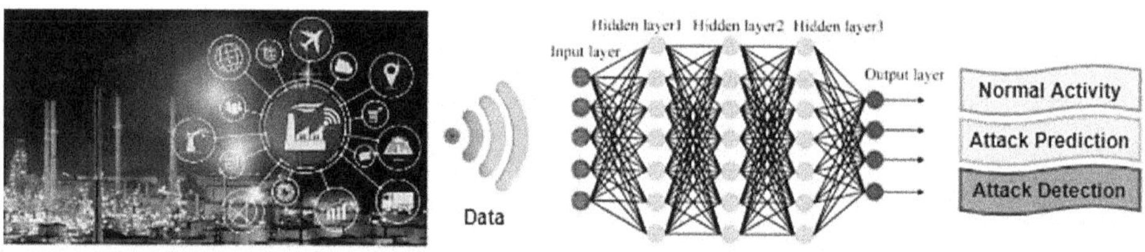

CONCLUSION

For a smart city, apart from what is available, advanced new sensors can be built in the future to provide a clean and categorized dataset before reaching the database based on previous results, thus reducing the computational cost. Billions of IoT devices will wirelessly communicate in the long-range and 5G technologies' restricted spectrum, as projected for the future smart cities. This requires the new agile technologies to have to overcome a large number of interferences while saving the power resources in the IoT nodes. Such issues are also in desire need to be looked at in order for existing technologies to improve. With the involvement of Big Data in the system, the computation cost has decreased, and efficiency has increased drastically. Although improved, there is still much scope for improvement. As our society is moving towards becoming driverless, it opens many possibilities for new improvements and implementation. Linking the vehicle software with the traffic control systems will make it easy to track and communicate and reduce human error possibilities. Linking them with the upcoming 5G technology and providing customized information as per the user's journey can also improve the comfort of the journey and help save tons of resources. With the study on various techniques explored through this chapter, it can be ascertained that the security of the IoT is protected by fullest at this moment but the possibility of new threats arising is always there. That's why researchers are working to learn and implement and retaliate on any such attacks but with the limit of being a human and being able to handle all the threats on their own. Thus, with the introduction of Deep Learning in the industry and making algorithms and technologies that learn and implement, the new age of security is been born, and with that creating it is the new war field.

REFERENCES

2019-2020 Australian Bushfires. (2019). *Center for Disaster Philanthropy (CDP)*. https://disasterphilanthropy.org/disaster/2019-australian-wildfires/

Abeshu, A., & Chilamkurti, N. (2018). Deep learning: The frontier for distributed attack detection in fog-to-things computing. *IEEE Communications Magazine*, *56*(2), 169–175.

Albino, V., Berardi, U., & Dangelico, R. M. (2015). Smart cities: Definitions, dimensions, performance, and initiatives. *Journal of Urban Technology*, *22*(1), 3–21.

Aminanto, M. E., Choi, R., Tanuwidjaja, H. C., Yoo, P. D., & Kim, K. (2017). Deep abstraction and weighted feature selection for Wi-Fi impersonation detection. *IEEE Transactions on Information Forensics and Security*, *13*(3), 621–636.

Azmoodeh, A., Dehghantanha, A., & Choo, K. K. R. (2018). Robust malware detection for internet of (battlefield) things devices using deep eigenspace learning. *IEEE Transactions on Sustainable Computing, 4*(1), 88-95.

Bacher, P., Madsen, H., Nielsen, H. A., & Perers, B. (2013). Short-term heat load forecasting for single family houses. *Energy and Building*, *65*, 101–112.

Bellini, E. (2020). World now has 583.5 GW of operational PV. *PV Magazine*. https://www.pv-magazine.com/2020/04/06/world-now-has-583-5-gw-of-operational-pv/

Boukerche, A., & Wang, J. (2020). Machine Learning-based traffic prediction models for Intelligent Transportation Systems. *Computer Networks*, *181*, 107530.

Brownlee, J. (2016). *Classification And Regression Trees for Machine Learning*. https://machinelearningmastery.com/classification-and-regression-trees-for-machine-learning/

Chang, Y. S., Chiao, H. T., Abimannan, S., Huang, Y. P., Tsai, Y. T., & Lin, K. M. (2020). An LSTM-based aggregated model for air pollution forecasting. *Atmospheric Pollution Research, 11*(8), 1451–1463.

Chatterjee, B., Das, D., Maity, S., & Sen, S. (2018). RF-PUF: Enhancing IoT security through authentication of wireless nodes using in-situ machine learning. *IEEE Internet of Things Journal, 6*(1), 388–398.

Chauhan, J., Seneviratne, S., Hu, Y., Misra, A., Seneviratne, A., & Lee, Y. (2018). Breathing-based authentication on resource-constrained IoT devices using recurrent neural networks. *Computer, 51*(5), 60–67.

Chen, X., Fang, H., Lin, T. Y., Vedantam, R., Gupta, S., Dollár, P., & Zitnick, C. L. (2015). *Microsoft coco captions: Data collection and evaluation server.* arXiv preprint arXiv:1504.00325.

Chen, X., Wang, H. H., & Tian, B. (2019). Visualization model of big data based on self-organizing feature map neural network and graphic theory for smart cities. *Cluster Computing, 22*(6), 13293–13305.

Classification and Regression. (n.d.). *Apache Spark.* https://spark.apache.org/docs/latest/ml-classification-regression.html#generalized-linear-regression

Crittin, F. (2004). *New algorithmic methods for real-time transportation problems* (Thesis). EPFL.

Diro, A., & Chilamkurti, N. (2018). Leveraging LSTM networks for attack detection in fog-to-things communications. *IEEE Communications Magazine, 56*(9), 124–130.

Drewil, G. I., & Al-Bahadili, R. J. (2021). Forecast Air Pollution in Smart City Using Deep Learning Techniques: A Review. *Multicultural Education, 7*(5).

Everingham, M., Van Gool, L., Williams, C. K., Winn, J., & Zisserman, A. (2010). The pascal visual object classes (voc) challenge. *International Journal of Computer Vision, 88*(2), 303–338.

Fan, C., Xiao, F., & Wang, S. (2014). Development of prediction models for next-day building energy consumption and peak power demand using data mining techniques. *Applied Energy, 127*, 1–10.

Ferdowsi, A., & Saad, W. (2018). Deep learning for signal authentication and security in massive internet-of-things systems. *IEEE Transactions on Communications, 67*(2), 1371–1387.

Gandhi, R. (2018). *Support Vector Machine — Introduction to Machine Learning Algorithms*. Towards Data Science. https://towardsdatascience.com/support-vector-machine-introduction-to-machine-learning-algorithms-934a444fca47

Gao, Z., Xu, C., Zhang, H., Li, S., & de Albuquerque, V. H. C. (2020). Trustful internet of surveillance things based on deeply represented visual co-saliency detection. *IEEE Internet of Things Journal*, *7*(5), 4092–4100.

Garg, S., Kaur, K., Kumar, N., Kaddoum, G., Zomaya, A. Y., & Ranjan, R. (2019). A hybrid deep learning-based model for anomaly detection in cloud datacenter networks. *IEEE eTransactions on Network and Service Management*, *16*(3), 924–935.

Giebel, G., Draxl, C., Brownsword, R., Kariniotakis, G., & Denhard, M. (2011). *The state-of-the-art in short-term prediction of wind power. A literature overview*. Academic Press.

Global Wind Energy Council. (2015). Global wind report: Annual market update. Global Wind Energy Council.

Guo, S., Dai, Y., Xu, S., Qiu, X., & Qi, F. (2019). Trusted cloud-edge network resource management: DRL-driven service function chain orchestration for IoT. *IEEE Internet of Things Journal*, *7*(7), 6010–6022.

Gupta, M., Abdelsalam, M., & Mittal, S. (2020). *Enabling and enforcing social distancing measures using smart city and its infrastructures: A COVID-19 Use case*. arXiv preprint arXiv:2004.09246.

Harrison, C., Eckman, B., Hamilton, R., Hartswick, P., Kalagnanam, J., Paraszczak, J., & Williams, P. (2010). Foundations for smarter cities. *IBM Journal of Research and Development*, *54*(4), 1–16.

Harrison, O. (2018). *Machine Learning Basics with the K-Nearest Neighbors Algorithm*. Towards Data Science. https://towardsdatascience.com/machine-learning-basics-with-the-k-nearest-neighbors-algorithm-6a6e71d01761

He, X., Jin, R., & Dai, H. (2018). Deep PDS-learning for privacy-aware offloading in MEC-enabled IoT. *IEEE Internet of Things Journal*, *6*(3), 4547–4555.

Hussain, T., Muhammad, K., Del Ser, J., Baik, S. W., & de Albuquerque, V. H. C. (2019). Intelligent embedded vision for summarization of multiview videos in IIoT. *IEEE Transactions on Industrial Informatics*, *16*(4), 2592–2602.

Hussain, T., Muhammad, K., Ullah, A., Cao, Z., Baik, S. W., & de Albuquerque, V. H. C. (2019). Cloud-assisted multiview video summarization using CNN and bidirectional LSTM. *IEEE Transactions on Industrial Informatics*, *16*(1), 77–86.

Kopf, D. (2020). *The countries with the worst traffic congestion - and ways to reduce it.* World Economic Forum. https://www.weforum.org/agenda/2020/07/cities-congestion-brazil-colombia-united-kingdom/#:~:text=Traffic%20analytics%20company%20Inrix%20puts,to%20traffic%20each%20year%2C%20respectively

Kursa, M. B., & Rudnicki, W. R. (2010). Feature selection with the Boruta package. *Journal of Statistical Software, 36*(11), 1–13.

Kuznetsova, A., Rom, H., Alldrin, N., Uijlings, J., Krasin, I., Pont-Tuset, J., ... Ferrari, V. (2020). The open images dataset v4. *International Journal of Computer Vision, 128*(7), 1956–1981.

Li, F., Shi, Y., Shinde, A., Ye, J., & Song, W. (2019). Enhanced cyber-physical security in internet of things through energy auditing. *IEEE Internet of Things Journal, 6*(3), 5224–5231.

Li, F., Shinde, A., Shi, Y., Ye, J., Li, X. Y., & Song, W. (2019). System statistics learning-based IoT security: Feasibility and suitability. *IEEE Internet of Things Journal, 6*(4), 6396–6403.

Liao, R. F., Wen, H., Chen, S., Xie, F., Pan, F., Tang, J., & Song, H. (2019). Multiuser physical layer authentication in Internet of Things with data augmentation. *IEEE Internet of Things Journal, 7*(3), 2077–2088.

Liu, Y., Wang, H., Peng, M., Guan, J., Xu, J., & Wang, Y. (2019). DeePGA: A privacy-preserving data aggregation game in crowdsensing via deep reinforcement learning. *IEEE Internet of Things Journal, 7*(5), 4113–4127.

Luong, N. C., Hoang, D. T., Gong, S., Niyato, D., Wang, P., Liang, Y. C., & Kim, D. I. (2019). Applications of deep reinforcement learning in communications and networking: A survey. *IEEE Communications Surveys and Tutorials, 21*(4), 3133–3174.

Lynggaard, P. (2020). Using neural networks to reduce sensor cluster interferences and power consumption in smart cities. *International Journal of Sensor Networks, 32*(1), 25–33.

Magaia, N., Fonseca, R., Muhammad, K., Segundo, A. H. F. N., Neto, A. V. L., & de Albuquerque, V. H. C. (2020). Industrial Internet-of-Things Security Enhanced With Deep Learning Approaches for Smart Cities. *IEEE Internet of Things Journal, 8*(8), 6393–6405.

Manimuthu, A., & Dharshini, V. (2020). *Framework for load power consumption in Han using machine learning and IoT assistance. IEEE Design & Test.*

Maran, P. S., Reddy, B. S., & Saiharshavardhan, C. (2021). Air Quality Prediction (IoT) Using Machine Learning. In *Advances in Electronics, Communication and Computing* (pp. 583–591). Springer.

Misbahuddin, S., Zubairi, J. A., Saggaf, A., Basuni, J., Sulaiman, A., & Al-Sofi, A. (2015, December). IoT based dynamic road traffic management for smart cities. In *2015 12th International conference on high-capacity optical networks and enabling/ emerging technologies (HONET)* (pp. 1-5). IEEE.

Mobile air quality monitoring with IoT. (2021). *Manx Technology Group*. https:// www.manxtechgroup.com/mobile-air-quality-monitoring-with-iot/

Morales, J. M., Minguez, R., & Conejo, A. J. (2010). A methodology to generate statistically dependent wind speed scenarios. *Applied Energy*, *87*(3), 843–855.

Mrinal. (2021). *Top 10 Open-Source Datasets for Object Detection In 2021*. Analytics Vidhya. https://www.analyticsvidhya.com/blog/2021/05/top-10-open-source-datasets-for-object-detection-machine-learning-in-2021/

Muhammad, K., Hussain, T., Del Ser, J., Palade, V., & De Albuquerque, V. H. C. (2019). DeepReS: A deep learning-based video summarization strategy for resource-constrained industrial surveillance scenarios. *IEEE Transactions on Industrial Informatics*, *16*(9), 5938–5947.

Muhammad, K., Hussain, T., Tanveer, M., Sannino, G., & de Albuquerque, V. H. C. (2019). Cost-effective video summarization using deep CNN with hierarchical weighted fusion for IoT surveillance networks. *IEEE Internet of Things Journal*, *7*(5), 4455–4463.

Nam, T., & Pardo, T. A. (2011, June). Conceptualizing smart city with dimensions of technology, people, and institutions. In *Proceedings of the 12th annual international digital government research conference: digital government innovation in challenging times* (pp. 282-291). Academic Press.

Pai, A. (2020). *Your Social Distancing Detection Tool: How to Build One using your Deep Learning Skills*. Analytics Vidhya. https://www.analyticsvidhya.com/ blog/2020/05/social-distancing-detection-tool-deep-learning/

Patient Exposure and the Air Quality Index. (2021). *United States Environmental Protection Agency (US EPA)*. https://www.epa.gov/pmcourse/patient-exposure-and-air-quality-index

Pawłowicz, B., Salach, M., & Trybus, B. (2018, September). Smart city traffic monitoring system based on 5G cellular network, RFID and machine learning. In *KKIO Software Engineering Conference* (pp. 151-165). Springer.

Punn, N. S., Sonbhadra, S. K., Agarwal, S., & Rai, G. (2020). *Monitoring COVID-19 social distancing with person detection and tracking via fine-tuned YOLO v3 and Deepsort techniques.* arXiv preprint arXiv:2005.01385.

Rezaei, M., & Azarmi, M. (2020). Deepsocial: Social distancing monitoring and infection risk assessment in covid-19 pandemic. *Applied Sciences (Basel, Switzerland)*, *10*(21), 7514.

Robertson, D. I., & Bretherton, R. D. (1991). Optimizing networks of traffic signals in real time-the SCOOT method. *IEEE Transactions on Vehicular Technology*, *40*(1), 11–15.

Russakovsky, O., Deng, J., Su, H., Krause, J., Satheesh, S., Ma, S., ... Fei-Fei, L. (2015). Imagenet large scale visual recognition challenge. *International Journal of Computer Vision*, *115*(3), 211–252.

Salama, A. S., Saleh, B. K., & Eassa, M. M. (2010, November). Intelligent cross road traffic management system (ICRTMS). In *2010 2nd International conference on computer technology and development* (pp. 27-31). IEEE.

Santana, M. C., Passos, L. A., Moreira, T. P., Colombo, D., de Albuquerque, V. H. C., & Papa, J. P. (2019). A novel siamese-based approach for scene change detection with applications to obstructed routes in hazardous environments. *IEEE Intelligent Systems*, *35*(1), 44–53.

Savjani, R. (2019). *Understanding firmware updates: what, why and when.* Einfochips. https://www.einfochips.com/blog/understanding-firmware-updates-the-whats-whys-and-hows/

Sodhro, A. H., Pirbhulal, S., & De Albuquerque, V. H. C. (2019). Artificial intelligence-driven mechanism for edge computing-based industrial applications. *IEEE Transactions on Industrial Informatics*, *15*(7), 4235–4243.

Sun, P., Aljeri, N., & Boukerche, A. (2020). Machine learning-based models for real-time traffic flow prediction in vehicular networks. *IEEE Network*, *34*(3), 178–185.

Surya, L., & Yarlagadda, R. T. (2020). AI Economical Smart Device to Identify Covid-19 Pandemic, and Alert on Social Distancing WHO Measures. *International Journal of Creative Research Thoughts.*

Tascikaraoglu, A. (2018). Evaluation of spatio-temporal forecasting methods in various smart city applications. *Renewable & Sustainable Energy Reviews*, *82*, 424–435.

Tian, Z., Luo, C., Qiu, J., Du, X., & Guizani, M. (2019). A distributed deep learning system for web attack detection on edge devices. *IEEE Transactions on Industrial Informatics*, *16*(3), 1963–1971.

Turton, W., & Mehrotra, K. (2021). *Hackers Breached Colonial Pipeline Using Compromised Password*. Bloomberg.https://www.bloomberg.com/news/articles/2021-06-04/hackers-breached-colonial-pipeline-using-compromised-password

Ullah, F., Naeem, H., Jabbar, S., Khalid, S., Latif, M. A., Al-Turjman, F., & Mostarda, L. (2019). Cyber security threats detection in internet of things using deep learning approach. *IEEE Access: Practical Innovations, Open Solutions*, *7*, 124379–124389.

Umer, M. A., Jilani, M. T., Junejo, K. N., Naz, S. A., & D'Silva, C. W. (2021). Role of Machine Learning in Weather Related Event Predictions for a Smart City. In *Machine Intelligence and Data Analytics for Sustainable Future Smart Cities* (pp. 49–63). Springer.

VE, S., Shin, C., & Cho, Y. (2021). Efficient energy consumption prediction model for a data analytic-enabled industry building in a smart city. *Building Research and Information*, *49*(1), 127–143.

Yang, D., Ye, Z., Lim, L. H. I., & Dong, Z. (2015). Very short-term irradiance forecasting using the lasso. *Solar Energy*, *114*, 314–326.

Yang, D., Yurtsever, E., Renganathan, V., Redmill, K. A., & Özgüner, Ü. (2021). A vision-based social distancing and critical density detection system for covid-19. *Sensors (Basel)*, *21*(13), 4608.

Zhang, X., Chen, X., Liu, J. K., & Xiang, Y. (2019). DeepPAR and DeepDPA: Privacy preserving and asynchronous deep learning for industrial IoT. *IEEE Transactions on Industrial Informatics*, *16*(3), 2081–2090.

Chapter 10
Machine Learning in Automobiles

Preety Khatri
Institute of Management Studies, Noida, India

Rohit Kumar
Institute of Management Studies, Noida, India

ABSTRACT

Maximum manufacturing operations in automotive industries are none the less largely dependent on experience-based human choices. The emergence of huge information, alongside gadget learning in automotive organizations, is leading to an increased degree of accuracy in choice making and advanced overall performance. It is pointless to mention that the automotive industry is getting ready for a revolution. The car is getting transformed by means of technologies. AI and system learning algorithms have found a growing degree of applicability in this industry. In this chapter, the authors have analyzed and study about how machine learning has been used for automobiles. Here they discussed autonomous vehicle test and development. In this chapter, they analyzed country-wise interest in autonomous cars and driverless cars. Motion planning, path planning, legal issues, and various applications related to machine learning for autonomous vehicles have been discussed in this chapter.

INTRODUCTION

One of the most thrilling era breakthroughs inside the previous few years has been the upward push of deep studying. Ultra-modern deep mastering models are being widely deployed in academia and enterprise, across a spread of areas, from photo

DOI: 10.4018/978-1-7998-9710-1.ch010

evaluation to herbal language processing. These models have grown from Fledgling research topics to mature strategies in real-global use. The growing scale of records, computational Strength and the associated algorithmic improvements are the main drivers for the development we see in this field. Those traits actually have a huge potential for the automobile industry and therefore the hobby in deep Mastering-based technology is growing. Lots of the product improvements, inclusive of self-using motors, parking and Lane-alternate assist or protection functions, which includes autonomous emergency braking, are powered by deep studying Algorithms (Gao et al., 2017). Deep mastering is poised to provide profits in overall performance and functionality for maximum ADAS (advanced driver assistance machine) solutions. Virtual sensing for automobile dynamics software, vehicle Inspection/heath tracking, automatic riding and records-driven product development are key regions which might be anticipated to get the most attention (Gang et al., 2017). This text gives an overview of the recent advances and some related challenges in deep learning techniques in the context of automobile packages (Zhou et al., 2017).

STATEMENT OF THE PROBLEM

Autonomous Vehicle Test and Development

A self-driving automobile, additionally known as an self-sufficient automobile (AV), related and autonomous car (CAV), driverless vehicle, robo-vehicle, or robotic automobile, is a car this is capable of sensing its surroundings and moving effectively with little or no human enter. Self-using cars combine a selection of sensors to perceive their environment, such as radar, lidar, sonar, GPS, odometry and inertial measurement gadgets (Gao et al., 2017). Advanced control systems interpret sensory data to identify suitable navigation paths, as well as limitations and applicable signage .Long-distance trucking is visible as being at the leading edge of adopting and enforcing the generation (Gang et al., 2017).

PURPOSE OF THE STUDY

Deep Learning in Automobiles

From voice assistants to self-riding automobiles, deep studying (DL) Is redefining the manner we engage with machines. DL has been Capable of reap breakthroughs in historically hard areas Of machine gaining knowledge of along with textual content-to-speech conversion, Image category,

Figure 1. Country-wise interest in autonomous car and driverless car

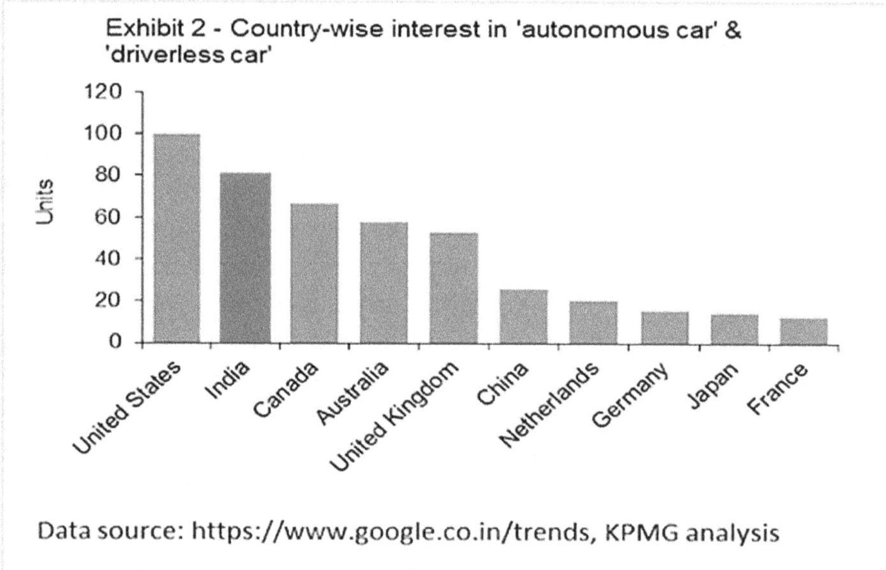

and speech reputation (Escobar & Morales-Menendez, 2018). DL commonly Refers to a category of models and algorithms primarily based on deep Artificial neural networks (ANN). Perceptron, the building Block of an ANN was first presented within the Fifties. ANNs had been a chief area of research in each neuroscience And laptop science till the late 1960s and the approach Then enjoyed a resurgence in the mid-1980s. At Bell Labs, Yann Le-Cun developed a number of DL algorithms within the Past due Eighties, together with the convolutional neural network (CNN). Pioneering deep neural networks via Yann Le-Cun ought to classify handwritten digits with properly speed and Accuracy and have been extensively deployed to read over 10% of all the cheques inside the U.S. inside the late 1990s and early 2000s. Despite the fact that the primary theories associated with ANN have existed For the reason that Nineteen Fifties, the field of DL has matured loads in the ultimate Decade and changed loads in the previous few years. The "deep" in Deep studying is not a connection with any kind of deeper Information accomplished through the approach; as a substitute, it stands for The many layers inside the neural community that make a contribution to a Version of the records (Ji et al., 2017).
There are 4 key factors that are riding the development and uptake of DL.

1. Extra Compute energy: e.g. Photographs processing unit (GPUs), tensor processing unit (TPUs) and so on.
2. Prepared large Datasets: e.g. ImageNet and so forth.

3. Higher Algorithms: e.g. Activation features like RELU, optimization schemes and so forth. Four. Software & Infrastructure: e.g. Git, robot operating System (ROS), Tensor Flow (Men et al., 2017), keras and so on (Men et al., 2017).

SIGNIFICANCE OF THE STUDY

Perception Systems for Object Detection

The belief system of a self-reliant car is answerable for mapping sensor observations into a semantic Description of the automobile's surroundings. 3D item detect a common feature inside this gadget and outputs a listing of 3-D bounding boxes around gadgets of interest. Diverse 3-D Object detection techniques have depended on fusion of different Sensor modalities to conquer barriers of person sensors (Wang et al., 2017). However, occlusion, limited subject-of-view and low-factor density of the sensor facts cannot be reliably and value-correctly addressed through multi-modal sensing from a single factor of view. Alternatively, Cooperative belief carries information from spatially various sensors distributed across the surroundings as a manner to mitigate those obstacles. This paper proposes two schemes for cooperative three-D item detection. The early fusion scheme combines point clouds from a couple of spatially diverse sensing factors of view before detection. In assessment, the late fusion scheme fuses the independently envisioned bounding boxes from more than one spatially diverse sensors. We evaluate the performance of each schemes using an artificial cooperative dataset created in two complicated using scenarios, a T-junction and a roundabout. The evaluation show that the early fusion approach outperforms overdue fusion via a good sized margin at the fee of better communication bandwidth (Men et al., 2017). The outcomes demonstrate that cooperative belief can take into account greater than ninety five% of the items as antagonistic to 30% for unmarried-point sensing within the maximum challenging scenario to offer practical insights into the deployment of such device, we document how the variety of sensors and their configuration impact the detection overall performance of the device.

Figure 2. Analysis of automated driving from 2000 to 2025 in future

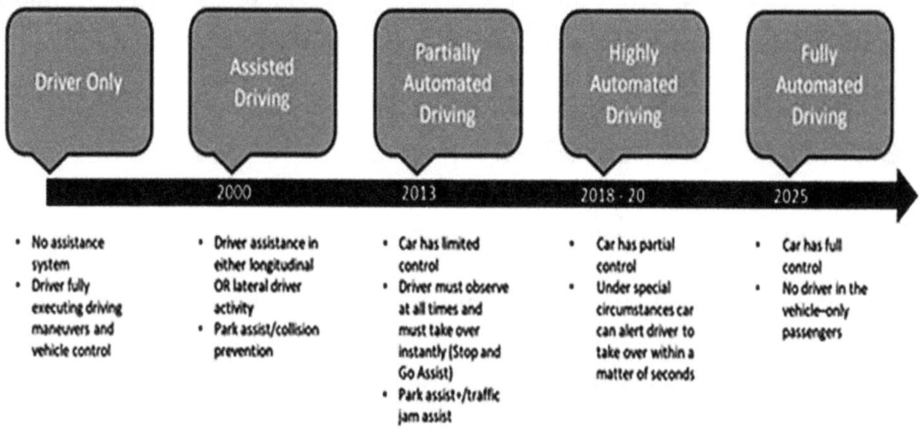

Motion Planning System

ARIA is an item-oriented automobile control utility that has a programming library (SDK) for C++ programmers who need to access their P3-DX platform and add-ons (Hu et al., 2018). However, the automobile firmware does no longer carry out any high-level robotic responsibilities. As an alternative, it's miles the activity of a clever purchaser walking on a connected pc to carry out this utility-level robotic control techniques and obligations, such as impediment detection and avoidance, localization, mapping, clever navigation and digicam manipulate. Indoor movement making plans of mobile robot deals with the troubles of localization, route making plans and the movement manage hows the motion making plans system used in this observe (Wang et al., 2017). Localization is the willpower of the positions of mobile robotic, boundaries and the target. Normally positions are given via a user or recognized by way of a digital camera. In the case of actual-time and dynamic working, consisting of moving obstacles or goals are used inside the environment, localization via a digital camera ought to be favored (Gao et al., 2017).

Because actual-time and dynamic packages are made in this study, a digital camera and photo processing techniques are used for localization (Ji et al., 2017). The digital camera is installed on the ceiling and it sends the actual-time pix of the surroundings to a pc. Each item is labeled with a extraordinary coloration on it; inexperienced for the boundaries, blue for the goal and yellow for the mobile robotic. So as to decide the heading angle of the mobile robotic, it's also categorized with a purple color. Course of the road that connects the centers of yellow and purple colored circles at the robotic equals to the heading attitude of it.

Figure 3. Pre-2010 Intelligent vehicle technologies

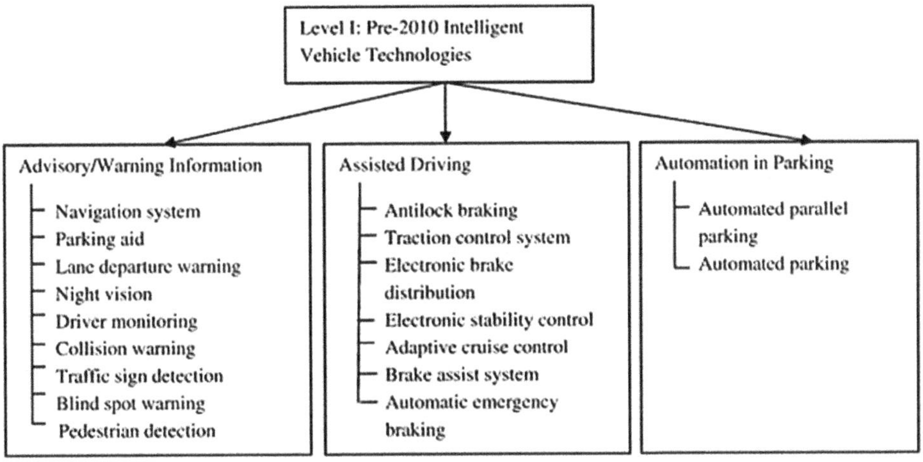

Photograph processing program going for walks on the pc, which is written in MATLAB, takes the immediately pix from the digicam. The program determines the coordinates of each item which has a coloration extraordinary from the bottom color, as proven within the higher proper photograph of the figure. Coordinate willpower is a complex hassle, due to the fact the real environment and its picture do not overlap because of the path of the digital camera. Therefore, coordinate transformation must be done with the aid of the usage of corner coordinates of the real surroundings and the photo. At the closing, photo processing software suits the coordinates of colours with a grid based totally map, as proven at the lower proper picture. On this way, localization is found out and the map is produced (Huang et al., 2017). The second one step of the motion making plans device is to send the map to the path planning technique. On this have a look at, genetic set of rules is used to decide the route which mobile robotic goes via it. Course planning with the genetic set of rules is explained in detail inside the following phase. Within the last step, with the intention to manipulate the robot platform, a patron software, which is written in MATLAB, sends command packets thru the robotic connection. This may be finished the use of direct instructions. Direct commands encompass -byte packet header, one-byte byte count, one-byte command wide variety, one-byte argument type, n-byte argument and two-byte checksum, as described by way of the robotic's running machine ARCOS. The direct command technique lets in sending any unusual or unique command immediately to the robot platform, without any intervening processing (Escobar & Morales-Menendez, 2018).

Path Planning and Optimization

Path making plans is a totally important task for the self -sustaining cell robotic. It is desired to discover a collision-loose motion in an obstacle inclined environment that allows you to navigate properly from the begin configuration to the aim configuration (Huang et al., 2017).

Figure 4. Path planning and view of automated vehicle

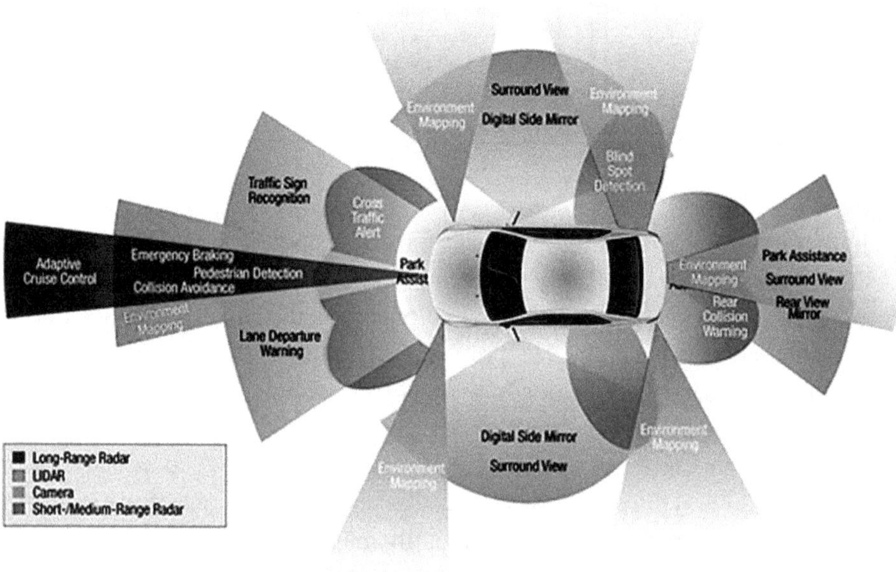

In lots of static and dynamic environments, cellular robots are more and more being hired. Typically, there are numerous possible paths for a robot to attain the goal from the begin region, however in situation, the first-rate feasible route is decided on in line with some guiding principle such as shortest distance, smoothness of the route, minimum energy consumption and many others. Or the most followed standards is the shortest distance with the minimal possible time. The path planning may be categorized into folds;

1. Neighborhood direction making plans and,
2. Worldwide direction making plans.

In neighborhood route making plans strategy, the robot has a restrained know-how (both in part-recognized or unknown) approximately the navigational environment. But, in worldwide course planning, the robotic has entire understanding approximately the navigational environment and thereby robotic can attain the goal by using following a predefined direction (Huang et al., 2017). However, worldwide direction planning techniques display constrained applications due to much less robustness in terrain uncertainty whereas, nearby route making plans techniques show greater flexibilities in partially known/unknown environments and presents an optimized course. It could be in addition categorized as classical technique and heuristic method (synthetic Intelligence method). The uses of cell robotic path making plans/arranging are included into restorative and surgical makes use of, person assist, safety, stockroom and movement packages, and moreover ocean and area investigation, robotized guided vehicles for shifting products in a plant, unmanned bomb switch robots, and planet investigation robots.

Addressing the Policy Issues Associated With Automated Driving Systems (ADS)

The convergence of records and communication technology (ICT) with automotive technology has already led to automation features in road vehicles and this fashion is expected to continue within the destiny owing to client call for, losing costs of additives, and stepped forward reliability (Gang et al., 2017). Even as the automation capabilities that have taken place thus far are specially in the form of data and driver warning technologies (labeled as level I pre-2010), destiny trends within the medium time period (level II 2010–2025) are anticipated to exhibit connected cognitive car features and embody increasing degree of automation inside the shape of advanced driving force assistance systems. Although autonomous automobiles were developed for research functions and are being tested in managed riding missions, the independent using case is most effective a long term (degree III 2025 +) state of affairs. It's on technological forecasts regarding automation, coverage challenges for each level of era improvement and alertness context, and the important device of fee-effectiveness for coverage analysis which enables policy choices at the automation systems to be assessed in a regular and balanced way (Gao et al., 2017). The fee of a machine in keeping with car is considered against its effectiveness in meeting policy objectives of enhancing safety, performance, mobility, convenience and lowering environmental results (Ji et al., 2017). Instance programs are supplied that illustrate the contribution of the technique in supplying facts for helping coverage selections. Given the uncertainties in device costs in addition to effectiveness, the device for assessing policies for destiny generation features probabilistic and utility-theoretic analysis capability. The coverage problems described and the evaluation framework permit the decision of policy challenges while allowing worth revolutionary automation in driving to enhance future road transportation.

Figure 5. On-going development of Autonomous vehicles (R&D)

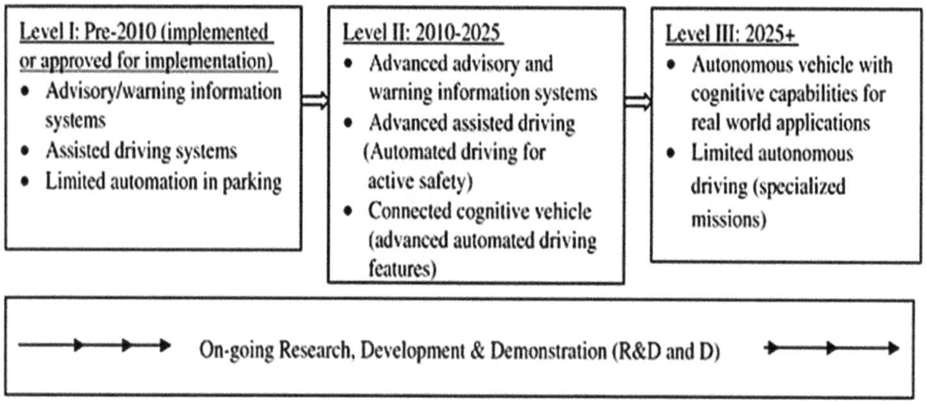

In recent years, fast trends in automobile technology have positioned public coverage in the capture-up mode. Advances in records and verbal exchange generation (ICT) have enabled the profession to go beyond the unique motive of the smart automobile and toll road machine (IVHS) initiative of many a long time ago and now we're within the technology of developing era for related cognitive cars (Gao et al., 2017). Further, experimental independent vehicle era has these days been tested efficaciously. The development and the scenario of huge-unfold applications of more and more automated vehicles in public street networks pose coverage demanding situations. Despite the fact that an economically possible self-sustaining vehicle is not possibly to be within the marketplace for decades, self-sufficient using as a public coverage problem has already emerged.

Legal Issues Surrounding Cyber Security and Privacy

Cyber security and facts safety are buzzwords in the meanwhile and for top motive. Banks and different financial establishments face constantly evolving cyber threats. The nature of the threat and the means through which cyber attacks are perpetrated are developing ever more sophisticated and the capacity fallout from a main cyber protection breach may be big.

The branch for business and skills' 2015 statistics protection Breaches survey observed that ninety percent of large firms had suffered a safety breach in the previous 12 months. The average price of the worst unmarried breach suffered by way of a big agency changed into a watch-watering £1.46m to £three.14m. Statistics protection breaches bring about loss of patron agree with, that can have an immediate impact on sales. However the fees of a security breach will even include enterprise disruption costs, compensation bills and regulatory fines (Wang et al., 2017).

The survey also observed that the character of cyber attacks experienced by using establishments has shifted, with a decrease number of denial of service assaults and a growth in attacks regarding malicious software. Possibly especially with the developing awareness of cyber security risks, inadvertent human error turned into mentioned because the principal cause of the worst security breaches, up to 50 percentage from 31 percentage in the preceding year.

Felony requirements with regards to cyber protection in the U.K rise up commonly from the statistics safety Act 1998, which requires organizations to take "appropriate technical and organizational measures" to guard non-public facts from unauthorized get admission to, damage, loss or disclosure. Such measures need to ensure a degree of protection that is suitable, taking into account the harm that may be precipitated to people in the event of a facts security breach and the character of the information. While determining which safety features to install area, the Act in addition specifies that companies ought to keep in mind the kingdom of technological improvement and the fees of enforcing the measures.

Human Factors of Automated Driving Systems

Automated riding can essentially exchange road transportation and improve nice of existence. However, at gift, the function of humans in automated automobiles (AVs) isn't always honestly established. Interviews had been conducted in April and might 2015 with twelve professional researchers inside the discipline of Human elements (HF) of computerized using to identify commonalities and exclusive views regarding HF challenges within the improvement of AVs. The experts indicated that an AV as much as SAE degree four have to tell its driver about the AV's abilities and operational repute, and ensure safety whilst converting between automated and manual modes (Gao et al., 2017). HF research should mainly address interactions between AVs, human drivers, and susceptible road users. Moreover, motive force education packages may additionally need to be modified to make sure that human beings are capable of the use of AVs. Subsequently, a mirrored image on the interviews is furnished, showing discordance among the interviewees' statements—which seem like consistent with an extended records of work on human factors studies, and the speedy improvement of automation era. We count on our angle to be instrumental for stakeholders worried in AV improvement and instructive to other parties. Automated using should essentially change road transportation and improve first-rate of lifestyles. But, the position of the human driver inside the computerized automobile isn't always yet without a doubt established. This work affords the results of an interview examine among 12 HF scientists worried in automated driving studies. A consensus become revealed many of the researchers regarding the HF demanding situations that need to be resolved previous to the deployment of AVs on public roads. Such demanding situations consist of the synergy among the human driving force and automation, capacity changes in driving behaviour due to automation, and the form of records that the human drivers will be receiving from the automatic riding device. Alternatively, a disparity was recognized between the researchers' concerns regarding the AVs improvement and deployment and the AVs technological advances: even though the researchers expressed that AVs should no longer be delivered except proven secure, reality indicates that enterprise is now near the advent of level three and stage 4 AVs on public roads (Huang et al., 2017).

Figure 6. Comparison of ownership cost & operating cost

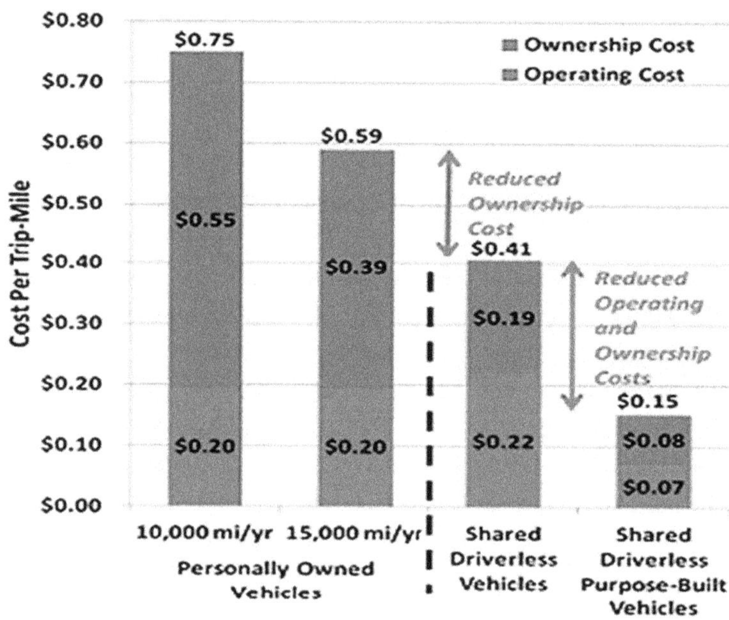

AUTONOMOUS VEHICLE APPLICATIONS

Automatic cars are an increasing number of present in modern-day society. Already, prototype motors had been automatic and judged dependable enough to force autonomously under widespread using situations. Beyond motors, there is a range of motors suitable for automation and throughout vehicle type there are not unusual hints recognized to standardize the procedure of making an autonomous car. In this text we overview what the commonplace elements are and illustrate with a few examples of automated floor vehicles developed with Freelance Robotics Pty Ltd.

Automatic Motors

The DARPA urban project did provide an incentive in the quest towards a totally computerized vehicle. DARPA has been at once related to a success development of independent motors via many studies college groups in addition to private sector. Industrial availability of a completely automatized car seems to be just around the corner. The remaining hurdles to business cognizance may be the important updates to road safety laws and, possibly greater tough to define, the overcoming of individual and network fears to accept as true with inside the wheels of an automated device on our roads (Gang et al., 2017). However, checks on present automated vehicles are proving their reliability, with take a look at outcomes higher than human judgments made on the road. Inside the cutting-edge marketplace, hooked up car agencies upload autonomous parts while they comply with the regulation. Commonplace examples of this emergence of automation in industrial cars encompass automated parking, computerized correction of the automobile's trajectory if the driving force crosses a continuous line, and alerts sounding to warn while other vehicles get too close (Shen et al., 2017).

Car Automation in Different Fields

The enterprise of vehicle automation is a whole lot broader than business vehicle automation. Freelance Robotics has worked on automation throughout many types of motors. As an example, we've evolved automation components for farming cars together with irrigators, tractors, and buggies, mining automobiles inclusive of drilling rigs, and also industrial vehicles like forklifts and car crash checking out automobiles. Civil engineering is an extra place of utility, with a success robots having been evolved for pipes inspection. Within the wider context of the automatic marketplace, those applications are only a few examples. Given the ability range and utility of applications, automatic vehicles include a clear increase marketplace.

Landmark Help in Nearby Positioning Structures

Any other tool for automated car localization is the usage of landmarks. Landmarks can be natural or artificially delivered to the automobile's environment. Whilst herbal, the vehicle identifies shapes which might be purported to be there, along with bushes, limitations, posts, et cetera.normally herbal landmarks can be complicated to pick out, reducing machine reliability. For example, a tree can alternate form, or boundaries appear exceptional from distinctive angles. In evaluation, artificial landmarks provide an clean, steady way for the automated vehicle to become aware of its function. Examples of appropriate artificial landmarks encompass reflectors, RF IDS, bar codes, traces on the floor, panels with geometric shapes and colours, lighting, wi-fi or IR beacons. When the automobile has correctly recognized positioning thru landmarks, it is able to update and music its position over time. In this way, dependable correction of positioning on its map turns into possible when the vehicle is a) in motion or b) moved in a regarded surroundings.

CONCLUSION

New era cars have severe tendencies and car brands are in opposition. Due to this opposition, sensible motive force assistance systems are playing a key role at the same time as automobile enterprise is being more automated. Researches show that finding a solution to parallel parking is one of the maximum wanted enhancements for drivers. Because parking is a very difficult subject matter for novice drivers. In particular in crowded cities this problem is getting larger due to the fact number of the automobile is increasing every day. (Gang et al., 2017) The purpose of the park assistant system is to assist the drivers have more fun and more efficient riding stories. Additionally some other intention is decreasing damages for the duration of the parking operation. Parking damages have very horrific effect on international's economic system. Due to the fact international locations and coverage organizations can pay cash for easy parking damages and also proprietors of the cars' sell their motors below its value due to harm records. System needs ultrasonic sensors that placed on the corners of the automobile. These sensors wanted now not only for doing parking moves, however additionally scanning the park area. There are a few problems about ultrasonic sensors (Shen et al., 2017). For instance, thin objects are now not seen through this sensors and some environmental adjustments as temperature, pressure, humidity, air turbulence, airborne particles and so on. Effect on ultrasonic reaction. In spite of those risks ultrasonic sensors are the most suitable sensor for these systems.

In this paper, we have discussed about machine learning in automobiles (Escobar & Morales-Menendez, 2018). This paper discussed about autonomous vehicle test and development, deep learning in Automobiles. In this paper we discussed about legal issues surrounding cyber security and privacy and autonomous Vehicle Applications also.

REFERENCES

Escobar, C., & Morales-Menendez, R. (2018). Machine learning techniques for quality control in high conformance manufacturing environment. *Advances in Mechanical Engineering*, *10*(2), 1–16. doi:10.1177/1687814018755519

Gang, L. H., Zhang, M. H., Zhang, L. F., & Hu, J. (2017). Automatic road vehiclemarking recognition for intelligent vehicle systems application. *Advances in Mechanical Engineering*, *9*(5), 1–11. doi:10.1177/1687814017706267

Gao, J. J., Xie, Y. N., Gu, F., Xiao, W., Hu, J., & Yu, W. (2017). A hybrid optimization approach to forecast automobile sales of China. *Advances in Mechanical Engineering*, *9*(8), 1–10. doi:10.1177/1687814017719422

Hu, Y. W., Wang, Y. P., Zhang, J. F., & Yu, H. (2018). Correlation degree analysis of arterial adjacent intersections for coordinated control subunit partition. *Advances in Mechanical Engineering*, *10*(1), 1–12. doi:10.1177/1687814017748748

Huang, H., Tang, Q. R., & Zhang, G. C. (2017). A novel approach to integrate potential field and interval type-2 fuzzy learning for the formation control of multiple AUVs. *Advances in Mechanical Engineering*, *9*, 1–11. doi:10.1177/1687814017737447

Ji, Y. F., Hu, X., Jiang, L. Y., Sun, J., & Wang, H. (2017). Algorithmic design andapplication of feedback control for coiling temperature inhot strip mill. *Advances in Mechanical Engineering*, *9*(9), 1–7. doi:10.1177/1687814017728647

Men, Y. Z., Yu, H. T., & Yu, H. B. (2017). Development of blockloading spectrum for car powertrain rig test correlatedwith customers' usage. *Advances in Mechanical Engineering*, *9*(9), 1–10. doi:10.1177/1687814017727472

Shen, X. P., Chen, F., Su, B. J., Chen, Q., & Yao, J. (2017). Optimization of park-and-ride system: A case study of Shunyi in Beijing. *Advances in Mechanical Engineering*, *9*(8), 1–8. doi:10.1177/1687814017714987

Wang, B., Gao, L. J., & Juan, Z. C. (2017). A trip detection model for individual smartphone-based GPS records with anovel evaluation method. *Advances in Mechanical Engineering*, *9*, 1–10.

Zhou, C. R., Jia, H. F., Gao, J. X., Yang, L., Feng, Y., & Tian, G. (2017). Travel mode detection method based on big smartphone global positioning sys-tem tracking data. *Advances in Mechanical Engineering*, *9*(6), 1–10. doi:10.1177/1687814017708134

Chapter 11
Object Detection Methods for Improving Smart City Safety

Kavita Srivastava
Institute of Information Technology and Management, Guru Gobind Singh Indraprastha University, India

ABSTRACT

The safety measures in a city are of major concern to its people. A smart surveillance system is necessary to make the safety measures robust and reliable. Currently, there are many object detection methods available for image analysis. Image analysis is only possible when we can detect the number of objects in the image, type of objects, as well as their location. Image analysis methods include Fast RCNN, YOLO, region-based convolution neural network (RCNN), and so on. These object detection methods are based on deep learning (DL) techniques. Video analysis requires the detection of moving objects. These methods involve background subtraction and extracting the foreground objects for motion analysis. There are several deep learning (DL)-based methods for video analysis such as the YOLO algorithm for detecting objects in a video frame and identifying their location. This chapter describes significant object detection methods in images and video that can be used in surveillance systems to improve the safety measures in a city.

INTRODUCTION

The safety measures in a city are of major concern to its people. A smart surveillance system is necessary to make the safety measures robust and reliable. Currently, there are many object detection methods available for image analysis. Image analysis is only possible when we can detect the number of objects in the image, type of objects, as

DOI: 10.4018/978-1-7998-9710-1.ch011

well as their location. Image analysis methods include Fast RCNN, YOLO, Region-Based Convolution Neural Network (RCNN), and so on. These object detection methods are based on Deep Learning (DL) techniques. Video analysis requires the detection of moving objects. These methods involve background subtraction and extracting the foreground objects for motion analysis. There are several Deep Learning (DL) based methods for video analysis such as the YOLO algorithm for detecting objects in a video frame and identifying their location. This chapter describes significant object detection methods in images and video that can be used in surveillance systems to improve the safety measures in a city.

A smart city aims to improve the quality of life of its citizens with the use of technology. The innovative solutions make the operations efficient and bring sustainability to mankind as well as the environment. Currently, cities across the world are on their way to become smarter by implementing smart solutions such as smart governance, smart energy, smart sanitation, and waste management, smart transportation, smart infrastructure, and smart healthcare to name a few. However, the cities overlook safety and security parameters. The use of technology helps people in many possible ways. But it has its drawbacks too. Evidently, the crimes in the cities are on rise. There are frequent incidents of traffic rule violations. So, the need for better security, and surveillance methods also arise.

The development of smart cities must ensure the safety needs of the people along with preserving the environment and ecosystem. Accordingly, a Smart City should also be a Safe City. A Safe City must adopt necessary measures to protect life, and property of people. At the same time, it should also ensure the privacy and security of data. Therefore, a smart and safe city needs to implement security and surveillance mechanisms by installing CCTV cameras. GPS-enabled vehicles help people in tracking the route they are traveling. The use of connected vehicles also enhances safety. A smart traffic management system can help in reducing traffic rule violations and prevent road accidents.

The smart security and surveillance system can help in the detection and identification of criminals, detecting suspicious activities and objects so that prompt action can be initiated. An efficient mechanism of CCTV footage analysis is also required. Smart security solutions also enable crisis management by providing early warning signs. For example, an indication of natural calamities such as floods, and cyclones can save the lives of millions of people.

Smart homes make use of many smart devices that collect large volumes of user's data daily. The security and privacy of user's data is also a concern that needs to be addressed. Hence, smart security solutions must also employ mechanisms to ensure the security and privacy of user's data.

Most of the security enhancement methods rely on CCTV footage and video surveillance. Therefore, Image analysis and video analysis is an integral component of security solutions. For instance, face detection methods play important role in many smart city applications such as smart door locks, biometric attendance, and criminal identification. Similarly, video analysis is required to detect traffic light violations, smart home security, and monitoring the security in schools, malls, and other public places. Self-driving vehicles should also have sophisticated mechanisms for image and video analysis. This is because these vehicles need to recognize the objects and also their relative distance so that they can apply the brake if necessary. Object detection methods also help in preserving resources by identifying the presence of people in a room so that the appliances such as air-conditioners, lights, and heaters can be automatically turned off and on.

There are object detection methods for image analysis as well as object analysis. Image analysis is only possible when we can detect the number of objects in the image, type of objects, as well as their location. Image analysis methods include Fast RCNN, Faster RCNN, YOLO, Region-Based Convolution Neural Network (RCNN), Region-Based Fully Convolution Neural Network (R-FCN), Single Shot Detector, (SSD), Histogram of Oriented Gradients (HOG), Spatial Pyramid Pooling (SPP-Net), and so on. These object detection methods are based on Deep Learning (DL) techniques. In particular, most of the object detection methods use Convolution Neural Networks (CNN).

Video analysis requires the detection of moving objects. These methods involve background subtraction and extracting the foreground objects for motion analysis. There are many Deep Learning (DL) based methods for video analysis. For instance, we can use the YOLO algorithm for detecting objects in a video frame and identifying their location. Further, we can use VGG-16 Network, VGG-19 Network, and Long Short Term Memory for object detection in a video.

Role of Object Detection in Smart City Security

The significance of Computer Vision and image analysis can't be underestimated in enhancing the security and surveillance. A smart city typically has lots of CCTV cameras installed on various places for monitoring. The images and video captured by these CCTV cameras must be analysed in real time in order to avoid dangerous situations.

The efficient methods of image and video analysis can help in averting many life-threatening incidents. Connected technologies can make it possible to fulfil the demands of operating smart cities. The technologies like Edge and Cloud enabled AI and Machine Learning are required to handle large amount of visual data. In fact, smart city solutions involve the use of technologies such as Internet of Things (IoT), AI, Machine Learning, and Deep Learning.

The following list provides a number of applications that are important for safety and security of smart cities. These applications require the use of computer vision and deep learning.

1. Detection of People at Restricted Areas

The real time video analysis can identify the presence of persons on a restricted area such as railway track. It can help prevent accidents.

2. Robberies and Theft detection in a home surveillance system.
3. Live mob analysis to prevent any dangerous behaviour by the crowd.
4. Preventing violence
5. Monitoring the workers' safety conditions at job sites.
6. Monitoring the city infrastructure such as roads, flyovers, and bridges to prevent accidents
7. Real time traffic rule violation detection to make the city safer for its people by reducing road accidents
8. Detecting suspicious activities such as detecting the people carrying weapons
9. Locating the parking space
10. Searching the missing persons

Apart from above applications, there are several other applications that are significant during pandemic. For example, detecting the people violating social distancing norms, mask detection, healthcare monitoring, and so on. These applications also require the use of deep learning for image and video analysis.

LITERATURE REVIEW

The research in object detection method started in 1990s. The earlier methods of object detection used handcrafted-features. Histogram of Oriented Gradients (HOG) is a significant method of object detection used before the emergence of Deep Learning.

Dalal et al., 2005 described a method of human detection using HOG. This paper demonstrates that the grids of Histogram of Oriented Gradients descriptors outperform the existing feature sets of human detection significantly. In this research work, a dataset of 1800 annotated images of persons was used. The images in the dataset comprises of wide range of pose variations and backgrounds.

He et al., 2008 presented a method for detecting humans in still images. A new feature descriptor called Scale Space Histogram of Oriented Gradients (SS-HOG) is proposed by combining Scale Space theory with Histogram of Oriented Gradients (HOG).

Shu et al., 2011, described the application of HOG in face recognition. In comparison to other existing methods, this approach achieved almost the same recognition rate. However, the computational time is significantly reduced. Hoang et al., 2014 discusses the application of HOG in Multi-Size Car detection. This study specifies two important improvements over the standard HOG. It solves the scaling problem by introducing Scalable Histogram of Oriented Gradients (SHOG). Further, it uses the laser data information instead of sliding window method. It achieves higher accuracy than standard HOG. In order to classify an object as car or non-car, it uses an SVM classifier.

Guennouni et al., 2014 presented an application of multiple object detection. The proposed method makes use of OpenCV libraries.

Girshick et al., 2015 presented the research paper entitled "Region-based convolutional networks for accurate object detection and segmentation", an object detection algorithm resulting in more than 50% higher accuracy of detection. They combined region proposals with CNN.

Zhao et al., 2015 presented an enhanced HOG for pedestrian detection. This approach combines the HOG features and the enhanced features. The detection results outperformed those obtained by applying the standard HOG.

Tang et al., 2017 discusses object detection methods based on Deep Learning. The study highlighted the differences between the traditional methods and Deep Learning based methods for object detection. Also, the challenges faced by deep learning methods are discussed.

Akcay et al., 2017 demonstrated the solutions of the problem of object detection in X-ray security imagery using the transfer learning approach. They have used a number of R-CNN and R-FCN variants. Faster R-CNN and R-FCN proved to be superior in comparison to the traditional sliding window driven CNN approach.

Kang et al., 2017 proposed an object detection system for advanced driver assistance systems in order to prevent accidents. The proposed object detection method is based on FCN with improved performance and realizes the real-time processing.

Yan et al., 2017 employed the R-FCN method for detecting vertical corner lines on buildings. Evidently. R-FCN performed better in comparison to traditional feature detection algorithms.

Zhu et al., 2017 introduced a fully convolution neural network called CoupleNet that couples the global structure information of the objects with their local parts. The proposed method obtains the proposals by RPN and feds them into the coupling module.

Li et al., 2018 addressed the problem of object detection in indoor scenes. The proposed object detector is based on the deep learning of small samples. In this study, the synthetic samples are augmented with available few samples. In addition, deep supervision learning along with deep prediction structure have been used in the Deep Convolution Neural Network.

Yustiawati et al., 2018 presented a method based on the Haar Cascade Classifier to identify the presence of objects.

Surasak et al., 2018 presented another application of HOG in detecting the presence of persons in a video. In this study, HOG is applied on every frame to detect and maintain the count of people in video. In their application, the three main tasks were performed – detection of people, counting people, and histogram generation.

Kanimozhi et al., 2019 proposed a faster object detection method based on Single Shot Multi-Box detection (SSD). The study reveals that the use of MobileNet along with SSD resulted in increased accuracy of real time detection of household objects.

Bhagya et al., 2019 presented a survey on various state of the art deep learning techniques for object detection. Also, it provides an extensive comparison of these techniques.

Wang et al., 2019 presented the design and implementation of an object detection system using Faster R-CNN. The proposed method shares full-image convolution features with the detection network and achieves higher accuracy.

Zhao et al., 2019 presents a review of deep learning based object detection methods. The study starts with a brief introduction on the history of deep learning. Further, generic object detection architecture along with some modifications are discussed. Along with it some prominent applications of object detection such as face detection and pedestrian detection are discussed.

Bratu et al, 2019 presents a comparison between two different object detection techniques. While Haar Cascade Classifier is a traditional object detection method, Deep Learning is the modern approach. The study presented a comparison of both approaches when applied on IoT devices.

Guo et al., 2020 proposed a cross-supervised learning approach for large-scale object detection. The proposed method termed as CS-R-FCN, utilizes the data flow of annotated images. A semantic aggregation strategy is also used. The proposed approach outperforms the previous related works.

Li et al., 2020 presented a review of object detection techniques. The paper discussed the traditional object detection methods such as HOG (Histogram of Oriented Gradients), SIFT (Scale Variant Feature Transform), SURF (Speeded Up Robust Features), and ORB (Oriented FAST and rotated BRIEF). Also, the methods based on Convolution Neural Network are discussed. Performance and features of various object detections methods are highlighted in this study. Some typical datasets such as INRIA, PASCAL VOC, ILSVRC, and COCO have been used in this study.

Zhang, J. et al., 2020 proposed a method called accurate R-FCN in order to improve the accuracy of R-FCN. In the proposed method a relational module is used to make the interaction between the appearance and features of the objects.

Zhang, L. et al., 2020 proposed HOG feature extraction without the use of normalization. In the proposed scheme, the normalization feature is merged with the gradient generation step. Instead of using linear difference based gradient, the logarithmic gradients are used. As a result, the speed of feature extraction is significantly improved.

Koçi et al., 2020 proposed a method for detecting threat objects such as explosives in X-ray images. For this purpose, Single Shot Detector (SSD), Region-based Fully Convolutional Neural Network (R-FCN), and Faster R-CNN were applied. Among these models, Faster-RCNN outperformed.

Zhou et al., 2020 proposed an approach where the HOG features are extracted directly from Bayer Pattern images with proper gamma compression. It significantly reduces the computational complexity and power consumption thereby making HOG more efficient.

Xu et al., 2020 proposed a method of improving the existing R-FCN method for object detection. In order to improve the speed and accuracy Whale Optimization algorithm based on SVM and novel anchor boxes are used.

Girshick et al., 2014 proposed a remarkable approach for object detection that combines high-capacity Convolution Neural Networks (CNN) to bottom up region proposals. It is basically done to localized and segment objects. Since, region proposals are combined with CNN, therefore the proposed approach is called as R-CNN (CNN with Region Proposals). With this approach mean Average Precision (mAP) is significantly improved. RCNN has three important parts. The first one is category-independent region proposals. The second one is a large Convolution Neural Network (CNN). While, the third part is a set of class-specific linear SVMs. Girshick et al., 2015, proposed another improvement over RCNN, known as Fast RCNN. Fast RCNN has shown improvement in both accuracy and speed of object detection. Gavrilescu et al., 2018 discussed an approach for real-time object detection known as Faster RCNN. In this study an example of using this approach is discussed that detects traffic indicators while driving a car. In Faster RCNN, the two techniques of Region Proposal Networks (RPN), and Fast RCNN are combined. Faster RCNN resulted in further improvement in mean Average Precision (mAP). Cao et al., 2019 described an approach for small object detection. It is basically the improved Faster RCNN. In this approach an improved loss function is proposed base on Intersection over Union (IoU) for bounding box regression. Bilinear interpolation is used to improve the regions of interest (RoI) pooling. It solves the problem of positioning deviation. The proposed approach is used to detect traffic signs. Liu et al., 2017 presented a study based on Faster RCNN. In this work, different neural networks are used to achieve the classification with Faster RCNN. Roh et al., 2017 discussed a method of refinement of Faster RCNN. The refining block and Faster RCNN are merged to get RF-RCNN. The proposed method shows improved performance over Faster RCNN. Abbas et al., 2018 discussed working of Faster RCNN. The research work mainly focuses on training the Faster RCNN. The paper discussed the training of network using custom based datasets. Qian et al., 2018 discussed the neural network inference. The proposed approach merges RCNN with NVIDIA Deep Learning hardware accelerator platform. Chen et al., 2019 describes the design and implementation of an object detection system using Faster RCNN. It is based on the earlier work on Faster RCNN. Yin et. al, 2020 discussed an enhanced Faster RCNN algorithm. The proposed work focuses on object detection in Aerial images. This study discusses the improvement of Faster RCNN by combining the multi-scale feature fusion method. Further, the residual module, and pooling layer are used to process the fused features.

Chen et al., 2019 discussed a method for object detection using deep learning. This paper discusses the Single Shot Detector for object detection with a refined feature fusion structure.

Sarda et al., 2021presented a method to detect objects appearing on the road using YOLO algorithm. Further, it classifies the objects using bounding boxes.

INTRODUCTION TO OBJECT DETECTION

Basically, object detection is more than the simple image classification. In case of image classification the outcome of classification provides a confidence score that tells about the class of the image. Whereas, object detection performs image classification as well as localization. Hence, object detection also determines the position of object in the image along with the information that whether the object is present in the image or not. Hence, we can say that image classification is a component of object detection. Moreover, object detection is not limited to identify a single object in the image. Rather, it can be used to detect multiple objects in the image along with their classes. Further, as a result of applying an object detection algorithm on an image, we get bounding boxes. Basically, a bounding box is a rectangle surrounding the detected object and serves the point of reference for the object. Fig. 1 shows the bounding boxes indicating the detected objects.

Figure 1. Bounding boxes indicating multiple objects in the image

Role of Deep Learning in Object Detection

Using Deep Learning techniques gives us several benefits over the use of handcrafted features for object detection. In brief, Deep Learning provides following functionality in object detection.

1. Deep Learning allows learning the distinguishing features in order to improve the accuracy of object detection and classification.
2. It is also possible to share the computation results while the training is still going on. It results in the improvement in speed.
3. Using Deep Learning the object detection model can be well-trained so that it can be used to detect new unseen images. Deep Learning allows learning enough feature distributions so that it can generalize well.

Object Detection Methods

The object detection task starts by taking an image or video as input and identifies the objects present in that image or video. Further, it classifies the objects by labelling them. In addition, object detection also identifies the location of objects in the image or video. Object Detection has been the topic of research since past two decades.

As a matter of fact, the research in this direction before the year 2014 comprises of research on traditional methods of object detection. After that, there was a drastic change in object detection methods with the advent of deep learning. Therefore, after the year 2014, research involves deep learning based object detection methods. Hence, the object detection methods are broadly classified into *Traditional Methods of Object Detection* and *Deep Learning Based Methods of Object Detection.*

- **Traditional Methods of Object Detection:** Earlier object detection methods use handcrafted features. Since they lack an effective representation of the image. Some of the early object detection methods are given below.
    - Viola Jones Detectors
    - Histogram of Oriented Gradients (HOG) detector
    - Deformable part-based model

Viola Jones Detectors

This particular detector was introduced by P. Viola and M. Jones for real-time detection of human faces. This detector uses the Haar wavelet for the feature representation of an image. For face detection, it uses the AdaBoost algorithm to select a small set of features. For reducing the overhead it uses the detection cascades.

Histogram of Oriented Gradients (HOG) Detector

Although, the concept of Histogram of Oriented Gradients (HOG) is first presented by Robert K. McKonell, in 1986, but it was used in applications later. In fact, its most prominent application is the Pedestrian Detection which was proposed in 2005 by N. Dalal and B. Triggs. Other applications of HOG detector are face detection, and people detection. Basically, this method divides the image into small squared cells. Further, the histogram of oriented gradients is computed in each cell. It normalizes the result and returns a descriptor for each cell.

Deformable Part-based Model (DPM)

Basically, DPM is an object detection method that is based on learning. Deformable part models are trained using a discriminative procedure. This procedure only needs bounding boxes for objects in images. These models make use of filters and form an effective system of object detection. These models represent highly variable object classes.

Problems With Handcrafted Features

Ultimately, the performance of hand-crafted features become saturated. Further, the advancement in Deep Learning techniques offered new opportunities in object detection techniques. As a result, learning becomes an essential component in object detection and Convolution Neural Network (CNN) has been used. With the help of CNN, the new object detection techniques have not only increased the accuracy of detection but also the speed of detection. Accordingly, the object detection has been performed with the help of neural networks.

- **Deep Learning Based Object Detection Methods:** The Deep Learning based object detection methods are broadly classified into two categories.
    - One Stage Object Detectors like YOLO, SSD, Transformer Based Neural Networks, and so on.
    - Two stages object detectors such as RCNN

The following section discusses different object detection methods in these two categories.

Two Stage Object Detectors

1. **RCNN:** R. Girshick proposed an object detection method that uses regions along with CNN. This method is known as RCNN. This model works in two stages. First of all, a set of object proposals are extracted by a selective search. Then, each proposal is transformed into a fixed size image. After that the image is fed as an input to a pre-trained network such as ImageNet or AlexNet. As an output of this task, we get the extracted features. Further a classifier such as an SVM classifier is applied. It determines the presence of an object within each region of the image. In addition, it also identifies the image category. The following procedure describes how to implement RCNN algorithm in Python.

    a. Import necessary libraries such as tensorflow and keras.
    b. Start selective search
    c. Calculate Intersection over Union (IoU)
    d. In the image dataset, each image should become the base for selective search. In order to do that, loop over image in the dataset.
    e. Retrieve the proposed regions
    f. Calculate IoU for all those images that pass the selective search.

g. Find IoU of the proposed region and the annotated region. Also include both negative and positive samples in the image dataset.
h. Now, since the training is complete, we can perform transfer learning
i. For transfer learning, we can further create training and test datasets and call the fit_generator() function of keras.
j. Once training is complete, we can perform predictions.
k. For this purpose, we pass the input image to selective search and its output to the trained model.
l. The trained model will predict whether, the image is an object or a background.

Although RCNN resulted in substantial performance improvement in terms of mean Average Precision (mAP), still there was one drawback. It performs lots of redundant computation that makes it slow. In fact, after RCNN, many variations of RCNN were implemented in order to improve the speed of object detection.

2. **SPPNet:** Spatial Pyramid Pooling (SPPNet) was also proposed in the year 2014. It doesn't require rescaling of the image. Instead it adds an additional layer called as Spatial Pyramid Pooling (SPP) layer in the CNN. This layer allows the generation of fixed-length representations regardless of the size of the image. Therefore, SPPNet avoids many repetitive computations. Although, it has increased some speed of detection, still it follows the two-stage approach. Also, it improves only the fully connected layer of CNN and ignores all previous layers.
3. **Fast RCNN:** Fast RCNN was proposed by R.Girshick in the year 2015. The improvement offered by Fast RCNN is that the training of the detector and bounding box regressor is done simultaneously. In other words, a single model is used for feature extraction from regions, classification, and returning bounding boxes. As a result both the performance and the speed of detection is increased. Still the speed is limited by proposal generation.
4. **Faster RCNN:** In the same year, R. Girshick introduced another deep learning based object detector known as Faster RCNN. Faster RCNN takes the advantage of Region Proposal Network (RPN). In fact selective search is replaced by region Proposal Network (RPN) in Faster RCNN. It is the first end-to-end, and the first near-real time object detection method based on deep learning. Hence, the individual blocks of the object detection system are integrated into a single unified framework. That becomes an end-to-end learning framework. However, the computation redundancy still remains there and forms the bottleneck in the speed of detection.

In short, first of all, the input images are fed to the convolution layers. It returns the feature map of images. When we get the feature maps, at the same time some anchor boxes are assigned to the image. Basically, anchor boxes are pre-defined bounding boxes having certain width and height. The reason of using anchor boxes is that the network predicts the probability of these anchor boxes rather than predicting bounding boxes directly. Further, the Region Proposal Network (RPN) is applied on the feature maps returned by the convolution layers. Hence, we get the region proposals in return. A classifier is applied on the returned proposals. Feature maps pass through the RoI pooling layer. This is done in order to make all proposals to the same size. Further these proposals are passed through fully connected layers to perform the classification. Hence, Faster RCNN can now predict the bounding boxes. The convolution layers present in the Faster RCNN are also termed as backbone. This backbone could be any pre-trained Convolution Neural Network such as VGG16, Inception v1, Inception v2, Inception v3, MobileNet, or ResNet.

RCNN is very accurate. But one of the drawback of RCNN is that it is very slow. This because, we have two NNs working together. However faster RCNN sometimes works much better than other approaches.

One Stage Object Detectors

The following section discusses some of the significant one-stage object detectors.

1. **You Only Look Once (YOLO):** In 2015, R.Joseph proposed a single-stage deep learning based object detector known as YOLO. In this approach, the proposal detection and verification task is completely eliminated. In this approach, the input image is divided into regions. Also, the bounding boxes and probability of each region is predicted simultaneously.

Since the emergence of YOLO, its five different versions are published. Hence YOLO can be implemented in five different ways.

Overview of YOLO

Basically, YOLO consists of a bunch of convolutions. Like RCNN and SSD, YOLO also utilizes anchor boxes. Based on the input image, we split the image into grids or cells. Each cell is having the same dimension. Each cell may contain an object. Therefore, each cell individually is responsible for detecting the object that it contains. Hence each cell predicts the bounding box and classifies the object contained in it.

YOLO also defines anchor boxes. Hence, it is similar to RCNN. But in terms of architecture, YOLO is more similar to Single Shot Detector (SSD). In fact, we get so many anchor boxes at different levels in YOLO.

YOLO v1

YOLO v1 divides the image into 13 X 13 cell. Further, each cell predicts 5 bounding boxes. As discussed before, the single neural network is applied to the full image. It is very fast and having a processing speed of 45 frames per second. However, it has one drawback in terms of large number of localization errors.

YOLO v2

YOLO v2 has many improvements over YOLO v1. First of all, it performs batch normalization. Batch normalization not only results in significant performance improvement, but also allow us to eliminate a dropout without overfitting. It also uses a high resolution (448 X 448) classifier that further improves performance. YOLO v2 performs convolution with anchor boxes. It applies k-means clustering on bounding boxes during training. Image size need not be fixed. Rather, we can vary it in every 10 batches. This improvement is termed as multi-scale training.

YOLO v3

It uses Darknet-53 having 59 convolution layers. Hence, it predicts more bounding boxes in comparison to YOLO v2.

YOLO v4

The main contribution of YOLO v4 is to have the CSPDarknet53 as the backbone network. In fact CSPDarknet53 has many features including Bag of Freebies (BoF) comprising of CutMix and Mosaic data augmentation, DropBlock, regularization, and Class label smoothing. YOLO v4 is divided into three parts – backbone, neck, and head.

The backbone part uses CSPDarknet53 with a feature called as Backbone. It has another feature called Bag of Specials (BoS).

BoS comprises of Mish activation, Cross stage partial connections (CSP), multi-input weighted residual connections (MiWRC).

Bag of Freebies (BoF) for detectors comprises of CIoU loss, CmBN, DropBlock regularization, Mosaic data augmentation, self-adversarial training, eliminate grid sensitivity, using multiple anchors for a single ground truth, Cosine annealing scheduler, optimal hyperparameters, random training shapes.

Bag of Specials (BoS) for detector comprises of Mish activation, SPP-block, SAM-block, PAN path-aggregation block, DIoU-NMS

The Neck part uses SPP, and PAN. While, the Head part comprises of YOLOv3.

YOLO v5

YOLO v5 is just YOLO v4 with native PyToch implementation.

SSD (Single Shot Multibox Detecter)

Basically, Single Shot Detector is a type of one stage Neural Network. SSD works as follows. It starts by having an image as input. It is passed to a backbone CNN. We get a feature map as a result. Then we have set of convolution layers. CNN layers subsequently become smaller and smaller. SSD is much more straight forward in comparison to faster RCNN.

So, we start with an image that passes through backbone CNN layers. Then we get the feature maps. We actually assign or generate anchor boxes from those feature maps. This is how SSD is different from the faster RCNN. So we get the anchor boxes for each cell or each pixel of the image. Then we use those generated anchor boxes in regression and classification so that we get the class of the object as well as bounding boxes.

Also we pass those feature maps to other convolution layers. The backbone convolution layers are mostly belong to pre-trained networks such as VGG or Inception. In contrast, the next set of convolution layers are basically ReLU layers. We do it several times with smaller layers. Then we pass the output from these layers to a function called non-maximum suppression. Finally, we get the output. There may be several bounding boxes for an object in the image. So we use an operation called IoU (Intersection over Union). We keep the bounding box with highest confidence. If IoU is more than 4.5 then we can say that the object that these bounding boxes surround will be of the same class.

Choosing the Right Neural Network for Object Detection Task

We need to consider following points while choosing a neural network for object detection:

1. Accuracy vs Inference Time
2. Accuracy Stratified by Object Size
3. Deploying the Model – on the cloud, or on the edge

Accuracy vs. Inference Time

Accuracy represents how precise the neural network is. While time represents the inference time taken by the neural network. In fact the accuracy of backbone networks determines the accuracy and precision of the overall model. The more accurate the backbone is, the better your object detector be.

Accuracy Stratified by Object Size

It means how our model will perform with respect to the object size. When the objects are large, the faster RCNN works best. It has the best mAP (mean average precision) for large objects. For medium sized objects, the faster RCNN still works best. For smaller objects, faster RCNN is best. So, if you have enough time, then you can go for Faster RCNN, since it is most accurate for objects of all sizes. However, for faster prediction, you can use SSD.

How the Image Resolution Affects the Object Detection Model?

For high resolution images (600X600), Faster RCNN is more accurate. Whereas for low resolution images (300X300) SSD works better. Although Faster RCNN is still the best. Further, YOLO is little bit slower than SSD, but its accuracy is better.

Overall YOLO is better than Faster RCNN in terms of precision and it is also better in terms of speed when compared to Faster RCNN.

REFERENCES

Abbas, S. M., & Singh, S. N. (2018, February). Region-based object detection and classification using faster R-CNN. In *2018 4th International Conference on Computational Intelligence & Communication Technology (CICT)* (pp. 1-6). IEEE.

Akcay, S., & Breckon, T. P. (2017, September). An evaluation of region based object detection strategies within x-ray baggage security imagery. In *2017 IEEE International Conference on Image Processing (ICIP)* (pp. 1337-1341). IEEE.

Bhagya, C., & Shyna, A. (2019, April). An overview of deep learning based object detection techniques. In *2019 1st International Conference on Innovations in Information and Communication Technology (ICIICT)* (pp. 1-6). IEEE.

Bratu, D. V., Moraru, S. A., & Guşeilă, L. G. (2019, August). A Performance Comparison between Deep Learning Network and Haar Cascade on an IoT Device. In *2019 International Conference on Sensing and Instrumentation in IoT Era (ISSI)* (pp. 1-6). IEEE.

Cao, C., Wang, B., Zhang, W., Zeng, X., Yan, X., Feng, Z., Liu, Y., & Wu, Z. (2019). An improved faster R-CNN for small object detection. *IEEE Access: Practical Innovations, Open Solutions*, *7*, 106838–106846.

Chen, S., Hong, J., Zhang, T., Li, J., & Guan, Y. (2019, August). Object detection using deep learning: Single shot detector with a refined feature-fusion structure. In *2019 IEEE International Conference on Real-time Computing and Robotics (RCAR)* (pp. 219-224). IEEE.

Dalal, N., & Triggs, B. (2005, June). Histograms of oriented gradients for human detection. In 2005 IEEE computer society conference on computer vision and pattern recognition (CVPR'05) (Vol. 1, pp. 886-893). IEEE. doi:10.1109/CVPR.2005.177

Gavrilescu, R., Zet, C., Foşalău, C., Skoczylas, M., & Cotovanu, D. (2018, October). Faster R-CNN: an approach to real-time object detection. In *2018 International Conference and Exposition on Electrical And Power Engineering (EPE)* (pp. 165-168). IEEE.

Girshick, R. (2015). Fast r-cnn. In *Proceedings of the IEEE international conference on computer vision* (pp. 1440-1448). IEEE.

Girshick, R., Donahue, J., Darrell, T., & Malik, J. (2014). Rich feature hierarchies for accurate object detection and semantic segmentation. In *Proceedings of the IEEE conference on computer vision and pattern recognition* (pp. 580-587).

Girshick, R., Donahue, J., Darrell, T., & Malik, J. (2015). Region-based convolutional networks for accurate object detection and segmentation. *IEEE Transactions on Pattern Analysis and Machine Intelligence*, *38*(1), 142–158.

Guennouni, S., Ahaitouf, A., & Mansouri, A. (2014, October). Multiple object detection using OpenCV on an embedded platform. In *2014 Third IEEE International Colloquium in Information Science and Technology (CIST)* (pp. 374-377). IEEE.

Guo, Y., Li, Y., & Wang, S. (2020, May). Cs-r-fcn: Cross-supervised learning for large-scale object detection. In *ICASSP 2020-2020 IEEE International Conference on Acoustics, Speech and Signal Processing (ICASSP)* (pp. 2553-2557). IEEE.

He, N., Cao, J., & Song, L. (2008, December). Scale space histogram of oriented gradients for human detection. In *2008 International Symposium on Information Science and Engineering* (Vol. 2, pp. 167-170). IEEE.

Hoang, V. D., Kurnianggoro, L., & Jo, K. H. (2014, November). Scalable histogram of oriented gradients for multi-size car detection. In *2014 10th France-Japan/8th Europe-Asia Congress on Mecatronics (MECATRONICS2014-Tokyo)* (pp. 228-231). IEEE.

Kang, M. S., & Lim, Y. C. (2017, November). High performance and fast object detection in road environments. In *2017 seventh international conference on image processing theory, tools and applications (IPTA)* (pp. 1-6). IEEE.

Kanimozhi, S., Gayathri, G., & Mala, T. (2019, February). Multiple Real-time object identification using Single shot Multi-Box detection. In *2019 International Conference on Computational Intelligence in Data Science (ICCIDS)* (pp. 1-5). IEEE.

Koçi, J., Topal, A. O., & Ali, M. (2020, December). Threat Object Detection in X-ray Images Using SSD, R-FCN and Faster R-CNN. In *2020 International Conference on Computing, Networking, Telecommunications & Engineering Sciences Applications (CoNTESA)* (pp. 10-15). IEEE.

Li, C., Zhang, Y., & Qu, Y. (2018, March). Object detection based on deep learning of small samples. In *2018 Tenth International Conference on Advanced Computational Intelligence (ICACI)* (pp. 449-454). IEEE.

Li, K., & Cao, L. (2020, May). A Review of Object Detection Techniques. In *2020 5th International Conference on Electromechanical Control Technology and Transportation (ICECTT)* (pp. 385-390). IEEE.

Liu, B., Zhao, W., & Sun, Q. (2017, October). Study of object detection based on Faster R-CNN. In *2017 Chinese Automation Congress (CAC)* (pp. 6233-6236). IEEE.

Qian, H., Xu, J., & Zhou, J. (2018). Object detection using deep convolutional neural networks. In *2018 Chinese Automation Congress (CAC)* (pp. 1151-1156). IEEE.

Qian, Y., Zheng, H., He, D., Zhang, Z., & Zhang, Z. (2018, August). R-CNN Object Detection Inference With Deep Learning Accelerator. In *2018 IEEE/CIC International Conference on Communications in China (ICCC Workshops)* (pp. 297-302). IEEE.

Roh, M. C., & Lee, J. Y. (2017, May). Refining faster-RCNN for accurate object detection. In *2017 fifteenth IAPR international conference on machine vision applications (MVA)* (pp. 514-517). IEEE.

Sarda, A., Dixit, S., & Bhan, A. (2021, February). Object Detection for Autonomous Driving using YOLO [You Only Look Once] algorithm. In *2021 Third International Conference on Intelligent Communication Technologies and Virtual Mobile Networks (ICICV)* (pp. 1370-1374). IEEE.

Shu, C., Ding, X., & Fang, C. (2011, April). Histogram of the oriented gradient for face recognition. *Tsinghua Science and Technology*, *16*(2), 216–224. doi:10.1016/S1007-0214(11)70032-3

Surasak, T., Takahiro, I., Cheng, C. H., Wang, C. E., & Sheng, P. Y. (2018, May). Histogram of oriented gradients for human detection in video. In *2018 5th International conference on business and industrial research (ICBIR)* (pp. 172-176). IEEE.

Tang, C., Feng, Y., Yang, X., Zheng, C., & Zhou, Y. (2017, July). The object detection based on deep learning. In *2017 4th international conference on information science and control engineering (ICISCE)* (pp. 723-728). IEEE.

Wang, C., & Peng, Z. (2019, June). Design and implementation of an object detection system using faster R-CNN. In *2019 International Conference on Robots & Intelligent System (ICRIS)* (pp. 204-206). IEEE.

Xu, C., Fan, J., & Liu, L. (2020, December). A novel object detection algorithm based on enhanced R-FCN and SVM. In *2020 IEEE Conference on Telecommunications, Optics and Computer Science (TOCS)* (pp. 363-367). IEEE.

Yan, L., Zhong, B., & Song, W. (2017, November). Region-based fully convolutional networks for vertical corner line detection. In *2017 International Symposium on Intelligent Signal Processing and Communication Systems (ISPACS)* (pp. 159-163). IEEE.

Yin, X., Yang, Y., Xu, H., Li, W., & Deng, J. (2020, December). Enhanced Faster-RCNN Algorithm for Object Detection in Aerial Images. In *2020 IEEE 9th Joint International Information Technology and Artificial Intelligence Conference (ITAIC)* (Vol. 9, pp. 2355-2358). IEEE.

Yustiawati, R., Husni, N. L., Evelina, E., Rasyad, S., Lutfi, I., Silvia, A., Alfarizal, N., & Rialita, A. (2018, October). Analyzing of Different Features Using Haar Cascade Classifier. In *2018 International Conference on Electrical Engineering and Computer Science (ICECOS)* (pp. 129-134). IEEE.

Zhang, J., Chen, S., & Hou, Y. (2020, November). Accurate Object Detection with Relation Module on Improved R-FCN. In *2020 Chinese Automation Congress (CAC)* (pp. 7131-7135). IEEE.

Zhang, L., Zhou, W., Li, J., Li, J., & Lou, X. (2020, December). Histogram of Oriented Gradients Feature Extraction Without Normalization. In *2020 IEEE Asia Pacific Conference on Circuits and Systems (APCCAS)* (pp. 252-255). IEEE.

Zhao, Y., Zhang, Y., Cheng, R., Wei, D., & Li, G. (2015). An enhanced histogram of oriented gradients for pedestrian detection. *IEEE Intelligent Transportation Systems Magazine*, *7*(3), 29–38.

Zhao, Z. Q., Zheng, P., Xu, S. T., & Wu, X. (2019). Object detection with deep learning: A review. *IEEE Transactions on Neural Networks and Learning Systems*, *30*(11), 3212–3232.

Zhou, W., Gao, S., Zhang, L., & Lou, X. (2020). Histogram of oriented gradients feature extraction from raw bayer pattern images. *IEEE Transactions on Circuits and Wystems. II, Express Briefs*, *67*(5), 946–950.

Zhu, Y., Zhao, C., Wang, J., Zhao, X., Wu, Y., & Lu, H. (2017). Couplenet: Coupling global structure with local parts for object detection. In *Proceedings of the IEEE international conference on computer vision* (pp. 4126-4134).

Chapter 12
Progression of Digital Healthcare With Deep Learning and Blockchain Methods for Smart Cities

Abeda Begum Mahammad
Bhagwant University, India

Rajeev Kumar
https://orcid.org/0000-0002-4141-1282
Chandigarh University, India

ABSTRACT

Healthcare is a fundamental feature of smart cities. Artificial intelligence-based systems usage during COVID-19 pandemic like virtual doctors, online consultations, telemedicine has enhanced the capability of reaching the treatment to patients on time, with high efficiency in smart cities. Governments have implemented patient monitoring and disease control systems during the COVID pandemic to reduce the spread of infection. The dynamics of healthcare provisions are changed exceptionally with the advent of deep learning and blockchain methods. Advanced levels of image diagnosis, centralized EHRs, emergency care, and intelligence-based recommendation systems are enhanced using deep learning and blockchain technologies. Integration of IoT devices with smartphones has enabled us to receive alerts to monitor the elderly or needy people with chronic conditions. The adaptation of AI in clinical diagnosis and predictive care systems is remarkable. High precision robotic systems based on AI technology are used in delicate and complicated surgeries in many city hospitals across the globe.

DOI: 10.4018/978-1-7998-9710-1.ch012

INTRODUCTION

With the advent of artificial intelligence, Blockchain and IOT systems the Smart Cities are getting developed with high efficiency and growth to improve the quality of human life in terms of health, finance, security, travel, and tourism. Digital Healthcare system is globalized, COVID-19 pandemic is the biggest example the way things are unfolded and speeded up from diagnosis, essential medicine provision, treatment procedures, monitoring patients at quarantine to vaccination roll out and implementation across the globe. It's evident that many developed and developing countries put their resources together to find solutions for COVID-19 vaccination. COVID-19 is a challenging situation for many countries, especially in cities as the population is dense, which could increase the spread of infections. Deep Learning Techniques such as Neural Networks are playing a prominent role in human disease diagnosis, emergency care systems, mobilization of medical facilities to reduce the time taken for diagnosis and treatment processes involved increasing the human life span and quality of life. In urban cities, population growth is enormous due to migrations after industrial revolution because of increased demands of human services. Processing large data sets into informative source using Deep Learning algorithms enabled huge developments in healthcare system. Artificial Intelligence and e-commerce systems enabled many services like virtual consultations, speedy and efficient diagnostic systems at an early stage of the disease, emergency care services, mobile health services and online medicine delivery systems to facilitate health services in Smart Cities.

AI, Blockchain and IOT Integration in Smart Cities

Smart Cities facilitate high end technology with well-established infrastructure with IOT device integration to provide a better quality of life in cities. Government and private organizations are collaborating in various projects to collect huge amounts of data and analysing this data to build strategies to increase healthcare systems. The key areas are economic growth, the mobility of services, communication & transport mode availability, better public service provision including immigration and citizenship services, environmental sustainability, co-existing of different nationalities & ethnicity people, modern healthcare systems, smart housing facilities & tourism opportunities. (Rocha, 2019)

Figure 1. Smart City model view

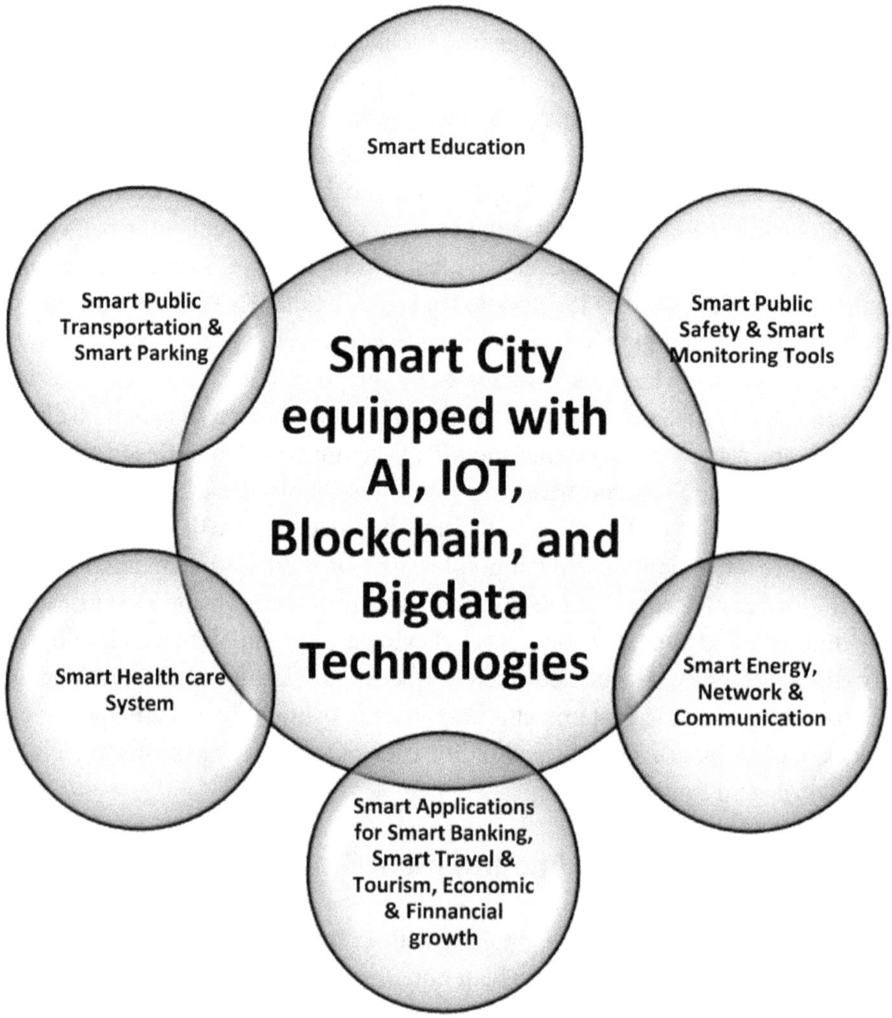

Smart cities are the solution for growing population needs in cities as the human intellectual grows to facilitate easy management in personal and professional time to lead a better quality of life. Rapid technology growth enhances the development of systems that enable automated systems such as IOT, Robotic devices for the individual service at home and workplace. With the augmentation of Artificial Intelligence, many human tasks are assigned to perform by the machines such as Robotic Vacuum Cleaner, Alexa. With the integration of these devices with IOT enabled much more safety and security of systems from Home Monitoring to Automatic Alerting Devices for older adults and kids at home such as Switching of household electronic items.

Smart Cities facilitate high end technology with well-established infrastructure integrated latest technology like AI, Blockchain and IOT systems.

With the integration of IOT systems with AI has provided many facilities in Smart Cities such as elderly people monitoring at home, medicine delivery management systems, handling emergency care situations. The electronic monitoring devices from BP monitoring systems, Diabetic monitoring systems to Intensive Care Units systems have made an enormous improvement in healthcare. Medical Practitioners can concentrate on patient care to provide better treatment and recommendations by offloading their monotonous work to machines. High population in smart cities causes the roads with heavy traffic congestion and jams, which put emergencies in trouble, hence big data and AI is helping come out of these problems and establish a peaceful urban environment lifestyle (Wang, 2020). Mobile health services are great help in seniors and needy people to avail the medical attention and procure safety from health hazards.

Machine Learning Algorithms in Disease Diagnosis

Machine Learning models are prevalent in Predictive analytics for disease diagnosis in high-risk patients to save their lives with early detection and preventive mechanisms. Data collection, processing, storing, and analysing have become the biggest source of healthcare technological advancement. Electronic Health Records are stored and centralized to retrieve the records in case of emergencies across the healthcare centres irrespective of their medical insurance providers. The governments are spending financial and human resources to address the needs of patients to improve the healthcare systems with EHRs storage and centralization. For example, Dubai Government has started a project on Electronic Medical Records to store and centralize EHRs for the patients' quick recovery and ease of use by medical professionals to save the patients with timely diagnosis and treatment. Governments like Dubai are strategizing the city to Smart City, investing and collaborating to create e-health services relentlessly to provide the higher quality of services to public and use of the latest technologies. In this process, electronic health records are created to collect the data of patients with patient details such as diagnostic information, treatment records, and medical prescriptions. These patient medical records can be shared across health providers to provide quick and efficient services to the patients in case of any emergencies, knowledge transfer and research purpose. These records were exceedingly helpful in COVID-19 pandemic situations. Dubai Health Authority is concentrating on 3 key factors, they are patient care redesign (healthcare design model), payment incentives (reward systems), and patient engagement systems. Even though, these strategies benefit healthcare systems like DHA provide financial benefits over the long term, but they provide high-quality services to patients (El-Hassan, 2017).

Critical diseases such as cardiovascular and cancer diseases are diagnosed and treated at an early stage of the disease with Deep Learning algorithms. With the invention of smartphone technology, common people are aware of using health devices to reduce the chances of entering diseases by adopting healthy lifestyle standards like eating nutritious food, following an exercise regimen, managing stress with various techniques that has increased the average lifetime of human.

Few decades ago, diseases like cancer, CVD, and psychological disorders such as anxiety disorders, depressions are rate to find, but nowadays these cases are common especially in the younger generation. Irrespective of the reasons for these disorders, we must tackle these problems and save the humanity to have a better quality of life. Many people will not be able to express these conditions due to the lack of understanding and societal denial factors, but we as a human community need to get a solution. In urban areas, especially these conditions are prevalent for which new technologies such as smart applications, online consultations, virtual clinics are big help to guide lifestyle changes like exercise, meditation, and eating nutritious food, providing psychological consultations to treat the patients for better community healthy. Youth are biggest sufferers of psychological disorders due to peer pressure, online games, lacking the motivation to lead a healthy life, we needed to address these critical issues to save our society and lead a healthy and happy life. Fortunately, many Governments and organizations are looking into this area to fix the issues in Smart Cities to provide high-quality services.

Healthcare applications enormously grown popularity nowadays due to developments in mobile and information technology. Individualized and personal care health apps are storing user data, providing timely alerts and recommendations in terms of health, exercise, meditation, and lifestyle in this busy era of urban population. Interactive applications are become popular. Public and private sector is investing in these apps to explore their commercial success and returns and promote their products to consumers, which is helpful for both consumers and companies. Targeted health behaviour and maintenance applications are big helpful tools for healthcare providers and organizations in addition to the consumers whether they are patients or healthy people who are willing to maintain their health (Boudreaux, 2014).

Data collected with the help of IOT devices such as public space usage, traffic movements, citizen mobility, drainage systems used for effective resource management in Smart Cities. The huge amount of data by IOT sensor devices is stored and analysed by researchers effectively using Deep Learning algorithms to better facilitate and use the resources in Smart Cities. Smart Cities are providing an ideal environment to live and work comfortably with the invention of new technology and infrastructure. For example, Parking spaces availability-checking systems (Smart parking system), airport-parking checkout procedures like consumer can make bill using smart machines, then the system will scrutinize whether vehicle can leave the parking space or should pay the bill, this procedure is avoiding the manual parking card payment system associated with delays and long queues. The sensor systems can read the vehicle number plate then decides whether bill has paid for the vehicle or not automatically, otherwise the bill is cleared at the exit gate to exit the vehicle from parking space. In this automation process, the vehicle owners no need to wait in the long queues to clear the bill at counter and exit points. Power supply and distribution are the biggest criteria in Smart Cities, which is managed efficiently with the help of high infrastructure technology such as smart street polls and solar system deployment in various public spaces like parks, bus, and metro stations to best use natural resources to preserve ecosystems inexpensively. Healthcare in Cities would benefit by DL routing mechanisms for better handling of network bottlenecks for data sourcing or transferring, autonomous self-driving refuse trucks for safer society by preventing accidents with automatic sensor systems, AI-based pedestrian detection systems to reduce accidents. Deep Learning technology is used extensively in the construction of buildings in Smart cities to facilitate home automation tools integrated with AI (Bhattacharya, 2020).

AI and Deep Learning Techniques in Image Analysis

In Smart cities, automated X-Ray, MRI, CT, PET scanning, image-screening systems are in place for quicker detection with minimal human efforts to reduce the cost of human resources and provide faster processing mechanisms using AI. Any healthcare provider can view these digital reports anytime in quick and safe method. Especially, DL techniques from CNN to autoencoders have helped to create solutions for medical imaging analysis and these methods are way better performed than human resources especially task-specific applications. AI-based applications are used heavily in the healthcare industry for patient monitoring systems remotely, discovery of new medicine, medical imaging and diagnostics systems, wearable systems like smart watches & gadgets integrated with mobiles, online consultations, virtual assistants, hospital management software. Many individuals use these devices nowadays due to affordability and ease of usage. Especially, in case of patients suffering from chronic conditions must be constantly aware of the potential risks for not following healthy diet intake and healthy lifestyle, for example cardiovascular diseases being one of the most widespread diseases across the globe. Patients must follow certain lifestyle to lead a better quality of life, which can be achieved with smart applications and gadgets. DL models are exceptionally used in detecting and classifying high-risk diseases like cancer, cardiovascular with high precision and accuracy compared to manual methods that reduced the time to detect anomalies with better accuracy rate. AI methods are used extensively for image-based technology in various departments such as radiology, pathology, dermatology, and ophthalmology. Complex image patterns are detected, segmented, and categorized with AI-based technology for better accuracy than humans to provide quantitative assessment. Physicians identify patterns based on their own experience which can vary the output, but the AI-based applications are trained with huge data to detect complex images with high accuracy and precision. In many hospitals, physicians are allowed to use these automated tools to quicken the process and provide better results to patients. Such imaging data can be used for research in scientific world and useful for feeding large and complex DL models to train for feature-engineering purposes. Radiologists are getting good support from AI-based systems to detect such as lung cancers, tumour classifications in the diagnosis of diseases, for better image quality, data visualization, assessment and generating reports. Medical images are stored in digital form with consistency in systems such as Picture Archiving and Communication System (PACS). (Hosny, 2018).

Many issues associated with computer vision, robotics and language processing are solved by DL models like Artificial Neural Networks. Many famous and top organizations such as Microsoft and IBM are investing in AI to invent new applications in healthcare systems. Machine Learning models are trained with the data as input and test these models after validation to get better accuracy and precision, the larger the data available to train, the better the accuracy of results. ANN models are used extensively in ML due to the advancements in Bigdata and other AI related technologies. DL is used in radiotherapy, radiomics, for theragnostic in neurosurgical imaging to detect CLE images, in prostate biopsy, and in image-guided therapy, radio genomics, neuroimaging, neuroradiology, neuropsychiatric disorders, breast cancer, mammography, chest imaging, imaging in oncology, medical ultrasound, and medical image analysis. DL is used in Image segmentation, in predicting disease based on images and text reports in various organs like the brain, kidney, prostate and spine and in image synthesis. For example, 3D images are used to differentiate cysts, polyps, or fibroids in a uterus, it's a remarkable improvement in healthcare, many times the physicians unable to differentiate these and end up with repeated D&Cs to remove them, but they keep showing in the scanning, if fibroid is there it must be operated with laparoscopy (Lundervold, 2019).

68% of the population will reside in urban areas by 2050 (Kundu, 2019), in the process of sustenance the ecological environment gets destroyed for constructing living places, providing uninterrupted power supply, and to provide better living systems including roads and transportation. Smart cities are providing a solution to these problems by using the resources to the best usage to provide a better quality of life in terms of health, economy, happiness by preserving the natural eco system. In this process, ML and Data Mining are playing a major role in creating Smart cities (Souza, 2019).

AI in Computer Vision

Deep Learning is used in Computer Vision. CV investigates digital images of radiology and pathology reports with high efficiency and specificity. CV used in applications such as screening in ophthalmology to detect diabetic retinopathy and automated detection of cancerous lesions in dermatology department. DL and CV jointly used in creating tools for radiation and magnetic exposure prediction. CV usage in surgery increasing day by day, the surgical procedure data are stored for analysing, using for new applications models' creation, educating other physicians, and for research. Surgeons are storing the data of Laparoscopy, endoscopy, and robotic camera procedures for future use, which is enhancing the healthcare industry growth in robotics and AI. Natural Language Processing combined with Computer Vision is a remarkable improvement in the field of healthcare. NLP is used for human language translations, unstructured data, to analyse and structure reports such as radiology, operative notes, for predicting the status of patient's health, for diagnosing cancer patient's risk levels, and to find operative site infections. To predict the risks associated with operation before it has taken place gives better understanding of the situations and provide an opportunity to patients' and for physicians to make better decisions about surgery. Especially, cardiac, and metabolic related events are important to know these risks. The latest ML algorithms are used to make these risk predictions for estimating the outcomes with better accuracy. Duke University performed an experiment with three different models using Lasso Regression, RF models, and extreme GBM, out of these three models Lasso Regression has shown better accuracy to predict post-operative mortality rates and comorbidity predictions. Electronic Medical Records are not only to store the patients' medical history, but they are also used for the training and educating physicians, for example AI systems are studying EMRs to evaluate the best surgical procedures and recommendation systems based on the surgeries taken place earlier on a patient. Massachusetts General Hospital launched risk predictor tool based on Machine Learning algorithm, it has scored the best performance compared to other tools. Combining Bigdata with ML techniques has led to easier data storage and evolution of new applications to use in clinical and surgical environments with much better accuracy and precision than before, by merging EHRs has evolved into higher accurate pre-operative and post-operative predictions. Computer Vision is playing a major role in guiding and assisting in Operative Rooms to Physicians, which will enhance their experience and easy accessibility of the areas in the operation, which is difficult for humans to reach out. AI is playing critical role in intraoperative decision making as well, such as port placing, confirmation of critical view, by providing better camera systems. In the future, ML models will be trained with a greater number of cases, and with ample of data to feed, which will learn and provide guidance to surgeons to perform critical operations to improve the success rates in surgeries. EMRs not only used for calculating risk and managing the resources, but these have also been used to integrate preoperative information with intraoperative incidents and postoperative results of patients. With this information, AI systems can be trained more efficiently to acquire better recommendations to any physician in the world to learn and to operate virtually. Surgeons and Clinical Practitioners should work closely with clinical and surgical equipment engineers and with data scientists to improve the technology further (Hashimoto, 2020).

AI and Robotics in Operating Rooms

AI in the department of surgery made a remarkable change in the healthcare world with advanced developments in imaging, navigation, surgical instruments, and robotics. Surgeons use AI technology-based tools in various surgeries related to gynaecology, cardiology, oncology, orthopaedics, neurology, ophthalmology, nephrology, and gastroenterology in Operating Rooms to assist. Complex surgeries are made easy with approaches such as PET scan, Ultrasound, Computed Tomography (CT), and MRI. Nowadays, many procedures are operated with minimally invasive surgery methods, which are reducing the recovery period required for the patients and many gadgets are facilitating for the early discharge provisions to maintain post-operative care. Post-surgical recovery period is reduced with minimal invasive operations like laparoscopy surgeries, which has given better patient prognosis. Reduced complications like less pain and blood loss, lesser in-patient stay, fast recovery are few other benefits with robotic surgery methods. By detecting the post-operative complications early, the patient and doctor can work together to avoid complications at an early stage. AI is widely used in risk classification, genomics, image techniques, medical diagnosis, precision medicine, and new drug discovery. DL algorithms like Deep Convolutional Neural Network are used in analysing images for diagnosis. With the advent of sensor and motor features, surgeries are made easy and helping in early detection of diseases to prevent further damages in the body and provide targeted therapies to patients. These methods increase the accuracy of diagnosis and surgery such as constrained tendon-driven serpentine mechanism (CTSM), NeoGuide colonoscopy system, miniature robotic system (Endotics), and Neurosurgical Intracranial Robot (MINIR) used in removing tumours of brain. Robotics is used to get better view of the organs with high precision and efficiency during surgeries. Robots are playing a key role in decision support systems in the healthcare system as well. Nowadays, DL models such as CNN are heavily used in pre-operative plans and able to use successfully in segmentation and classification of cancers. RNNs are used in the prediction of renal failures, bleeding problems in cardiovascular disease patients' post-operative times with better accuracy. DL models are used to detect and categorize malignant tumours with better sensitivity and specificity. Convolutional autoencoder models trained to detect prostate cancer with PET images. Pulmonary nodules are detected with 3D CNNs. CNNs are used in orthopaedics to detect cartilage lesions. Deep Reinforcement Learning models are used to detect breast cancer (Zhou, 2020).

Robotics in Operating Room is playing a vital role along with Surgeon by managing trajectory, pace, and depth of the equipment motion precisely. AI technology with Robotics can learn complex surgical procedures to enhance the surgical skills to use at a great precision and accuracy. Nowadays, collaborative robots (cobot) are popular in Surgical Rooms to perform multiple tasks together to assist physicians during complex surgeries. Otorhinolaryngology is a study of diseases of the ear, nose, and throat. AI-based tools are widely used in otorhinolaryngology (head and neck surgery) and neurosurgery departments. AI-based applications are integrated with hearing aids to patients, in addition to assisting to hear the noises, these devices are helpful to remember the places, filter the environmental sounds, check the physiological variances, and to predict the language speaking abilities. 13 Machine Learning algorithms are used in the classification of otoneurological disease diagnosis with 77% accurate results. ML methods are used for rotational testing. Dizziness is diagnosed with 93% accuracy in vertigo patients using ML algorithms. AI-based treatments such as speech recognizable hearing aid tools are used for brain stem evoked response audiometry (BERA) tests. Tonsillitis, Pharyngitis and Rhinosinusitis of ENT related diseases studied with 90% accomplishment with ML and ANN algorithms. ML algorithms are used in Meniere's disease, respiratory tract infections and influenza, oral cancer, and thyroid disease diagnosis, to predict bleeding problems in patients with tonsillectomy and adenoidectomy, to detect vocal cord disorders with better accuracy. Robotic surgery proved to be better than normal laser surgery in head and neck cancers surgery. Rhinoplasty surgeries are better projected and Glial tumours are better segmented in brain surgeries with AI. ANNs seem better in detecting lesions, glial tumours, tumour categorization with better precision in the neurosurgery department. MRI-guided robotic arms facilitate the surgeons with greater mobility in complicated surgeries such as brain surgery, neurosurgery, oncologic surgery, cervical surgery, lumbar herniated disc surgery, arteriovenous malformation surgeries. Random Forest, SVM models are used in diagnosing various tumours according to the size, shape, and volume to categorize. AI is used for stroke studies based on CT image diagnosis with CNNs, which will speed up neuroradiology workflow. AI technology is imparted in smartphone applications, surgical equipment, and clinical decision support systems in radiology. Augmented and virtual reality approaches are used in surgeries training to make better decisions in pre-operative scenarios to reach the targets of successful completion of surgeries in complicated cases (Lee, 2021).

VR and AR in Smart Cities

Virtual Reality and Augmented Reality (VR and AR) enable the digital reports to be viewed in 3-dimensional. VR enables to simulate the results for the procedures using images and noises. Stereoscopic glasses on headset create a 3D image to view digital information. VR provides complete simulation such as look and feel, whereas AR is an interactive experience where we look with smartphone, digital contact lens or headset. VR and AR are going to be a big game changer for Physicians or Surgeons, who can check the treatment options on organs before surgery occurs and it can be used for teaching students as well. Surgeons can study anatomical complexity using VR in surgeries before operations occur to simulate the results such as Plastic surgeries, and cardiovascular related surgeries. AR is used in Doctor's education, for navigating hospitals, defibrillator station wayfinding, intravenous assistance, and vein detection, and for hospital automation (Vles, 2020).

Human resources like nurses, are the scarcity in city hospitals to attend patients and complete documentation process. The integration of applications to smartphones have enabled remote patient monitoring systems in place with Internet of Things and smart gadgets. Brain haemorrhage causes due to brain artery damage, which may burst and bleed, eventually effects the function of adjacent brain cells. If this condition is not diagnosed and treated on time, it may lead to death or permanent disability. Brain haemorrhage is 5 types: intraventricular, intraparenchymal, subarachnoid, epidural, and subdural. CT scans are used to investigate such cases. Deep Learning algorithms such as Support Vector Machine (SVM) and Feedforward Neural Network (FNN) are used in the investigation of Brain haemorrhage. Automated detection and screening tools to diagnose brain haemorrhage are commonly used in critical care facilities. Applications integrated with IOT are providing health facilities at hospitals without long queues and waiting period, home care diagnostic facilities, cancer care, continuous glucose monitoring (CGM) to detect blood glucose levels automatically. In smart cities, routine work performed by human resources in various departments the clinical environments are getting automated to reduce human efforts, increasing the work efficiency, and to perform things faster than humans. The development of IOT systems have greatly transformed health industry enabling home care services in various chronic conditions such as kidney disease, diabetes, cardiovascular diseases, cancer (Hang, 2020). Telehealth systems are in great help during the COVID-19 pandemic, many developed and developing countries are adapted these systems quickly to save the humanity from infections and get treatment being at home.

In Smart cities, people health at workspace is considered the most important factor for the financial growth of society. The absenteeism of employees may be associated with physical or psychological illness such as suffering from chronic conditions, depression, sleep disorders, weight gain and associated diseases. Larger organizations are focusing on employee's health to promote good health and healthy work environment by implementing health risk assessments. Various programs are designed to educate the employees for healthy nutrition, sleep, physical exercise, meditation techniques for stress management. These techniques implementation has become practical with the usage of digital devices with support groups like healthcare practitioners, professional development, volunteers, and online applications (Howarth, 2018).

AI in Image Analysis

Technological advancements in Internet of Things and sensor devices are inculcated smart healthcare system is an essential part of smart city amenities. Collecting information of people using wearable or sensing devices called crowdsensing. To acquire quality healthcare services, Deep Learning algorithms process and use the data or information acquired by crowdsensing. High level of computing resources such as bigger storage, CPU, GPU (processing power), and memory capacity, such as Bigdata environments enabled the processing and training of data with Neural Networks made easy and affordable in Smart Cities. These data are used for various developments in smart cities such as economic growth, to build better quality healthcare systems, ease traffic congestion, to provide public transportation, and public safety. Quality healthcare depends on social sustainability in smart cities, for example smart devices to measure air pollution levels. Machine Learning algorithms were unable to resolve some functions; hence Deep Learning algorithms are better implemented in sensor device relevance with data to develop new applications to resolve healthcare problems in smart cities. DL algorithms impersonate the human brain and create various perceptron layers to train the data and achieve results with high accuracy and specificity. Deep Learning algorithms need huge data sets required to train the models, but the results are more accurate than ML algorithms. DL models learn from experiencing rather than rule-based approach with huge datasets using bottom-up perspective to get the best approximation of function, they work better on non-labelled and unstructured data as well. The systems developed with DL algorithms can perform better in unforeseen events with the help of features such as back propagation. DL models could achieve 94.95% accuracy with ECG (electrocardiogram) reports of coronary diseases. DL algorithms could classify EEG (electroencephalogram) signals using Deep Neural Networks, such as in epileptic diseases, decision making and blood glucose levels monitoring. Convolutional Neural Networks (CNNs) are used to diagnose Parkinson's (neurodegenerative disorder) with data collected by patient's wearable sensor devices to find different stages of disease to help the patient's timely needed treatment. CNNs are used to find energy consumption details with the help of monitoring device, to detect when elderly people have a tendency of falling. By including CNNs with physiological biometrics and behavioural Biometrics has provided higher security and spoofing in digital healthcare systems. Autoencoder Neural Network works on three layers, and it follows the minimalist approach with encoder and decoder functions. Autoencoders not only used in feature learning, but they are also used in generative model purpose. Types of Autoencoders: Undercomplete, Regularized, Sparse, Denoising, and Contractive. In Convolutional Neural Networks each layer identifies the image partially, then the output is passed on to the next layer with multi-spectral resolution, this process is repeated with each layer so that the end output image quality is much better to calculate and project the image with better clarity, for example in image analysis (MRI, PET scan). Convolutional, Pooling and Fully connected layers are represented with Convolutional Neural Networks. Deep Belief Network builds like a tree, based on random variables connected between various layers to categorize the data, it learns layer by layer. Boltzmann Machine neural network has uniform nodes connected, in which neurons assist to make decisions. BM algorithms are suitable for learning and search-related issues to solve. Deep Learning algorithms on wearable devices with sensors such as wrist bands, VR goggles, EEG headsets, smart watches are playing an important role in personalized healthcare system, with the latest developments in technology the data is acquired with high-quality, which will be useful for patient prognosis and medical industry research (Obinikpo, 2017).

Figure 2. Workflow of IOT, AI, Blockchain, Bigdata, User data and Smart Applications

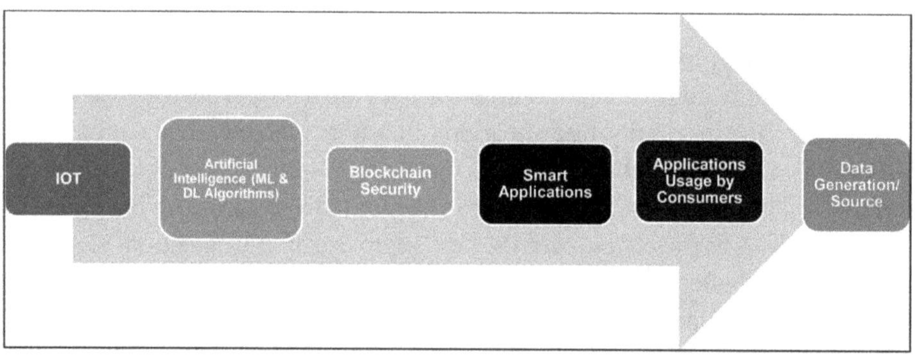

AI in COVID-19 Pandemic

Deep Learning algorithms of Artificial Intelligence, Bigdata Analytics, Blockchain, IOT devices are made revolutionary developments in healthcare system. These technologies are inter-related to play a vital role in gathering data, training the data to learn, patient care, and to automate the processes in hospitals. The outbreak of COVID-19 coronavirus disease is contagious, which made chaotic situation in the whole world due to the quick spread of infection. Governments have spent lot of resources to trace down the infection, isolating infected individuals with quarantine rules, safeguarding the elderly people and kids. IOT, Bigdata, Blockchain and AI technologies are played an important role in monitoring, investigating, diagnosis, preventing and reducing the impact of COVID-19 disease. IOT devices enable to access the data by health groups such as worldometer to know the COVID-19 patients' number, which include daily infected cases in various countries and extremity of the disease (Ting, 2020). Bigdata enabled to transfer, store, and process the data with analytics to study COVID-19 cases, disease symptoms, treatment options, to reduce infections in populated areas like urban areas, so that the healthcare authorities take appropriate actions to prevent spreading the disease. For example, United Arab Emirates government was able to take appropriate timely actions to restrict the spreading of infection by building emergency care centres, providing isolation wards, medicine distribution, guiding people across the country to follow COVID-19 rules in private and public premises. Enhancements in digital and telecommunication technologies enabled the individuals to aware and implement safeguard measures using Smartphone applications, Government health applications for example Dubai Health Authority application, thermal scanners to recognize people with high temperatures. Deep Learning algorithm developments in Artificial Intelligence enabled to find and diagnose COVID-19 cases by providing inexpensive and accurate tests to identify the disease was need of the hour. Many countries were not equipped with enough resources to test these cases, but DL algorithms with enormous data sets on these cases, enabled to train the models for screening symptomatic and asymptomatic cases so that the patients' will be given immediate treatment or send for an isolation. Even in the case of not severe cases, medical practitioners had to be extremely cautious and follow the treatment process, monitor, and isolate to avoid spreading infection. AI promoted drug discovery and vaccine for COVID-19 disease. In the initial months of COVID-19 many hospitals have stopped routine services such as non-covid cases doctor's visits, non-critical surgeries temporarily. Digital tools and e-commerce facilitated with virtual doctors, online appointments, home care systems during COVID-19 pandemic. Tele Medicine with blockchain technology has enabled facilities such as uploading the reports to hospital websites to diagnose, treat patients and deliver medicine at the doorstep. Many Artificial Intelligence based applications reduced the workload of medical practitioners such as chatbot for online help to identify symptoms, to educate individuals on hygiene measures to be followed, and recommend patients to medical attention in case of severe symptoms. Mobile applications such as Dubai Health Authority application has updated with current information and statistics on COVID-19, recording patient's information, and helped in preventing unnecessary hospital consultations in case of mild symptoms (Ting, 2020). AI, IOT, Blockchain technologies have contributed to developing, supply, and implement vaccination drive for COVID-19 across the globe successfully.

Blockchain is an essential component to provide trust to safeguard the data and sensitive information of people in a smart city. Blockchain enables trust factors with enormous developments in storing, processing and data across the globe to various organizations, individuals, agencies, and governments (Kundu, 2019).

AI in Pharmaceutical Industry

Machine Learning algorithms are playing an important role in drug discovery development with magnificent results and accommodating research opportunities with the advent of technology developments in Bigdata to store enormous data sets and process data with high level computing resources. Many pharmaceutical establishments are exploring ML techniques to adapt, for getting financial benefits with the implementation of successful business models by reducing the resource intensive methods, and to prevent the need for animal experimentation. Most of the developers and researchers can use free software and easy programming languages such as Python and R to develop ML algorithms effectively with huge datasets. ML techniques such as artificial neural networks, random forest, support vector machines, k Nearest Neighbour, k-means, decision tree, principal component analysis (PCA) are popularly used in drug discovery. These algorithms and tools facilitate scientists to research without computer science knowledge like programming background. The collaborated third-party applications are easy to use for working with ML algorithms such as linear regression, categorization. ML algorithms have some limitations like dimensionality, performance, big data sets, sample size, lack of autonomy, and transparency. Post implementation of these algorithms in various applications, if the data is changed in data set, the maintenance of the applications is required, which adds up a lot of additional work and expense (Elbadawi, 2021). Reinforcement Learning is another ML technique to address the autonomy issue, this algorithm learns dynamically based on actions adjusted with a feedback system to increase reward, it's suitable for dynamic fields such as gaming, finance trading and robotics. Transfer learning is a Deep Learning technique used to address a smaller sample size data where transfer knowledge from one model to another to solve tasks, these models (supervised and unsupervised) can be deployed faster specifically to use in image analysis in the healthcare system. Transfer learning was significantly useful in the prediction of blood cancer with minimal data availability. Multitask learning algorithms explored to increase the performance to run models because transfer learning models perform sequentially, which will take longer time to run.

This algorithm is useful with smaller data sets with significant noise and to study the effects of multi-target drugs. In patients with cancer or tuberculosis, multi-target drugs are more effective in treatments. Multitask models can be combined with gradient boosting training models on smaller datasets. Low-labelled dataset issue is addressed with Active learning models. Generative models can predict the next data sequence, they can generate new sample data which can be used for data augmentation in case of a small sample size available to do the predictions. Bayesian Neural Networks do not need a larger data set to train the models and they can avoid overfitting problems. Explainable algorithms are used in decision-making processes (Elbadawi, 2021). Generating large size of datasets in pharmaceutical industry is an expensive task, hence data augmentation techniques are going to be useful to predict drug discovery with better accuracy and specificity. To address these limitations Deep Learning algorithms have taken front seat to drive large data sets to train with the models on unstructured data as well as the best results in terms of efficiency and accuracy.

Artificial Intelligence algorithms are not only used in drug discovery and development but also in drug testing, drug repurposing, increasing the productivity of the pharmaceutical industry by providing the best business models. AI is targeted to reduce human workload and increase profits, economic growth of the pharmaceutical industry with inexpensive drug testing models. Many web-based applications are in use to reduce the cost of drug testing, for example, to find the chemical compound toxicity in drugs. AI is helpful in structure-based drug design with the prediction of protein structure of the target, to predict drug-protein interactions with protein of receptors with better accuracy. Artificial Neural Networks are used to develop capsules with decision support systems. Mathematical tools integrated with AI establish enormous help in quick production of new drugs in pharmaceuticals (Paul, 2021).

With wireless and telecom revolution Telemedicine able to provide medical services to rural areas, remote areas, nursing facilities, jails, individual's home, especially during COVID-19 pandemic. Otolaryngology department protocols could successfully implement with the help of telemedicine services. With advanced cameras, high-speed internet services, and smart applications telemedicine services are feasible to implement. The patients' survey proved that 78.9% improvement in 2021 year in recovery rate. Telemedicine services solved most of the ENT problems in patients without the need to travel to the hospital for treatment (Wu, 2021)

CONCLUSION

A smart city services are well facilitated amidst of Artificial Intelligence, Blockchain, Big data analytics and IOT to provide smart services efficiently. The massive increase in data volume has led to the challenges of data protection, security, threat, and privacy. Blockchain has solved many data security problems still security threats and ransom demands are common in cybercrime recently. Performance, Speed and Accuracy are three key factors in developing and maintaining smart applications, another challenge is that companies need to allocate huge budgets to maintain the data and skill set of employees. Organizations should spend noticeable time and money to improve the processes and maintain the digital systems. Big data and Cloud Computing have resolved the issue of expensive hardware cost and scalability of organization data capacity requirements to maintain data, yet data security, privacy and performance are debatable to this day. Even though there is lot of improvement in Machine Learning algorithms to serve the growing needs of various industry requirements, there is yet lot of improvements needed on algorithms to work better with a smaller sample size as the data collection is a big challenge and expensive task for some fields. Deep Learning has revolutionized various fields with automation, image analysis, natural language processing, and facial recognition yet there is lot of work to be done to improve the efficiency, precision and avoid mistakes to the level of human intelligence. The field of Artificial Intelligence has taken the lead in digital technology development to make the systems for easy access, reachability with higher accuracy and specificity. Continuous progress and evolution in Deep Learning Algorithms, with the help of open-source software to develop, analyse and research to acquire astonishing results and developments by relentless efforts of computer developers' society in digital healthcare systems and various other fields is remarkable for the future of smart cities.

REFERENCES

Bhattacharya, S., Somayaji, S. R. K., Gadekallu, T. R., Alazab, M., & Maddikunta, P. K. R. (2020). A review on deep learning for future smart cities. *Internet Technology Letters*, *187*(1), 1–6. doi:10.1002/itl2.187

Boudreaux, E. D., Waring, M. E., Hayes, R. B., Sadasivam, R. S., Mullen, S., & Pagoto, S. (2014). Evaluating and selecting mobile health apps: Strategies for healthcare providers and healthcare organizations. *Translational Behavioral Medicine*, *4*(4), 363–371. doi:10.100713142-014-0293-9 PMID:25584085

El-Hassan, O., Sharif, A., Al Redha, M., & Blair, I. (2017). Tracking the Implementation of Electronic Medical Records in Dubai, United Arab Emirates, Using an Adoption Benchmarking Tool. *Studies in Health Technology and Informatics*, *245*, 64–68. PMID:29295053

Elbadawi, M., Gaisford, S., & Basit, A. W. (2021). Advanced machine-learning techniques in drug discovery. *Drug Discovery Today*, *26*(3), 769–777, 1359–6446. doi:10.1016/j.drudis.2020.12.003 PMID:33290820

Hang, C., Khan, S., Kou, B., Nazir, S., Liu, W., & Anwar, H. (2020). A Smart Machine Learning Model for the Detection of Brain Hemorrhage Diagnosis Based Internet of Things in Smart Cities. *Complexity*, *2020*, 1–10. doi:10.1155/2020/3047869

Hashimoto, D. A., Ward, T., & Meireles, O. R. (2020). The Role of Artificial Intelligence in Surgery. *Advances in Surgery*, *54*, 89–101. doi:10.1016/j.yasu.2020.05.010 PMID:32713441

Hosny, A., Parmar, C., Quackenbush, J., Schwartz, L. H., & Aerts, H. (2018). Artificial intelligence in radiology. *Nature Reviews. Cancer*, *18*(8), 500–510. doi:10.103841568-018-0016-5 PMID:29777175

Howarth, A., Quesada, J., Silva, J., Judycki, S., & Mills, P. R. (2018). The impact of digital health interventions on health-related outcomes in the workplace: A systematic review. *Digital Health*, *4*, 2055207618770861. doi:10.1177/2055207618770861 PMID:29942631

Kundu, D. (2019). Blockchain and Trust in a Smart City. *Environment & Urbanization Asia*, *10*(1), 31–43. doi:10.1177/0975425319832392

Lee, K., Jang, J., Yu, Y., Heo, J. S., Han, H., Yoon, Y., Kang, C. M., Hwang, H. K., & Kang, S. (2021). Usefulness of artificial intelligence for predicting recurrence following surgery for pancreatic cancer: Retrospective cohort study. *International Journal of Surgery*, *93*(106050), 393–401. doi:10.1016/j.ijsu.2021.106050 PMID:34388677

Lundervold, A.S., & Lundervold, A. (2019). An overview of deep learning in medical imaging focusing on MRI. *Zeitschrift für Medizinische Physik*, *29*(2), 102-127. doi:10.1016/j.zemedi.2018.11.002

Obinikpo, A.A., Kantarci, B. (2017). Big Sensed Data Meets Deep Learning for Smarter Healthcare in Smart Cities. *Journal of Sensor and Actuator Networks*, *6*(4), 1-17. doi:10.3390/jsan6040026

Paul, D., Sanap, G., Shenoy, S., Kalyane, D., Kalia, K., & Tekade, R. K. (2021). Artificial intelligence in drug discovery and development. *Drug Discovery Today*, *26*(1), 80–93. doi:10.1016/j.drudis.2020.10.010 PMID:33099022

Rocha, N. P., Dias, A., Santinha, G., Rodrigues, M., Queirós, A., & Rodrigues, C. (2019). Smart Cities and Healthcare: A Systematic Review. *Technologies, 7*(3), 1-8. doi:10.3390/technologies7030058

Souza, J. T., Francisco, A. C., Piekarski, C. M., & Prado, G. F. (2019). Data Mining and Machine Learning to Promote Smart Cities: A Systematic Review from 2000 to 2018. *Sustainability*, *11*(4), 1–11, 1077. doi:10.3390u11041077

Ting, D. S. W., Carin, L., Dzau, V., & Wong, T. Y. (2020). Digital technology and COVID-19. *Nature Medicine*, *26*(4), 459–461. doi:10.103841591-020-0824-5 PMID:32284618

Vles, M. D., Terng, N., Zijlstra, K., Mureau, M., & Corten, E. (2020). Virtual and augmented reality for preoperative planning in plastic surgical procedures: A systematic review. *Journal of Plastic, Reconstructive & Aesthetic Surgery*, *73*(11), 1951–1959. doi:10.1016/j.bjps.2020.05.081 PMID:32622713

Wang, L., Fan, X., Chen, J., Cheng, J., Tan, J., & Ma, X. (2020). 3D object detection based on sparse convolution neural network and feature fusion for autonomous driving in smart cities. *Sustainable Cities and Society*, *54*, 1–5. doi:10.1016/j.scs.2019.102002

Wu, C. N., Luo, S. D., Lin, H. C., Huang, J. T., Lee, C. H., Liu, S. Y., Tsai, M. H., Wang, C. C., Fan, S., Wang, P. S., & Lan, K. C. (2021). Eligibility for live, interactive otolaryngology telemedicine: 19-month experience before and during the COVID-19 pandemic in Taiwan. *Biomedical Journal*. Advance online publication. doi:10.1016/j.bj.2021.07.012

Zhou, X., Guo, Y., Shen, M., & Yang, G. (2020). Artificial Intelligence in Surgery. *Frontiers of Medicine*, 1–28. doi:10.100711684-019-0733-5 PMID:32705406

Chapter 13
Smart City:
A Modernization for Uplifting the Standard of Living of the Citizens

Monika Maheshbhai Patel
Natubhai V. Patel College of Pure and Applied Sciences, CVM University, India

Ghanshyam Parmar
Natubhai V. Patel College of Pure and Applied Sciences, CVM University, India

ABSTRACT

In the era of IT, each and every city has become digitalized due to the high speed of internet and the world wide web. In today's scenario, each urban area has transformed to a "smart city." A smart city is a municipality that uses information and communication technologies to increase operational efficiency, share information with the public, and improve both the quality of government services and citizen welfare. The aim of smart cities is that citizens can easily access the available resources digitally. The main mission of smart cities is to promote economic growth of the nation and optimize the city services to the citizen by improving quality life for citizens. In smart cities, everything is connected to each other, and this is highly dependent on technology. Moreover, various technologies, applications, and challenges for implementing smart cities are discussed in this chapter.

INTRODUCTION

In this techno wise world, on account of rapidly improvement in information and technology (ICT), we've seen an expanding hybrid inside the use of this new innovation. As a result of innovation today, the whole world become a ton of more brilliant

DOI: 10.4018/978-1-7998-9710-1.ch013

and more astute. Most eminently inside the presentation and intuitiveness of urban services. This has become typically referenced as “Smart Cities”. Along with the Internet of Things (IoT), Artificial Intelligence, sensors, and various gadgets, smart cities gather data of specific area and utilize this information to oversee resources and assets with productivity. (Cabot, 2019; TWI Ltd, n.d.; University, n.d.)

A geographical region that utilizes differing types of electronic devices and sensors to collect information is named as smart city. On account of the progression of innovation, any city is converted into smart city by fixing some infrastructure as well as citizen of city can easily access digital information by single tip of finger. A savvy city uses information and communication technology (ICT) to reinforce operational proficiency, sharing of data with the general public, provide a far better quality of state service and lavish life style to its citizen. (Cabot, 2019; TWI Ltd, n.d.; University, n.d.)

Whenever city is considered then individuals are at core of the town. The cities are convergences of individuals. It provides places wherever people live, alternative ways to satisfy with people and exchange ideas with one another, earn livelihoods, access education, health and other services and luxuriate in a lifetime of good quality. Therefore, cities ought to work for people. Subsequently as a revolution, the cities are converted into smart cities. On a daily basis a smart city becomes higher version to provide basic desires of the citizens in smart manner. The core elements in a Smart City are shown in figure 1.(Kota city, n.d.)

To construct any city as a smart city, it is desirable whether the city is appropriate or not for considering it as a Smart City. A city’s smartness is set employing a set of characteristics, including:(TWI Ltd, n.d.)

- An infrastructure fulfills the need of advanced technology
- Environmental initiatives
- Effective and extremely purposeful public transportation
- Confident and progressive city plans
- People ready to live and work within the city, by accessing its resources

Figure 1. Core elements in smart city
(Kota city, n.d.)

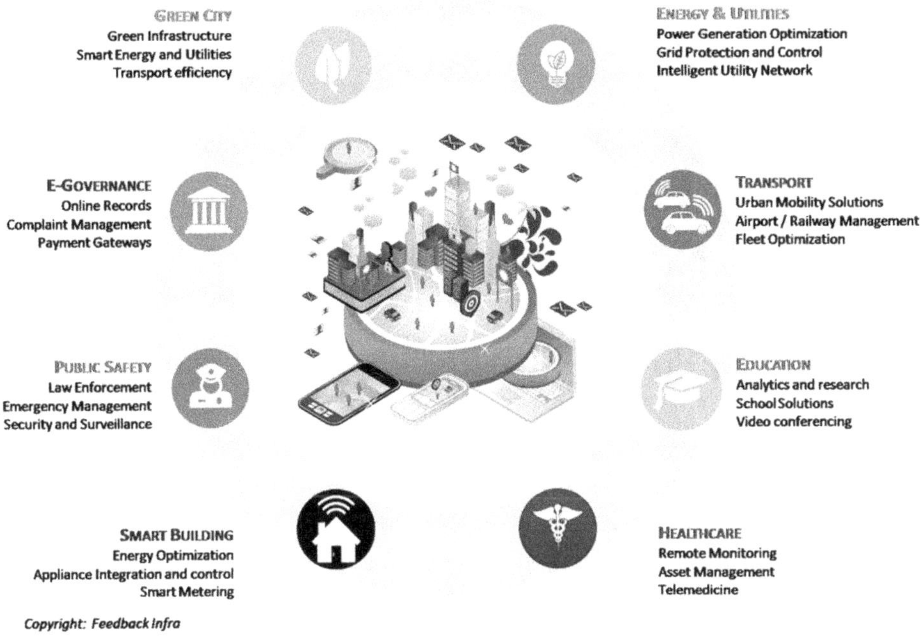

HOW SMART CITIES WORK

To make any urban city as smart city some ICT related technologies are required. ICT framework collects the data from connected devices and make it available for the citizens. By establishing connection between devices and data, it becomes easy to reduce the costs of accessing data, improve sustainability and streamline factors such as energy distribution and refuse collection, reduced traffic congestion, improve air quality, manage the waste management etc. To improve the quality of life and enable economic growth through a network of connected IoT devices and other technologies, smart cities has to follow the four steps given as shown in figure 2.(TWI Ltd, n.d.)

Figure 2. Four steps for smart city

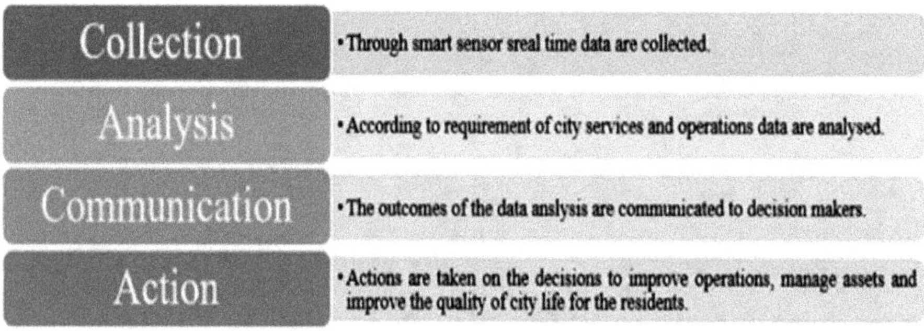

AIM OF SMART CITY

Smart cities use information gathered from sensor-empowered gadgets and Internet of Things innovation to foster understanding and insight which will change the manner in which districts oversee assets, perform administrations, secure public wellbeing and guarantee personal satisfaction for inhabitants. The aims of smart city are listed below: (Spectrum Enterprise, n.d.)

1. To build data-driven decisions that expansion the effectiveness and cost-efficiency of city programs.
2. To further develop city works and advance monetary development which consequently improve the personal satisfaction for residents by utilizing savvy innovations and information investigation.
3. Use trend setting innovation, information and investigation to further develop the executives of city assets and lives of residents.
4. To advance urban areas that give center framework and give a fair personal satisfaction to its residents, a spotless and reasonable climate.

VARIOUS APPLICATIONS

Figure 3 demonstrates the various applications used for Smart City.(ARC, 2021; Bismart, n.d.; Libelium, 2020; Stefanini, 2020; Thales, n.d.; TWI Ltd, n.d.; Wade, 2016)

Figure 3. Applications of smart city
(Libelium, 2020)

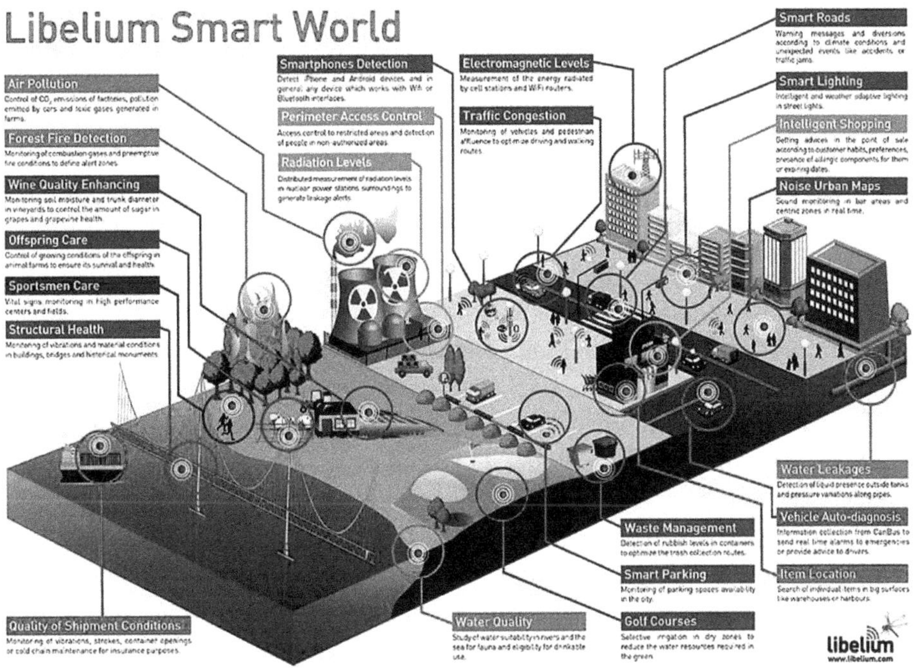

Smart Infrastructure

Smart Infrastructures include a few administrators from various areas of action, like energy, public vehicle, public wellbeing. Smart Infrastructures joins the genuine and computerized universes across energy frameworks, buildings and industries, upgrading the way individuals live and work and essentially further developing productivity and maintainability. Digital technologies are turning out to be progressively significant for urban areas to have the conditions for continuous development; structures and metropolitan foundations should be arranged all the more productively and economically. In smart infrastructure intelligent innovations are utilized to accomplish an energy-proficient and harmless to the ecosystem framework. For instance, CO2 discharges ought to be kept as low as possible for instance investing in electric vehicles and self-moved vehicles. In addition, to decrease the requirement for electrical force, smart lightening is one in every of the answer that provide light once somebody walks past them. One can likewise set the brightness levels. In expansion to this, Smart meters are utilized to follow energy utilization by diminishing use of energy which sway on monetary need.(Mohamed, 2018)

Air Quality Management

Smart cities are also implementing tools which can catch contamination information continuously and estimate outflows. Having the ability to predict pollution accurately allows cities to urge to the basis of their emissions problems and conceptualize vital approaches to restrict the measure of air contamination they put out.(Mohamed, 2018)

Traffic Management

Traffic congestion is rising in urban areas all throughout the planet. Contributing factors include extending urban populations, aging infrastructure, inefficient and uncoordinated traffic light timing and an absence of ongoing information. Because of progression of innovation, presently it turns out to be not difficult to limit the traffic of generally globe. Today, Smart Traffic Management Systems make it conceivable to build the limit of city roads without really adding new streets. With the approach of associated vehicle innovation, these frameworks can likewise straightforwardly control vehicles when required. In addition, asphalt incorporated sensors send constant updates of traffic stream to a focal traffic the management platform, which examinations the information and consequently changes traffic signals to the traffic circumstance within seconds. At an equivalent time, authentic information is utilized to anticipate where traffic can go without human involvement.(Mohamed, 2018)

Smart Parking

Smart parking may be a strategy that smart cities can use to extend parking efficiency and combat numerous issues associated with traffic. Utilizing smart parking system, for drivers it becomes easy to seek out a vacant parking lot specially in crowded public places. The mix of advanced mobile phones and detecting gadgets can make the smart parking structure. It likewise refined using cameras, counting utilities at the doorways or entryways of parking structures, sensors implanted in the asphalt of individual parking spots, and so forth. With the assistance of mobile app, drivers will get the alarm for vehicle parking spot effectively. It saves fuel just as lessen the pressure of driver for looking through parking spot. Today, smart Parking may be a reality and doesn't require complicated infrastructure and a high investment. Indeed, even it's an optimal application for a moderate sized smart city initiative. (Mohamed, 2018)

Smart Waste Management

The smart waste management system gives a sound climate to its residents by viably gathering the waste. Utilizing trend setting innovation and sensors it turns out to be not difficult to find vacant and full canisters for waste, waste recycling efficiency, locate space and route for trucks to discharge the waste. Also, through mobile app transporter gets the warning in regards to full bins. In the event that containers are half unfilled, the executives will get the alarm and canister's area for not gathering containers for reusing. With this framework, one can dynamically and efficiently manage the waste collection by predicting waste status, classifying garbage location, and checking the measure of waste.(Mohamed, 2018)

Smart Education

Education is spine of improvement. Smart cities need schooling facilities which ensure understudies obtain computerized education, innovative reasoning, effective communication, teamwork and therefore the ability to make high-quality projects. ICT enabled smart education provides better use of data, upgraded control and appraisal, higher support for life-long learning for all people. Smart learning ought to be adaptable for all, no matter gender, physical abilities, age, religion and other groupings. Smart classroom technology supports the professionalization of the teaching process, supporting instructors to more readily plan and enrich their lectures and to react flexibly to the requirements of scholars and conditions in classrooms, resulting in increased efficiency and better educating execution. Smart learning is that the best path to guarantee the most splendid future for advanced education. For Smart City, smart education has been given extraordinary significance in order that subsequent generation are often educated and trained with technology enabled smart learning programs.(Al Nuaimi et al., 2015)

Smart Traffic Lights

Smart traffic lights are one among many smart city initiatives that use advanced technology to enhance the management of city services and therefore the quality of life for citizens. The smart traffic lights are utilized to make a more intelligent traffic management system that permits smart cities to scale back congestion and travel times, to enhance safety and to diminish carbon emissions. The circumstance of red and green lights, "walk" and "don't walk" signs can be changed by smart traffic signals to control the progression of traffic and people on foot all the more effectively. Smart traffic signals use information from sensors, cameras, GPS, vehicles, cells phones and other devices to detect patterns of traffic and therefore the volume of vehicles, pedestrians and bicyclists. By dynamically responding to time period conditions, smart traffic lights will facilitate to move traffic through town more efficiently to scale back commute times, decrease congestion, minimize carbon emissions and improve safety for drivers, pedestrians and bicyclists.(Al Nuaimi et al., 2015; Stefanini, 2020)

Smart Grid

The smart grid is a crucial component of a sensible city. A smart grid is an electricity network supported digital technology that is utilized to supply electricity to consumers via two-way data communication. This technique allows for monitoring, analysis, control and communication within the availability chain to assist with further developing effectiveness, decrease energy utilization and cost, and maximize the transparency and reliability of the energy supply chain. This involves placing smart sensors and meters on production, transmission, and distribution systems additionally to consumers access points to urge granular near real-time data about the present power production, consumption, and faults.(Stefanini, 2020),(Al Nuaimi et al., 2015)

Smart Health

Medical services are a core facility necessary for each town across the planet. Smart healthcare utilizes innovations, like the Internet of Things (IoT), AI, wearable gadgets and sensors, Big Data, Business Intelligence, and so forth to smartly react to the requirements of the medical services environment. With the assistance of IoT empowered sensors it turns out to be not difficult to recognize blood and glucose levels in patients, and send the subsequent information to specialists for investigation and anticipation. Medical services experts can dissect each of the information gathered by sensors to endorse exceptionally customized therapies and meds for patients. Artificial intelligence is likewise applied to perform tasks like analyzing laboratory tests, x-rays, CT scans, and data entry. AI-based applications can be utilized to get to the current ailment of patients that can give help with clinical discussion. Technologies like Blockchain redefine the methods of maintaining and populating Electronic Health Records as well as they also help link them to other services like payments and insurance. Robotics is one of the extensively used technologies in smart healthcare. Surgical robots enable surgeons to perform robot- supported minimally invasive surgeries. Robots are decreasingly being used in radiology, patient monitoring in the ICUs, medical arrangements, and outfit running. Features and capabilities such as smart parking, automated patient enrollment, smart wards/ ICUs, online payment systems etc. ameliorate the patient experience. Smart healthcare has numerous benefits for smart metropolises, and these apply to the patients, governments, medical fraternity and sellers. (Al Nuaimi et al., 2015)

ESSENTIAL TECHNOLOGIES FOR SMART CITY

Smart cities are smart Metropolises where everything is associated with one another and this is exceptionally relied upon advancements. "Transmission technologies" and infrastructure are the bottom technologies that support the framework of a smart city. Without the structure of innovation, it's onerous to create a shift towards smart megacity.(Choudhary, 2019; Integra, n.d.; Paz, n.d.)

Technological knowledge is a key to transform a city into smart city which is all around associated, manageable and versatile, where data isn't simply accessible yet in addition findable. It isn't a new thing that smart megacity is each about furnishing smart services to its citizens which can save their time and ease their lives. It is likewise about connecting them to the administration where they can give their feedback to the public authority as of how they need their city to be. Furthermore, this point cannot be transformed into reality without innovation. Utilizing innovation, the authorities can assemble city knowledge and this insight when coordinated with the tasks, make the urban areas more astute and more secure. The essential technologies without which smartness of a city can never be enhanced are shown in below figure 4. (Choudhary, 2019; Integra, n.d.; Paz, n.d.)

Figure 4. Essential technologies used in smart city

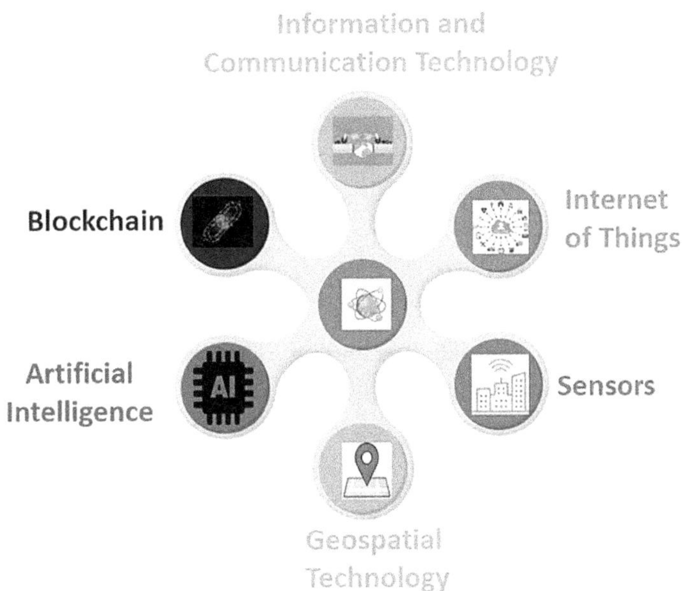

Information and Communication Technology

Making a two-way correspondence channel is vital for a city to be brilliant. ICT constructs an extension between the residents and the public authority where the residents can communicate with the public authority and consequently, the public authority can assemble a city which the selection of its voters. ICT assists the government to investigate the demand pattern of the state and therefore produce a pool of assets to address the equivalent on the web. The electronic medium of communication in a very community helps in making a collective intelligence which may be deployed for resource improvement with the assistance of analytics and deep learning. (Choudhary, 2019; Integra, n.d.; Paz, n.d.)

Internet of Things

Internet of things resembles veins of the town unfold all across and associating each spot. Each gadget that is important for a smart city should be connected with one another so that they can talk among and can take opinions for themselves which consequently permits overseeing assets of a megacity populace. On account of the IoT gadgets like connected sensors, lights, and meters can gather and break down information for the better working of smart metropolises by perfecting civic services, supportability, safety, mobility and transparency of the city. Hence, for smart city IoT is a " data collection technology ". (Choudhary, 2019; Integra, n.d.; Paz, n.d.)

Sensors

Sensors are hidden but ubiquitous factors of the metropolitan scene. Sensors are a pivotal element of any intelligent control system. Sensors are like transformers which convert parameters of a physical nature to an electronic signal, which can be interpreted by humans or can be fed into an independent system. These signals for conventional detectors, amongst others, include light, pressure, temperature, moisture, humidity and variety of other parameters. (Choudhary, 2019; Integra, n.d.; Paz, n.d.)

Geospatial Technology

Smart Metropolises use Geospatial technology to make civic surroundings more accessible and environmentally friendly or to descry disasters that affect the ecosystem. Depending on the requirements of citizens, this technology reacts to their demand by offering results in transport, power plants, water supply organizations, civil protection or public centers, among others. Geospatial data management and ecological checking in urban communities empowers the location of changes in the climate, further develops avoidance, speeds up the reaction to environment disasters and extreme natural phenomena and improves decision-making. In this sense, this technology makes it easier for public administrations to create higher choices within the ecological administration of their cities so as to make municipalities a lot of committed to the natural environment. (Choudhary, 2019; Integra, n.d.; Paz, n.d.)

Artificial Intelligence

Smart Megacity is an advanced transformation producing an enormous quantum of information. Those data are of no utilization until and except if they are handled, which creates data consequently. This gigantic quantum of data generation can be done by the use of Artificial intelligence that can make sense out of those data. Preventive support is one of the primary elements of this innovation and perhaps the main benefit. With the utilization of AI, districts can distinguish water spills and forestall occurrences. Traffic the executives and working on the wellbeing of residents on the streets is one more capacity of AI in brilliant urban areas. What's more, increasingly more AI-fueled surveillance cameras are being utilized that go past video reconnaissance capacities. (Choudhary, 2019; Integra, n.d.; Paz, n.d.)

Blockchain

Blockchain application is new to smart megacity conception. Blockchain technology secures data knowledge flow. Blockchain can corroborate individualities and authorize access. Its integration into good cities may higher connect all town services whereas boosting security and transparency. Blockchain records and stress deals in inflexible records and makes data exchanges between distributed gadgets seamless and cost-effective. In some ways, blockchain is anticipated to impact metropolises through smart contracts, which help with billing, processing deals and handling facilities management. Smart contracts are self-executing contracts with the terms of the agreement between client and vendor directly written into lines of code. They permit trusted deals and agreements to be carried out among distant parties without the need for an interceding third party, making the process safer, cheaper and quicker. (Choudhary, 2019; Integra, n.d.; Paz, n.d.)

CHALLENGES FACED BY SMART CITIES

For quite a whereas currently, sensible cities are among the dominant motifs in newsrooms and media. Creating metropolises smart, connected, and resilient has been the mission of developed and developing countries worldwide. By using digital technologies, town functions are efficient and optimized whereas keeping negative ecological impact to a minimum. At the side of this, sensible cities boost monetary gains and improve the citizen's quality of life. The government, technical experts, and town planners are collaboratively operating to rework this formidable plan into reality.(Joshi, 2019; Stone, 2018)

A smart town could be a novel answer to create economical use of natural resources, ameliorate the citizens' standard of living, and conjointly attain economic development. However, to attain success, challenges for good cities ought to be addressed first which are shown in below figure. Thus, within the future topic we have a tendency to mention a number of the common challenges facing Smart City solutions nowadays, and the way developers will begin to handle these barriers to success. (Joshi, 2019; Stone, 2018)

Insufficient Funds

For the implementation of good town funding is that the biggest challenge. Creating smart cities means that planting smart, complicated infrastructure for implementing digital technologies.

Besides, tons and heaps of smart devices ought to be integrated for knowledge assortment. Additionally, to make sure smart mega city success, governments ought to rent enough technical experts and town planners. Further, network needs have to be correctly met. Moreover, the hardware put in needs to be audited oftentimes for maintenance. All of those eat up loads of cash. Governments ought to think about fashioning a technique to make applicable revenue models for his or her good town initiatives. (Joshi, 2019; Stone, 2018)

Lack of Experienced Professionals

Another most-pressing challenge for smart metropolises is the lack of professed professionals. For preparing a strategy to achieve smart megacity design success, relating areas for perpetration of technologies, and operating these tools, tech experts are needed. The government and the concerned stakeholders should take the count of professionals needed and hire them before they start with their design plans. (Joshi, 2019; Stone, 2018)

Infrastructure With Network Connectivity

Smart Cities utilize sensor technology to gather and analyze information in an effort to improve the quality of life for residents. For the smart management of a municipality, several sensors, cameras, and actuators are installed everywhere. These sensors gather and send large volumes of data in real time. Analysis and processing of the collected data should happen almost instantaneously for efficient management of city operations. And for instant processing, high-speed Internet connectivity is mandatory. Complicated and costly infrastructure is involved in installing and maintaining these sensors. (Joshi, 2019; Stone, 2018)

Cybersecurity Risks

As IoT and sensor technology use expands, therefore will the threat level to security. This makes one wonder…is technology very thought of "smart" if hackers will be burgled it and stop working a whole city? These internet-connected devices can transmit vast chunk of data in real time. Although this information facilitates in providing potency at municipality functions, it presents serious security risks that cannot be unnoticed. Data from parking lots, CCTV cameras, EV charging stations, and GPS systems contains nonpublic information of citizens. Not each connected device is cyber-resilient as of now. If that's the case, criminals will simply gain access to the information and use it for outlawed intent. Hence, government and IT professionals should strengthen the safety borders of sensible devices and therefore the supporting infrastructure. (Joshi, 2019; Stone, 2018)

Privacy Concerns

In any significant city, there's a harmony between personal satisfaction and attack of security. While everybody needs to partake in a more helpful, serene, and sound climate, no one needs to feel like they are continually being checked by "Big Brother." Cameras introduced everywhere may assist with discouraging wrongdoing, yet they can likewise introduce dread and neurosis in reputable residents. Another substantial concern is the measure of information being gathered from every one of the smart sensor occupants come into contact with every day. (Joshi, 2019; Stone, 2018)

Educating and Engaging the Community

For a Smart City to truly live and thrive, it wants "smart" voters who are engaged and laboriously taking advantage of new technologies. With any new city-wide tech project, a part of the implementation method should involve educating the community on its advantages. This could be done through a series of in-person city hall-style conferences and email campaigns with citizen registration, moreover as an internet education platform that keeps electors engaged and up-to-date. At the point when a local area feels like it's having an impact in the general choices that influence day to day existence, and is being imparted to in an unmistakable and insightful way, it's more adept to utilize the innovation and urge others to utilize it also. This is critical to a Smart City's prosperity. The city is teaming up with residents, entrepreneurs, large companies and startups to form a 'city of tomorrow.' (Joshi, 2019; Stone, 2018)

Being Socially Inclusive

Smart transit programs that give riders constant updates are a good thought for a clamoring city. In any case, imagine a scenario where a large portion of the number of inhabitants in that city can't stand to take mass travel or Uber. What might be said about a developing senior population that doesn't utilize cell phones or apps? How might smart technology reach and advantage these groups of individuals? It's essential that Smart City arranging includes the consideration of all gatherings of individuals, in addition to the well-off and technologically advanced. Innovation should always be attempting to unite individuals, instead of isolating them further dependent on financial gain or education levels. (Joshi, 2019; Stone, 2018)

4 PILLARS OF SMART CITIES OF INDIA

The term Smart is infiltrating the Indian populace at a way quicker rate recently. We as a whole know at this point that an Indian Smart City is the modernization of the current urban areas to make them smart.

Every country has its own definition dependent on its goals and prerequisites, with, obviously, the normal target of each smart city being driving monetary development and up the standard of life. For India, a smart city is characterized as: "Urban areas that feature all core infrastructure required like assured water and electricity supplies, sustainable environment, efficient sanitation, solid waste management, and public transport, adequate healthcare and education facilities and affordable housing, especially for economically-weak sections of society."(Oizom, 2019)

Smart Metropolises focus to put individuals first in their methodology. They carry out changes by tapping metropolitan arranging best practices, digital and information technologies, public-private associations, and policy change. The Indian government dispatched a mission in 2015 called Smart Cities Mission to meet the objects characterized previously. 100 urban communities were chosen to be subsidized for carrying out their Smart City Plans (SCPs). The execution of the mission is extensively isolated into 4 pillars which are shown in below figure 5.(Oizom, 2019)

Figure 5. Pillars of smart cities of India

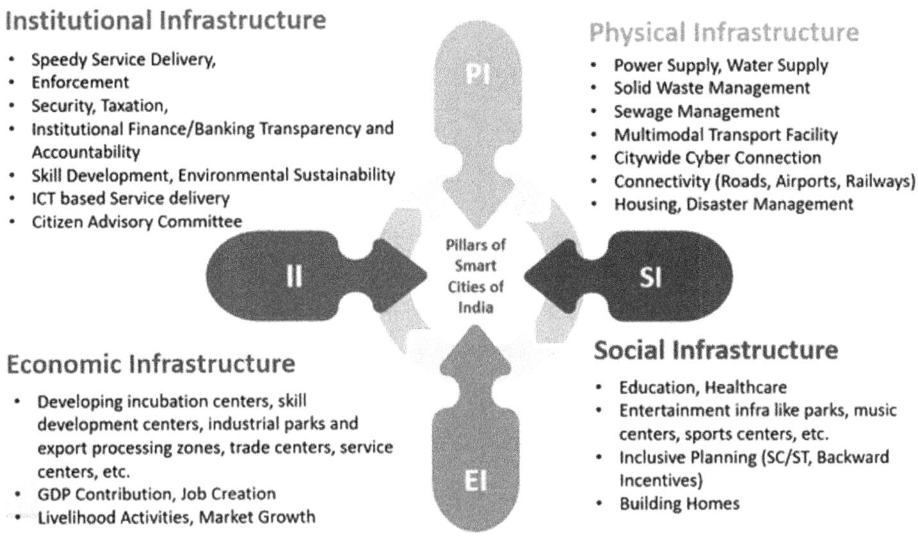

REFERENCES

Al Nuaimi, E., Al Neyadi, H., Mohamed, N., & Al-Jaroodi, J. (2015). Applications of big data to smart cities. *Journal of Internet Services and Applications*, *6*(1), 1–15. doi:10.118613174-015-0041-5

ARC. (2021). *Smart Cities*. https://www.arcweb.com/advisory-services/infrastructure-smart-cities

Bismart. (n.d.). *What Exactly is a Smart City?* https://blog.bismart.com/en/what-is-a-smart-city

Cabot, T. (2019). *An Introduction to Smart Cities*. https://www.arcweb.com/blog/introduction-smart-cities-0

Choudhary, M. (2019). *Six technologies crucial for smart cities*. https://www.geospatialworld.net/blogs/six-technologies-crucial-for-smart-cities/

Integra, N. (n.d.). *Smart Cities, how they work and how to tackle change*. https://nexusintegra.io/smart-city-what-is-how-it-works/

Joshi, N. (2019). *4 challenges faced by smart cities*. https://www.allerin.com/blog/4-challenges-faced-by-smart-cities

Kota City. (n.d.). *What is a Smart City?* https://smartcitykota.com/about/

Libelium. (2020). *50 Sensor Applications for a Smarter World*. https://www.libelium.com/libeliumworld/top_50_iot_sensor_applications_ranking/

Mohamed, Y. (2018). *What is the role of IoT in Smart Cities?* Academic Press.

Oizom. (2019). *4 Pillars of Smart Cities of India and the Role of Smart Environment*. Academic Press.

Paz, A. (n.d.). *New Technologies in Smart Cities*. https://www.e-zigurat.com/innovation-school/blog/new-technologies-smart-cities/

Spectrum Enterprise. (n.d.). *What is the aim of smart cities?* https://enterprise.spectrum.com/support/faq/smart-cities/what-is-the-aim-of-smart-cities.html

Stefanini. (2020). *Top 5 Applications of IoT in Building Smart Cities*. https://stefanini.com/en/trends/news/top-5-applications-of-iot-in-building-smart-cities

Stone, S. (2018). *Key Challenges of Smart Cities & How to Overcome Them*. https://ubidots.com/blog/the-key-challenges-for-smart-cities/

Thales. (n.d.). *Secure, sustainable smart cities and the IoT*. https://www.thalesgroup.com/en/markets/digital-identity-and-security/iot/inspired/smart-cities

TWI Ltd. (n.d.). *What is a Smart City? Definition and Examples*. https://www.twi-global.com/technical-knowledge/faqs/what-is-a-smart-city

University, T. O. (n.d.). *Introduction to smart cities*. https://www.open.edu/openlearn/ocw/mod/oucontent/view.php?id=67877

Wade, M. R. (2016). *What is a Smart City anyways?* https://www.imd.org/research-knowledge/articles/what-is-a-smart-city-anyways/

Chapter 14
Smart City:
Holistic Development Approach Using the ICT Paradigm – An Indian Perspective

Sandeep K. Budhani
Graphic Era Hill University, India

Mukesh Joshi
Graphic Era Hill University, India

Naveen Tewari
https://orcid.org/0000-0002-6104-5005
Graphic Era Hill University, India

ABSTRACT

In this chapter, the authors discuss advances like 5G, 6G, and 7G. They see that IoT has made our lives so natural, yet if we see it from an Indian point of view, things are completely changed. In India, there are various difficulties. The objective of smart urban communities could likewise be to advance financial development and improvement in regions to establish a perfect and stable climate, to fund-raise for all, to help the hindered, or to make city administration more straightforward. Provided that this is true, the smart metropolitan standards ought to extend and profoundly change our conventional framework by uniting new highlights, unwavering quality, versatility all through the city, and the effortlessness of day-by-day benefits in metropolitan regions. Smart city applications are presently an innovation that administrations match and a blend of numerous direct organizations for arrangements in a wide scope of conditions, for example, medical services, delivery, business and modern organization, and security.

DOI: 10.4018/978-1-7998-9710-1.ch014

INTRODUCTION

What Is a Smart City?

There is generous writing on the meaning of brilliant urban communities which could be ordered into three significant sources:

1. **Examination and scholastic view** - Puts supportability, as the essential plan to be accomplished, where personal satisfaction and economy arise as second-level need factors.
2. **Corporate area** - Looks at ICT as a panacea, accepting that the necessary results like city effectiveness, the board, foundation, climate, and personal satisfaction follow naturally. Prominently, there is ostensible accentuation on the general usefulness, strength, city structure and metropolitan plan of the smart city.
3. **Government area** - Letaifa S. (2015) mentioned that this area mirrors a bigger comprehension of the utilization of ICT in conveying administration, perceives the basic importance of HR, and puts accentuation on personal satisfaction just as climate. Nevertheless, an extremely predetermined number of definitions have risen up out of this area.

Smart City Framework

Zubizarreta I. et al. (2015) deliberated that Setting up a smart city is more than building a wilderness of concrete or redesigning an older style framework by essentially robotizing utilizing sensors, remote checking, and controls to significant city administrations. It is a finished change in outlook in daily existence as you apply new innovations, particularly new ICT prompting savvy results. Along these lines, we want to foster the framework by digitizing the significant region like banking, medical services and so on and for execution of the idea of smart city; we want to build the education rate. Just expanding the education rate in not the arrangement but rather we need to drift our public to utilize the ICT. Until and except if they will be constant of ICT, the idea of smart city can't be fruitful.

For infrastructural improvement, the most importantly step is to run mindfulness program in regards to the utilization of ICT then we need to give the fundamental equipment and programming. For this, web availability assumes an imperative part.

ROLE OF ICT AND CHALLENGES

Conventional arrangements are as of now not adequate to meet the present metropolitan issues. Over the long haul, plainly any huge metropolitan region that doesn't acknowledge to some extent part of Smart City's vision will have a troublesome future. The significance of applying the standards of Smart City is the way that these enhance the administrations given by all partners, increment the existence of the city, and at last work on the personal satisfaction of its residents. According to Raparthi K. (2014), the outcomes may not be pretty much as quick as the resuming of an expressway, for instance, yet will at last make the city more appealing to organizations and residents, and more aggressive on a worldwide scale. The vision of Smart City is on the way and therefore, Smart City ICT and IoT framework should think of a test to help these significant thoughts for the fate of our urban areas.

Smart City Design and Planning

The talks that are accessible in the public space on brilliant urban communities, just as its applications, are to a great extent drove by technologists and executives with optional contributions from modelers and metropolitan originators. Numerous utilizations of brilliant city advances depend vigorously on ICT. It is accepted by engineers and metropolitan planners that they don't zero in on the spot or the actual type of the city, or the sociocultural and world of politics it will possess. In the current discussion on smart urban areas the emphasis on what sort of socio-spatial results that it incites is forgotten about. This part attempts to address the huge hole in the current standard talk between the actual sign of urban areas of future and a distinct metropolitan vision for the spatial and typological setups of the smart city. The linkages between metropolitan structure and ICT are not immediate and rely upon numerous elements neighborhood attributes, needs and needs of the residents, worldwide market influences, public and global financial movements and so forth besides, the linkages are two-way, in that they are harmonious in affecting one another. By and large, while the connection between metropolitan structure and ICT are not immediate, they are more enthusiastically to comprehend and require long haul study. According to Mosannenzadeh F. and Vettorato D. (2014), the capability of ICT applications practically speaking can't be disregarded by anybody today. Smart city applications for certain areas, for example, transportation and water the board are surely known and effectively applied. Applications to the act of metropolitan plan and arranging are as yet arising yet offer some encouraging leads. Utilization of brilliant innovation as a guide to metropolitan plan and arranging practice can support to develop a city as a smart city. For example, the admittance to data made accessible by Google Earth has effectively changed the way to deal with spatial plan. Joined with instruments like Google Street view, it permits organizers and metropolitan planners to handily get across scales. This data can assist with adjusting metropolitan spaces and their qualities by understanding the footfalls, inclination designs, and client audits, computing porousness of huge domains, hydrological designs, populace movements, and thickness and even reproduce monetary investigation.

The second point would include making plans for brilliant metropolitan spaces. The flexibility and versatility of spaces to shifted, multi-disciplinary requests of a savvy worldwide society will be the need of things to come. It is crucial for ponder how changes in innovation will impact the utilization of room as well as the typologies of room in the city. A couple of inquiries that yield up:

- Will the coming of internet shopping render customary shopping regions repetitive?
- Will enormous coordinated server farms and control rooms become the new focuses of government?
- Will the utilizations of ICT change the financial aspects of metropolitan agglomerations? Through Urban Design we can adjust the type of metropolitan public spaces to take into account an arranged society, make dynamic gathering spaces and adaptable, multi-use spaces in the public domain. Thirdly, the potential outcomes made by the inescapability of intelligent innovation, permits resident investment in space making that can be uncommon.

Advantages of IoT and ICT

Advantages of IoT and ICT in Urban Design and Planning ICT benefits inside Urban Design, Urban Planning and the board processes are particularly clear today, as guides, information and evaluation models are progressively turning into a typical legacy: the reconciliation of sensors, web and wiki based technologies with GIS applications is an extremely productive way of working on the odds of helpful communication between residents, strategy creators and the abilities in question inside the metropolitan arranging processes. Cloud innovations, famous among expert and shoppers the same, permit normal updates straightforwardly from the source through a consistent incorporation of decentralized data sets. Geo-referred to frameworks are integral to dynamic cycles at the neighborhood and local level, working with choices of institutional and enterprising entertainers, for instance by sharing area information and empowering optimizing of authoritative strategies. Shared data sets can support public-private organizations and task financing by making information, data and achievability examines accessible to specialized workplaces or by guaranteeing multi-utilities commitments. Association of neighborhood accomplices connecting with worldwide channels through an inexorably shared administrative and arranging process is upgraded with ICT. Metropolitan Design and Urban Planning practices can profit from ICT apparatuses utilized for consistent and differed information assortment, information investigation, planning and recreation. With brilliant city advances, it is additionally conceivable to upgrade the power and extent of apparatuses that are as of now utilized. The concept of smart city shows a focal brilliant arranging convention for ICT applications that can be utilized across scales, areas and client bunches and so on. In utilizing innovation as an instrument for Urban Design and Urban Planning, the objective isn't to computerize the cycle or limit the job of organizers and creators, however to help the course of independent direction. Metropolitan arranging upheld by information on cities improvements; developments and its kin can create plans dependent on an intensive comprehension of urban communities and objective independent direction. ICT gives a chance to accomplish objectives in plan and arranging that essentially might not have been imaginable previously.

Different Concerns While Utilizing IoT for Urban Design and Urban Planning

With regards to IoT, the uses of advancements that make aloof articles dynamic, receptive and programmed as far as detecting and reaction can be countless. Figure 6 gives an undeniable level outline of the IoT system. IoT gives systems administration to associate individuals, things and applications, which empower controller, the executives, just as intelligent coordinated administrations. Subsequently, more things can be checked and controlled, more perplexing errands can be cultivated in lesser time and choices made can be more exact.

IoTs impact on Urban Design and Urban Planning is relied upon to upset and change the way all organizations, states, and residents cooperate with the actual world. This degree of disturbance will altogether affect the world in working on the personal satisfaction of each person. Be that as it may, city the board by incorporating IoT into city frameworks require countless contemplations like utilizing existing actual resources, connecting with neighborhood information eco-frameworks, clear information the executives systems, straightforwardness, security and protection concerns. The effect of uses of IoT on Urban Design ideas like

INNOVATION RELATED STANDARDS IN URBAN DESIGN

Smart city writing is ready with different definitions, however the basic responses to what exactly makes a city smart? Furthermore, when does it turn into a smart city? These are especially imperative to metropolitan type of both Greenfield and Brownfield urban communities since a smart city is an urbanization interaction and not an item. It is made and doesn't basically arise.

There are a few rules that spin around innovation and Urban Design. A portion of the significant ones include:

- Consider metropolitan life before metropolitan spot; think about metropolitan spot before innovation.
- **Urban people group:** Consultations on plans for new improvements ought to completely take advantage of the abilities of online media, virtual universes and different innovations to guarantee that networks impacted by them are given the largest, most vivid chance conceivable to add to their plan.
- **Transport:** Transport plans supporting new advancements ought to show that they have accommodated conventional vehicle interest as well as that which may be made by online plans of action and other social advances
- Identify brilliant machine works that give speedier and more precise outcomes inside and across city organizations, and build up foundation and cycles for information assortment, coordination, understanding and examination, and so forth
- Identify corresponding data works and rearrange city organizations to tackle robotization, for example consolidating information examination in arranging emotionally supportive networks and advancing arranging processes that illuminate team up and co-make with the local area
- Create development encouraging offices that get advancements, association and cycles of and across offices, to shape organizations and start projects that help city monetary, ecological and social destinations
- Anticipate and address potential difficulties in coordinated effort, regardless of whether emerging between accomplices or from innovation holes
- Incorporate various, multi-scale criticism circles to evaluate the viability of drives utilizing measurements, and apply criticism contributions to refine cross-office processes and long haul techniques
- Avoid lock-in using open information and open-source stages rather than restrictive information designs and through the organizing of associations where not a solitary innovation supplier or innovation overwhelms
- Sequence advancement and venture to painstakingly design the execution of essential spine foundation, or the utilization of explicit advances

Who Is the Recipient of Smart City?

Residents are the essential recipients of a Smart City. In any case, they will benefit from the arrangement of norms for city frameworks in a roundabout way. It will be for every city to gather the residents' concerns and treat them methodically as per this arrangement of norms for city frameworks, alongside the connected instruments and directing materials. Residents' concerns include:

- Satisfactory Water Supply.
- Guaranteed Electricity Supply.
- Disinfection, including Solid Waste Management.
- Effective Urban Mobility and Public Transport.
- Reasonable Housing, particularly for poor people.
- Boundless and extraordinary utilization of Data and Technology.
- Great Governance and Citizen Interest.
- Economic Environment
- Wellbeing and Security of Citizens, especially ladies and kids.
- Reasonable Healthcare for everybody.
- Current Education for youngsters and grown-ups.

IoT, BIG DATA, CLOUD, 5G: THE SUPPORTING TECHNOLOGIES OF ICT

Internet of Things

The Internet of Things is an environment of always expanding intricacy, and the jargon of its language is dynamic. At this moderately incipient stage, the IoT biological system is divided and scattered. Over the long haul, the IoT environment ought to go through a smoothing out and coordinating interaction and a weaving together of its singular pieces, in light of the fact that the IoT will assume an inexorably significant part by the way we live and maintain our organizations.

Different projects dispatched by Govt. of India, for example, Smart Cities, Digital India, Make in India, AMRUT, and Startup India is the vital drivers of IoT industry's development in the country. Further, numerous uber projects embraced by the Government of India will help in the powerful, and supportable use of assets by the use of IoT arrangements organizations. Nonetheless, on account of inescapable wide range of correspondence innovations, different heterogeneous guidelines bringing about absence of interoperability and harmonization, the environment is probably going to remain divided in not so distant future, if not tended to right away.

As repeated before, Internet of Things is about heterogeneous and mindful gadgets connecting to improve on people group's life somehow or another or the other. Henceforth, the heterogeneity of the IoT worldview has made it basic to have a new glance at the predominant designs and structures of the IoT and ICT Infrastructures being created or being conveyed.

The IoT esteem chain is maybe the most assorted and confounded worth chain of any industry or consortium that exists on the planet. Truth be told, the gold rush to IoT is unavoidable to the point that in the event that one joins a large part of the worth chain of most industry exchange affiliations, principles bodies, the environment accomplices of exchange affiliations and norms bodies, and afterward include the diverse innovation suppliers taking care of those businesses, one draws near to understanding the extent of the undertaking. In this totally heterogeneous situation, concocting normal fit principles is a significant obstacle.

Rejuvenating the Internet of Things requires a Comprehensive Systems Approach comprehensive of smart handling and detecting innovation, availability, programming, and administrations, alongside a main environment of accomplices. Harmonization in the Communication and Application Protocols, Messaging Middleware, Cloud, Big Data, and Data Analytics, Artificial Intelligence will acquire some similarity this different environment.

Network for M2M/IoT

Network is the way into any digitalization approach. It is the empowering agent for IoT, cloud, and information investigation as it gives the association between the things and the control, tasks, examination, and business applications. While availability is one of its significant parts, IoT is considerably more about offering the types of assistance and semantic augmentations to advance information to important data that can be deciphered and perceived by all applications permitting to develop own insight.

In light of the particular necessities of the different application regions worried, for instance, data transmission, reach, nature of administration, inactivity, and accessible assets various conventions, will be utilized at the various layers of the correspondence stacks created for the large number of correspondence innovations and organization designs.

The wide assortment of correspondence advances, conventions, and principles created over the most recent couple of many years, be it wired or remote, address all sort of utilization cases in correspondence prerequisites with fitting models and organization geographies needed for M2M/IoT applications. In remote correspondence Wi-Fi, ZigBee, 6LoWPAN, Bluetooth advancements might be utilized for short-range availability among device(s) to the door; while Cellular innovations like 2G,

3G, LTE, and wireline advances like Fiber, Fiber to the Home (FTTH), and so on, might be utilized relying upon the organization for interfacing the M2M/IoT passage to the ideal server.

Inside the IoT worldview, we have various classifications of partners Consumers, Enterprise, Industrial, and Infrastructure. Each category's limitations and prerequisites are very assorted and unique in relation to other people, which make the IoT worldview as one of the most complicated and heterogeneous. As a rule, we have the pattern to utilize monetarily generally accessible, off-the-rack advancements like Ethernet and Wireless LAN additionally for modern applications. In the last setting, care will be taken to guarantee that the particular necessities of the modern applications worried, for instance, inertness and accessibility are upheld.

Big Data and Analytics

Information examination, brilliant or huge information, is worried about dealing with and investigation of huge arrangements of information. This incorporates effective capacity, search, and perception. Prescient upkeep, process enhancement, supply arranging, and item quality control are models for the utilization of information examination. The information will be the information from the cycle and supply chains, item lifecycle data, and reproduction information from the turn of events and designing cycles.

The IoT environment is intensely reliant upon information assortment and transmission. Associated sensors gather a lot of information through the Internet, empowering M2M collaboration and handling of the information for specific administrations. Various kinds of information are sent and handled inside the IoT biological system. The information basically incorporates individual information and delicate individual information like monetary data, area, wellbeing related data, and so forth, that is ascribed to a person.

Edge Computing: Cloud Computing Paradox for IoT

Progress of IoT information handling to the organization edge was relied upon to occur in the early IoT advancement lifecycle stage. Notwithstanding, diminishing availability expenses and rising correspondence networks throughput prompted a sluggish speed of this pattern, which brought about a shift toward concentrated cloud handling. Presently falling costs and the expanding handling force of edge gadgets have launched the progress towards network edge. The change to edge processing might essentially affect associations IT and Operational Technology (OT) frameworks, and how the establishment of new age advanced items is laid.

Change of IoT information handling to the organization edge was relied upon to occur in the early IoT improvement lifecycle stage. Be that as it may, diminishing availability expenses and rising correspondence networks throughput prompted a lethargic speed of this pattern, which brought about a shift toward concentrated cloud handling. Presently falling costs and the expanding handling force of edge gadgets have launched the change towards network edge. The change to edge figuring might essentially affect associations IT and Operational Technology (OT) frameworks, and how the establishment of new age advanced items is laid.

5G Services

The beginning of the 5G time will reshape current remote specialized strategies utilized for IoT-based applications. IoT can't flourish without successful and reasonable remote network, interoperability, and normal principles. Consequently, 5G can possibly have a momentous effect in transit where future IoT biological systems are planned, particularly in the space of adaptability, dormancy, dependability, security, and the degree of individual control on network boundaries. With a guarantee of 20Gbps top information rate, under 1ms idleness and 90% decrease in network energy usage, 5G will spike the following round of telecom foundation ventures across the globe.

The sharp climb in purchaser information and the expansion of IoT gadgets will fuel the development of 5G. A vital prerequisite for 5G-network carry out is accessibility of a solid dependable backhaul, which is non-existent in India, as of now. As 5G organizations should uphold huge volume of information from arising applications like IoT, smart urban communities, necessity of a solid and dependable backhaul (from cell pinnacle to arrange administrators Point of Presence) turns into a basic concern. Further to help 5G necessity for inactivity decrease (from 50ms to 1ms) and speed from 100 Mbps to 20Gbps, the Fiber organization in India should be expanded from current market of 1618 million Fiber km each year to something like 50 million Fiber km each year. 5G will likewise require a multifold expansion in little cells sending, with every little cell having backhaul on Fiber. The level of pinnacle backhaul on Fiber for the administrators should increment fundamentally from 20% to 70%, 80% levels. Further, for high limit microwave backhaul, utilization of E-Band (7176 GHz combined with 8186 GHz) and V-Band (5764 GHz) should be started.

Wi-Fi Offload

Wi-Fi has ended up being colossally famous with cell phone clients as a minimal expense answer for worked on limited inclusion, and portable broadband experience, at scene explicit areas, particularly inside. Generous traffic on a cell phone today is persisted Wi-Fi. If we can accomplish high traffic volumes on Wi-Fi, it would be not difficult to achieve a high extent of traffic offloading from the cell organizations to Wi-Fi organizations, subsequently, opening up cell range for upgrading outside versatile organization limit.

The ability to convey enormous volumes of traffic over radio waves is the most difficult perspective. In a brilliant city climate, huge number of portable cells would be needed to give the network and convey traffic. Nonetheless, cost, ecological perspectives, and particularly radio impedance issues are to be tended to. Therefore, the administrators are acquainting tiny radio cells with convey the necessary limit and inclusion. These phones are described by moving enormous volumes of information over an exceptionally brief distance. Portable administrators worldwide have begun to execute a versatile information offloading procedure. All in all, they should observe correlative advances for conveying information initially focused on for 3G/4G organizations. Wi-Fi has arisen as the favored offloading innovations. A third innovation, WiMAX, is additionally in presence.

The India Perspective

In the way to deal with the Smart Cities Mission, the goal is to advance urban communities that give center foundation and give a nice personal satisfaction to its residents, a spotless and practical climate and utilization of Smart Solutions.

The Biggest Challenge

Occupations

1. India will have around 400 million extra people in the workforce constantly 2050.
2. Farming and related exercises that give means to around 220 million of the current labor forces of 500 million can't retain this extra work minus any additional lessening levels of profit.
3. There must be a huge exchange of individuals from essential to auxiliary and tertiary areas, and from provincial to metropolitan regions.
4. Enterprises and organizations are creating some distance from megacities into lower request urban areas or rustic areas, while the casual area is moving into the megacities.
5. It is the non-dirtying tertiary action and development of select casual area that are driving the restricted urbanization in million or more urban areas.
6. Casual vocations should be incorporated into metropolitan plans and drafting guidelines: disorderly labor force accesses markets and essential conveniences.

Different Difficulties

1. Comprehensive Thinking
2. Choosing fitting good examples
3. Advancing Innovation
4. Energy:
    a. Integrate Energy and Waste Management
5. Promote Low Carbon Mobility
6. Building Construction:
    a. Incorporate Building Systems and City Networks/Systems
    b. Advance Materials Sciences
7. Total Quality Management
8. Metropolitan Management:
    a. Use (clean) Big and Open information for recreation and displaying
    b. Revamp educational programs in design, arranging, designing
    c. Study Urban Networks and Systems
    d. Converge organizations and advancements through IoT, M2M, and normalization
    e. Command and Control Centers
9. India: Areas of concern Housing, Slums, Overcrowding, Public Safety, Garbage, Pollution, Infrastructure, Transportation, Disaster Management, Education, Healthcare, Governance, Accountability, Transparency.
10. Utilizing India's uniqueness: Socio-Economic Disparity, Religion, Coexistence, Interdependence, Non-requesting Citizens, Collective Voice, Tolerance, Coping Mechanism, Youth, Growth, and Entrepreneurial Spirit.

Smart Cities Need Brilliant Citizens

Generally, drives like the Smart Cities Mission are marks of created economies. An emerging nation will turn into an effectively evolved country when its residents additionally overhaul and update themselves. In India, the Mission is going full speed ahead, with the best 20 smart urban communities venturing into the following phase of execution. It centers around the innovation overlaying the fundamental framework that will be implicit perfect spots and in adequate amounts in the urban areas. The most crucial perspective the residents who live and work in these urban areas should be indispensable to the execution interaction too. The accomplishment of the Mission is immovably vested in smart residents. A smart resident is one who has city sense and regards the law. Some of unanswered inquiries regarding savvy urban communities, taking everything into account are:

1. I will submit to the traffic rules, drive inside speed cutoff points and halt from hopping signals.
2. I will put walkers first.
3. I Will regard elderly folks and give way to senior residents.
4. I will leave vehicles at assigned spots and not elsewhere.
5. I will keep up with cleanliness their pads as well as in the normal spaces of their apartment buildings.
6. I will toss trash just in receptacles and practice source isolation during waste disposal.

Mindfulness about smart arrangements assumes significant part in growing genuine brilliant residents. However, the nearby specialists of our savvy urban communities will make considerable interests in smart arrangements, they can't hold back on endeavors to raise resident mindfulness on the proficient use of these arrangements and administrations.

For instance, energy saving can't be accomplished only with brilliant meters in a home. To decrease energy utilization and get a good deal on charges, purchasers need to screen their energy use as well as put forth an attempt to change the entire family everyday energy use conduct. This would incorporate moving to energy-productive apparatuses, lessening TV time, and turning off electrical machines when not being used, particularly during top periods.

A smart city interfaces individuals with their current circumstance and city to make more productive and ideal connections between accessible assets, innovation, local area administrations, and occasions in the metropolitan texture. This association is an apparatus that interfaces the execution of the savvy city and the proposed innovation.

Citizen Participation

Most created urban communities guarantee that residents take an interest in each viewpoint, from cleaning to security prerequisites. Resident support guarantees resident fulfillment, which thus guarantees most extreme effectiveness of the proposed innovation.

Great administration is constantly estimated by the degree to which it includes its residents in the general dynamic interaction.

In an undeniably mind boggling world, residents inputs are a basic asset for strategy making. Great independent direction requires the information, encounters, perspectives and upsides of people in general. The investment of residents has become more straightforward through internet based government entryways.

Such cooperation lessens the contention of feelings and makes execution more straightforward. Smart residents should be completely comprehensive, imaginative and practical.

Joint Commitment of Residents and Government

Political will and the specialized ability to draw in residents in strategy making, or giving exact information on government execution, are the signs of created majority rules systems. However making strategies for a city's development and thorough advancement is significant, a brilliant city recognizes that approaches alone are adequately not to arrive at their objectives. It requests support from its residents and nearby partners to get this going.

Residents are called upon to together assume liability and take part simultaneously. Building social capital is fundamental to guarantee that smart residents obtain the capacities and abilities to address the difficulties of things to come. Really at that time does a city become future prepared.

Innovative Support

Innovation has given the world new aspects; worldwide, residents are becoming innovation situated in each viewpoint concerning their solace, comfort and security. Innovation support has turned into a fundamental element for the development of a city and its residents.

The utilization of imaginative data and correspondences innovation (ICT) applications, advanced cells and brilliant installations are all important for the most common way of making smart residents. At the point when city tenants utilize the Internet to make more brilliant, more educated decisions, urban areas become more astute as well. The right methodology towards the Mission includes a harmony among mechanical and non-innovative methodology.

The ICT stage is the best device to connect the current hole between the public authority and residents. The savvy city idea fundamentally accentuates the consideration of ICT arrangements as a feature of its whole proposition. Going ahead, such a framework will be hearty and reasonable. There are parcel of applications and online stage accessible to work with resident cooperation, some of which are the Swachh Bharat Clean India portable application, IPaidABribe.com, Safecity India, Next Bengaluru and Kumbhathon. Urban areas become more unique by the utilization of such stages.

As of now, the public authority has dispatched fundamental devices to assist residents with taking part proactively. These incorporate road battling, training programs in schools and universities, media ads and hoardings, interview programs with government authorities, and online support.

MyGov.in is a superb illustration of ICT coordinating and expanding the effectiveness of resident commitment. All arrangement level choices will straightforwardly include resident's suppositions. This stage will make the framework more straightforward and go about as an intelligent gathering among residents and the Government. It permits residents to post their remarks and ideas on any proposition and furthermore incorporates various sorts of support (i.e., casting a ballot, raising public mindfulness, pushing for an issue, checking political cycles) that will best advance popularity based improvement in various settings. It is an innovation that is made to deliberately interface resident's gatherings and enhance their voices.

What's more, private innovation engineers are constantly investigating shrewd advances from brilliant mobiles to savvy furniture and apparatuses, advertising them at reasonable costs to reach to all classes of individuals. This aides in cost saving as well as associating with around the world utilized brilliant advances.

Also, extraordinarily shaped social orders in private states and corporate organizations across India are presently taking up drives like cleaning the area, streets and water bodies, tree ranch, electronic waste, and reusing. This is the sort of inspiration expected of residents to take part in taking the brilliant city drive.

To guarantee a more noteworthy portion of online cooperation channels, for example, through cell phone applications and web-based media, districts needs to put resources into savvy individuals not simply in brilliant advancements. Really at that time will instruments like cell phones and portable applications can possibly reform city administration and add to the creation of individuals driven brilliant urban communities. These urban communities need to have comprehensive, imaginative, and maintainable shrewd residents. Without them, the savvy city proposition will hold great just for documentation, not execution

A city is an impression of how its residents see it, and a brilliant city really is the means by which a city acts as an inventive biological system. There is nothing of the sort as a standard layout or an enchanted across the board savvy city application. Permitting residents to become dynamic during the time spent city plan and developing empowering base advancement and community oriented methods of creating frameworks out of some inexactly joined parts will help in carrying out the Smart City Mission effectively.

SMART CITY ADVANTAGES AND DISADVANTAGES

(Madakam S. and Ramaswamy R. (2014) said that shrewd urban areas have used innovation to make their day to day environments more productive and utilized assets. Shrewd urban communities have the fundamental elements for an extended quick development rate in metropolitan populaces throughout the following not many years. The meaning of the brilliant city isn't consistent, and a few things keep on developing. These urban areas make the most out of the data and correspondence innovations to work fair and square of administrations, its resident's prosperity, manageability and generally monetary development.

Benefits of Smart City

There are a few benefits of Smart City, and finding out about these can assist you with gauging the advantage and the inconveniences from it.

Better Transportation Benefits

A smart city can possibly definitely work on the current degree of transportation all through a city. It will have better traffic the board, the capacity to follow public transportations and serve its residents better with steady data and low costs.

More Secure Communication

A smart city will have the most mechanical advances and organizations with the private area will help society since there will be lesser crime. The case of such innovation is tag acknowledgment; associated wrongdoing focuses, shooting identifiers, better crisis administrations, and body cameras.

Proficient Public Administration

Since there is a restricted measure of regular assets left to satisfy the need of individuals, shrewd urban communities will have advances and the vital apparatuses to eliminate our utilization of normal assets and reduction misuse of water, power, and so forth without eliminating any elements.

Diminished Natural Impression

According to Hall R. (2000), a shrewd city has large number of energy-proficient structures that can further develop the air quality, utilize environmentally friendly power sources, and reduction the reliance on non-environmentally friendly power sources. These will assist with diminishing the natural effect we have on the climate.

More Computerized Value

People should approach rapid internet providers at reasonable costs and gadgets. If they approach public Wi-fi in neighborhoods the city, all inhabitants will have equivalent freedoms.

Financial Advancement Openings

Investing in shrewd urban areas will prompt working on their local and worldwide seriousness and draw in new inhabitants and further develop business. Since the whole city will approach an open information stage, data, and so on organizations will prosper. They can settle on informed choices with the accessible innovations and lead to monetary turn of events.

Improvement of Framework

Old streets, structures, roadways, spans require huge speculations to keep up with their state and increment their helpful life. Yet, with the assistance of savvy innovations, urban communities will can systematically foresee and recognize the regions that can cause foundation disappointments before it happens.

Open Positions

Giffinger R. et al. (2008) discussed that a brilliant city will have numerous organizations and open positions since individuals will get equivalent admittance to fundamental assets like transportation, web association, and propositions for employment.

Disadvantages of Smart City

However Smart City enjoys many benefits, there are a few inconveniences. Knowing these can assist one with understanding the other side of the coin. The drawbacks are as per the following:

Restricted Protection

Since the specialists or the public authority will approach surveillance cameras and clever frameworks associated through various spaces, the residents will experience issues in keeping up with their secrecy. Facial acknowledgment and such things will definitely change the idea of protection or individual space.

Social Control

Individuals who can follow and unify the information they assemble with surveillance cameras will have more prominent power. It tends to be an administration, a private office, or different specialists. They will have the ability to control a residents information and can without much of a stretch control general feelings.

Abundance Network Trust

Since the resident of these shrewd urban areas will depend for the most part on hardware and organizations, they will lose independence in their navigation and could become awkward. They would not have the option to respond fittingly in a situation where these devices are not usable.

Trouble in the Pre-trade Stage

Cosgrave E. and Tryfonas T. (2012) discussed that though cash will be accessible, the shrewd innovations will be as yet in their pre-business stages. These urban areas will come up short on the innovation related abilities and limit.

Pre-preparing Is Required

If individuals of the city don't know about innovation, then, at that point, they cannot utilize it. Without preparing, they will see it insignificant to their regular routines and will see it hard to use it.

CONCLUSION

The Positive Impacts

As globalization turns out to be more conspicuous, it is getting simpler for individuals to speak with individuals of different countries and advantage from the worldwide effect of innovation. A Smart City will fuse innovation all through the city through savvy innovation, for example, Wi-Fi and modems giving individuals more admittance to data and mechanical use in the city. According to Tiwari A. and Jain K. (2014), there are some beneficial outcomes for the foundation of the nation, as there will be greater redevelopment plans around India, like created nations abroad. For example, the Khan River Redevelopment Plan could have positive effects as many individuals will utilize the space for entertainment and happiness, while utilizing the riverfront.

Numerous urban areas all throughout the planet have great riverfronts or waterfronts as open spaces, giving numerous conveniences and qualities to the overall population like the Navy Pier in Chicago, USA and Docklands in Melbourne, Australia. Savvy urban communities will help Indian urban areas contend and co-work with different urban areas around the world, giving comparable offices to individuals of India.

The Negative Impacts

As shown, there can be constructive outcomes of the Smart City plan in India. Notwithstanding, the broad development of IT with many designers being counseled can be an adverse result for organizers and draftsmen. Lombardia P. et al. (2012) conversed that the public authority ought to think about the perspectives of organizers, geographers and modelers, alongside IT experts to have an all-encompassing methodology towards Smart Cities. The manners by which individuals are affected in Smart Cities and how the foundation can help people is a center part of metropolitan preparation. Organizers should be counseled for Smart City development; in any case, the public authority has limitlessly focused on the Information Technology part of the Smart City disregarding the city as a design and base up techniques towards city arranging. An absolutely specialized city may have many negative ramifications, for example, absence of human contact and absence of eye to eye collaborations which are a need as indicated by the glad city file. The glad city record shows that one of the primary purposes behind bliss in urban areas is the need to convey and coincide as a local area in a city.

Although a portion of the Smart City foundation arranging in India may prompt local area commitment in a city, authorizing innovation rather than place making and city arranging approaches may have a general adverse consequence to the Smart City. Shrewd Cities ought to likewise zero in on monetary development and giving more open positions to individuals. Notwithstanding, with the approach of AI innovation, which is viewed as joined in Smart Cities, many positions will be performed by Artificial Intelligence supplanting people groups skill in them. This, once more, can contrarily affect the general public later on. Shrewd urban areas ought to be worked by having people as a top priority, rather than simply mechanical improvement. This is the reason organizers ought to likewise add to Smart City development, rather than specialists and specialized specialists.

As shown, there are numerous positive and adverse consequences of Smart City development in India. The emphasis on working on the framework and giving innovation to individuals will have colossal advantages on the Indian populace. In any case, who the recipients are and how much Smart City development helps individuals on a human scale is easily proven wrong. There should be a more prominent spotlight on building urban areas for people group's benefits and giving foundation that is favorable to individuals, rather than framework that diminishes people groups capacities. Innovation can do both, and it is the obligation of the public authority to examine these issues with specialists in city improvement to support individuals of India.

REFERENCES

Cosgrave, E., & Tryfonas, T. (2012). Exploring the Relationship between Smart City Policy and Implementation. *Smart 2012: The First International Conference on Smart Systems, Devices and Technologies*, 79-82.

Debnath, A. (2014). A methodological framework for benchmarking smart transport cities. *Cities-The International Journal of Urban Policy and Planning, 37*, 47-56.

Giffinger, R. (2008). The role of rankings in growing city competition. European Urban Research Association.

Gill, H. (2014). Smart Cities and Affordable Housing in India. *The International Journal of Business & Management, 2*(8), 67-73.

Hall, R. (2000). The Vision of A Smart City. *2nd International Life Extension Technology Workshop*, 1-7.

Letaifa, S. (2015). How to strategize smart cities: Revealing the Smart model. *Journal of Business Research, 68*, 1414–1419.

Lombardia, P. (2012). Modelling the smart city performance. *Innovation - The European Journal of Social Science Research, 25*(2), 137 - 149.

Madakam, S., & Ramaswamy, R. (2014). Smart Cities - Six Dimensions. *International Conference on Advances in Computing and Information Technology - ACIT*, 38-41.

Mosannenzadeh F., & Vettorato D. (2014). Defining smart city: a conceptual framework based on keyword analysis. *Tema - Journal of Land Use, Mobility and Environment*, 683 - 694.

Raparthi, K. (2014). Assessing Smart-Growth Strategies in Indian Cities: Grounded Theory Approach to Planning Practice. *ASCE - Journal of Urban Planning and Development*, 1-10.

Tiwari, A., & Jain, K. (2014). GIS Steering Smart Future for Smart Indian Cities. *International Journal of Scientific and Research Publications, 4*(8), 1-5.

Zubizarreta, I. (2015). Smart City Concept: What It Is and What It Should Be. *ASCE - Journal of Urban Planning and Development,* 1-8.

Compilation of References

Smart City Hub. (n.d.). https://smartcityhub.com/technology-innnovation/the-top-ten-companies-that-build-smart-cities/

Wikipedia contributors. (2022, March 15). *Humanoid robot*. Wikipedia. https://en.wikipedia.org/wiki/Humanoid_robot

Kosowatz, J. (2020, February 3). *Top 10 Smart Cities in the World*. ASME. https://www.asme.org/topics-resources/content/top-10-growing-smart-cities

Bisht, A. K., Singh, R., Bhutiani, R., & Bhatt, A. (n.d.). *An Investigation of Assessment and Modelling the Water Quality of Rivers Based on Artificial Neural Networks-An Initiative towards the River Ganga*. Academic Press.

Pandya, P., Saxena, A., & Bhatt, A. (n.d.). *The Human capacity assessment through time series prediction of Artificial Neural Network*. Academic Press.

What is a Smart City? Definition and Examples. (n.d.). *TWI*. https://www.twi-global.com/technical-knowledge/faqs/what-is-a-smart-city

Datt, G., Bhatt, A. K., & Saxena, A. (2010). An Investigation of artificial neural network based prediction systems in rain forecasting. *International Journal on Recent and Innovation Trends in Computing and Communication*, *3*(8), 5338–5349.

Bhatt, A., Dubey, S. K., & Bhatt, A. K. (2018). Analytical study on cardiovascular health issues prediction using decision model-based predictive analytic techniques. In *Soft computing: theories and applications* (pp. 289–299). Springer. doi:10.1007/978-981-10-5699-4_28

Saxena, A., Bhatt, A. K., Pandey, B., Tripathi, P., & Dutt, G. (2016). HSTL IO Standards Based Processor Specific Green Counter Design on 90nm FPGAAbhay Saxena. *International Journal of Control and Automation*, *9*(7), 331–342. doi:10.14257/ijca.2016.9.7.31

Johnson, B. (2008). Cities, systems of innovation and economic development. *Innovation*, *10*(2-3), 146–155. doi:10.5172/impp.453.10.2-3.146

2019-2020 Australian Bushfires. (2019). *Center for Disaster Philanthropy (CDP)*. https://disasterphilanthropy.org/disaster/2019-australian-wildfires/

Keenan, T. P., & Trotter, D. M. (1999). The changing role of community networks in providing citizen access to the Internet. *Internet Research*, *9*(2), 100–108. doi:10.1108/10662249910264855

Komninos, N. (2013). *Intelligent cities: innovation, knowledge systems and digital spaces*. Routledge. doi:10.4324/9780203857748

Rosa, L., Silva, F., & Analide, C. (2021). Mobile Networks and Internet of Things Infrastructures to Characterize Smart Human Mobility. *Smart Cities*, *4*(2), 894–918. doi:10.3390martcities4020046

Thite, M. (2011). Smart cities: Implications of urban planning for human resource development. *Human Resource Development International*, *14*(5), 623–631. doi:10.1080/13678868.2011.618349

United Nations. (2012). *World Urbanization Prospects. United Nations, Department of Economic and Social Affairs, Population Division: the 2011 Revision*. Highlights.

Washburn, D., Sindhu, U., Balaouras, S., Dines, R. A., Hayes, N., & Nelson, L. E. (2009). Helping CIOs understand "smart city" initiatives. *Growth*, *17*(2), 1–17.

Saxena, A., Chang, A. Y., Kulathuramaiyer, N., & Bhatt, A. (2019). The Conceptual Framework of Articulated Human Intelligence in Perspective of Industry 4.0. *Indian Journal of Science and Technology*, *12*(6), 6. doi:10.17485/ijst/2019/v12i6/141147

Stone, S. (2019, March 6). *Key Challenges of Smart Cities & How to Overcome Them*. Ubidots Blog. https://ubidots.com/blog/the-key-challenges-for-smart-cities/

Abbas, S. M., & Singh, S. N. (2018, February). Region-based object detection and classification using faster R-CNN. In *2018 4th International Conference on Computational Intelligence & Communication Technology (CICT)* (pp. 1-6). IEEE.

Abeshu, A., & Chilamkurti, N. (2018). Deep learning: The frontier for distributed attack detection in fog-to-things computing. *IEEE Communications Magazine*, *56*(2), 169–175.

Agniel, D., Kohane, I. S., & Weber, G. M. (2018). Biases in electronic health record data due to processes within the healthcare system: Retrospective observational study. *BMJ (Clinical Research Ed.)*, 361. doi:10.1136/bmj.k1479 PMID:29712648

Ahmad, M. A., Eckert, C., & Teredesai, A. (2018, August). Interpretable machine learning in healthcare. In *Proceedings of the 2018 ACM international conference on bioinformatics, computational biology, and health informatics* (pp. 559-560). ACM.

Akcay, S., & Breckon, T. P. (2017, September). An evaluation of region based object detection strategies within x-ray baggage security imagery. In *2017 IEEE International Conference on Image Processing (ICIP)* (pp. 1337-1341). IEEE.

Al Nuaimi, E., Al Neyadi, H., Mohamed, N., & Al-Jaroodi, J. (2015). Applications of big data to smart cities. *Journal of Internet Services and Applications*, *6*(1), 25. Advance online publication. doi:10.118613174-015-0041-5

Albino, V., Berardi, U., & Dangelico, R. M. (2015). Smart cities: Definitions, dimensions, performance, and initiatives. *Journal of Urban Technology*, *22*(1), 3–21.

Al-Hader, M., Rodzi, A., Sharif, A. R., & Ahmad, N. (2009). Smart city components Architicture. *2009 International Conference on Computational Intelligence, Modelling and Simulation*. 10.1109/CSSim.2009.34

Alliance, L. (2021). *What is LoRaWAN® Specification*. Retrieved from www.lora-alliance.org: https://lora-alliance.org/about-lorawan/

Alsaadi, M., Lisitsa, A., Khalaf, M., & Qasaimeh, M. (2019, August). Investigating the capability of agile processes to support medical devices regulations: The case of xp, scrum, and FDD with EU MDR regulations. In *International Conference on Intelligent Computing* (pp. 581-592). Springer.

Alzubaidi, L., Zhang, J., Humaidi, A. J., Al-Dujaili, A., Duan, Y., Al-Shamma, O., Santamaría, J., Fadhel, M. A., Al-Amidie, M., & Farhan, L. (2021). Review of deep learning: Concepts, CNN architectures, challenges, applications, future directions. *Journal of Big Data*, *8*(1), 1–74. doi:10.118640537-021-00444-8 PMID:33816053

Aminanto, M. E., Choi, R., Tanuwidjaja, H. C., Yoo, P. D., & Kim, K. (2017). Deep abstraction and weighted feature selection for Wi-Fi impersonation detection. *IEEE Transactions on Information Forensics and Security*, *13*(3), 621–636.

Apps, B. M. (2021). *Bluetooth Low Energy (BLE) Technology*. Retrieved from www.blemobileapps.com: https://www.blemobileapps.com/ble-mobile/

Araromi, O. A., Graule, M. A., Dorsey, K. L., Castellanos, S., Foster, J. R., Hsu, W. H., & Wood, R. J. (2020). Ultra-sensitive and resilient compliant strain gauges for soft machines. *Nature*, *587*(7833), 219–224. doi:10.103841586-020-2892-6 PMID:33177670

ARC. (2021). *Smart Cities*. https://www.arcweb.com/advisory-services/infrastructure-smart-cities

Azmoodeh, A., Dehghantanha, A., & Choo, K. K. R. (2018). Robust malware detection for internet of (battlefield) things devices using deep eigenspace learning. *IEEE Transactions on Sustainable Computing, 4*(1), 88-95.

Bacher, P., Madsen, H., Nielsen, H. A., & Perers, B. (2013). Short-term heat load forecasting for single family houses. *Energy and Building*, *65*, 101–112.

Bean. (2021). Transforming The Insurance Industry With Big Data, Machine Learning And AI. *Forbes*. https://www.forbes.com/sites/randybean/2021/07/06/transforming-the-insurance-industry-with-big-data-machine-learning-and-ai/?sh=542a599ff8a6

Bellini, E. (2020). World now has 583.5 GW of operational PV. *PV Magazine*. https://www.pv-magazine.com/2020/04/06/world-now-has-583-5-gw-of-operational-pv/

Benbasat, I., & Wang, W. (2005). Trust in and Adoption of Online Recommendation Agents. *Journal of the Association for Information Systems*, *6*(3), 72–101. doi:10.17705/1jais.00065

Bennett, C. C., & Šabanović, S. (2014). Deriving minimal features for human-like facial expressions in robotic faces. *International Journal of Social Robotics*, *6*(3), 367–381. doi:10.100712369-014-0237-z

Bhagya, C., & Shyna, A. (2019, April). An overview of deep learning based object detection techniques. In *2019 1st International Conference on Innovations in Information and Communication Technology (ICIICT)* (pp. 1-6). IEEE.

Bhattacharya, S., Somayaji, S. R. K., Gadekallu, T. R., Alazab, M., & Maddikunta, P. K. R. (2020). A review on deep learning for future smart cities. *Internet Technology Letters*, *187*(1), 1–6. doi:10.1002/itl2.187

Bibri, S. E., & Krogstie, J. (2017). The core enabling technologies of big data analytics and context-aware computing for smart sustainable cities: A review and synthesis. *Journal of Big Data*, *4*(1), 38. Advance online publication. doi:10.118640537-017-0091-6

Bismart. (n.d.). *What Exactly is a Smart City?* https://blog.bismart.com/en/what-is-a-smart-city

Boudreaux, E. D., Waring, M. E., Hayes, R. B., Sadasivam, R. S., Mullen, S., & Pagoto, S. (2014). Evaluating and selecting mobile health apps: Strategies for healthcare providers and healthcare organizations. *Translational Behavioral Medicine*, *4*(4), 363–371. doi:10.100713142-014-0293-9 PMID:25584085

Boukerche, A., & Wang, J. (2020). Machine Learning-based traffic prediction models for Intelligent Transportation Systems. *Computer Networks*, *181*, 107530.

Bratu, D. V., Moraru, S. A., & Guşeilă, L. G. (2019, August). A Performance Comparison between Deep Learning Network and Haar Cascade on an IoT Device. In *2019 International Conference on Sensing and Instrumentation in IoT Era (ISSI)* (pp. 1-6). IEEE.

Briggs, P., Jeske, D., & Coventry, L. (2017). The Design of Messages to Improve Cybersecurity Incident Reporting. In *International Conference on Human Aspects of Information Security, Privacy, and Trust*. Springer. 10.1007/978-3-319-58460-7_1

Brownlee, J. (2016). *Classification And Regression Trees for Machine Learning*. https://machinelearningmastery.com/classification-and-regression-trees-for-machine-learning/

Brown, P., Christensen, K., & Schuster, D. (2016, September). An Investigation of Trust in a Cyber Security Tool. *Proceedings of the Human Factors and Ergonomics Society Annual Meeting*, *60*(1), 1454–1458. doi:10.1177/1541931213601333

Butavicius, M., Parsons, K., Lillie, M., McCormac, A., Pattinson, M., & Calic, D. (2020). When believing in technology leads to poor cyber security: Development of a trust in technical controls scale. *Computers & Security*, *98*, 102020. doi:10.1016/j.cose.2020.102020

Cabot, T. (2019). *An Introduction to Smart Cities*. https://www.arcweb.com/blog/introduction-smart-cities-0

Cao, C., Wang, B., Zhang, W., Zeng, X., Yan, X., Feng, Z., Liu, Y., & Wu, Z. (2019). An improved faster R-CNN for small object detection. *IEEE Access: Practical Innovations, Open Solutions*, *7*, 106838–106846.

Chang, Y. S., Chiao, H. T., Abimannan, S., Huang, Y. P., Tsai, Y. T., & Lin, K. M. (2020). An LSTM-based aggregated model for air pollution forecasting. *Atmospheric Pollution Research*, *11*(8), 1451–1463.

Chatterjee, B., Das, D., Maity, S., & Sen, S. (2018). RF-PUF: Enhancing IoT security through authentication of wireless nodes using in-situ machine learning. *IEEE Internet of Things Journal*, *6*(1), 388–398.

Chauhan, J., Seneviratne, S., Hu, Y., Misra, A., Seneviratne, A., & Lee, Y. (2018). Breathing-based authentication on resource-constrained IoT devices using recurrent neural networks. *Computer*, *51*(5), 60–67.

Chen, X., Fang, H., Lin, T. Y., Vedantam, R., Gupta, S., Dollár, P., & Zitnick, C. L. (2015). *Microsoft coco captions: Data collection and evaluation server.* arXiv preprint arXiv:1504.00325.

Chen, S., Hong, J., Zhang, T., Li, J., & Guan, Y. (2019, August). Object detection using deep learning: Single shot detector with a refined feature-fusion structure. In *2019 IEEE International Conference on Real-time Computing and Robotics (RCAR)* (pp. 219-224). IEEE.

Chen, X., Wang, H. H., & Tian, B. (2019). Visualization model of big data based on self-organizing feature map neural network and graphic theory for smart cities. *Cluster Computing*, *22*(6), 13293–13305.

Choudhary, M. (2019). *Six technologies crucial for smart cities.* https://www.geospatialworld.net/blogs/six-technologies-crucial-for-smart-cities/

Cianchetti, M., Laschi, C., Menciassi, A., & Dario, P. (2018). Biomedical applications of soft robotics. *Nature Reviews. Materials*, *3*(6), 143–153. doi:10.103841578-018-0022-y

Classification and Regression. (n.d.). *Apache Spark.* https://spark.apache.org/docs/latest/ml-classification-regression.html#generalized-linear-regression

Clavel, C., Plessier, J., Martin, J. C., Ach, L., & Morel, B. (2009, September). Combining facial and postural expressions of emotions in a virtual character. In *International Workshop on Intelligent Virtual Agents* (pp. 287-300). Springer. 10.1007/978-3-642-04380-2_31

Cloud computing trends for 2019. (2019, February 22). *Networks Unlimited.* https://www.networksunlimited.com/cloud-computing-trends-for-2019/

Columbus, L. (2019). State Of AI And Machine Learning In 2019. *Forbes Magazine, 8.*

Cosgrave, E., & Tryfonas, T. (2012). Exploring the Relationship between Smart City Policy and Implementation. *Smart 2012: The First International Conference on Smart Systems, Devices and Technologies*, 79-82.

Creed, C., & Beale, R. (2008). Psychological responses to simulated displays of mismatched emotional expressions. *Interacting with Computers*, *20*(2), 225–239. doi:10.1016/j.intcom.2007.11.004

Crittin, F. (2004). *New algorithmic methods for real-time transportation problems* (Thesis). EPFL.

Dairi, A., Harrou, F., Sun, Y., & Senouci, M. (2018). Obstacle detection for intelligent transportation systems using deep stacked autoencoder and $ k $-nearest neighbor scheme. *IEEE Sensors Journal*, *18*(12), 5122–5132. doi:10.1109/JSEN.2018.2831082

Dalal, N., & Triggs, B. (2005, June). Histograms of oriented gradients for human detection. In 2005 IEEE computer society conference on computer vision and pattern recognition (CVPR'05) (Vol. 1, pp. 886-893). IEEE. doi:10.1109/CVPR.2005.177

de Visser, E. J., Pak, R., & Shaw, T. H. (2018). From "automation" to "autonomy": The importance of trust repair in human-machine interaction. *Ergonomics*, *61*(10), 1409–1427. Advance online publication. doi:10.1080/00140139.2018.1457725 PMID:29578376

Debnath, A. (2014). A methodological framework for benchmarking smart transport cities. *Cities-The International Journal of Urban Policy and Planning, 37*, 47-56.

Deloitte. (2015). *Industry 4.0—Challenges and solutions for the digital transformation and use of exponential technologies*. Study of Deloitte Consulting. https://www2.deloitte.com/content/dam/Deloitte/ch/Doc uments/manufacturing/ch-en-manufacturing-industry-4-0-24102014.pdf

Diro, A., & Chilamkurti, N. (2018). Leveraging LSTM networks for attack detection in fog-to-things communications. *IEEE Communications Magazine*, *56*(9), 124–130.

Drewil, G. I., & Al-Bahadili, R. J. (2021). Forecast Air Pollution in Smart City Using Deep Learning Techniques: A Review. *Multicultural Education*, *7*(5).

Du, W., Rao, N., Liu, D., Jiang, H., Luo, C., Li, Z., Gan, T., & Zeng, B. (2019). Review on the applications of deep learning in the analysis of gastrointestinal endoscopy images. *IEEE Access: Practical Innovations, Open Solutions*, *7*, 142053–142069. doi:10.1109/ACCESS.2019.2944676

EETimes. (2021). *India's LPWAN Networks Expand to Meet Smart Cities IoT Demand*. Retrieved from www.eetimes.com: https://www.eetimes.com/indias-lpwan-networks-expand-to-meet-smart-cities-iot-demand/#

Elbadawi, M., Gaisford, S., & Basit, A. W. (2021). Advanced machine-learning techniques in drug discovery. *Drug Discovery Today*, *26*(3), 769–777, 1359–6446. doi:10.1016/j.drudis.2020.12.003 PMID:33290820

El-Hassan, O., Sharif, A., Al Redha, M., & Blair, I. (2017). Tracking the Implementation of Electronic Medical Records in Dubai, United Arab Emirates, Using an Adoption Benchmarking Tool. *Studies in Health Technology and Informatics*, *245*, 64–68. PMID:29295053

Elizabeth, B. (2009). Clustering of Internet risk behaviors in a middle school student population. *The Journal of School Health, 79*(11), 547–553. doi:10.1111/j.1746-1561.2009.00447.x PMID:19840232

Everingham, M., Van Gool, L., Williams, C. K., Winn, J., & Zisserman, A. (2010). The pascal visual object classes (voc) challenge. *International Journal of Computer Vision, 88*(2), 303–338.

Fan, C., Xiao, F., & Wang, S. (2014). Development of prediction models for next-day building energy consumption and peak power demand using data mining techniques. *Applied Energy, 127*, 1–10.

Fan, W., & Bifet, A. (2013). Mining big data. *SIGKDD Explorations, 14*(2), 1–5. doi:10.1145/2481244.2481246

Ferdowsi, A., & Saad, W. (2018). Deep learning for signal authentication and security in massive internet-of-things systems. *IEEE Transactions on Communications, 67*(2), 1371–1387.

Figueroa, E. (2021). *Long range, low power Wi-Fi® for IoT.* Retrieved from www.wi-fi.org: https://www.wi-fi.org/discover-wi-fi/wi-fi-certified-halow

FLASHNET. (2021). *Everything you need to know about smart street lighting.* Retrieved from www.intelilight.eu: https://intelilight.eu/smart-street-lighting/

Frontiers in massive data analysis. (2013). doi:10.17226/18374

Gandhi, R. (2018). *Support Vector Machine — Introduction to Machine Learning Algorithms.* Towards Data Science. https://towardsdatascience.com/support-vector-machine-introduction-to-machine-learning-algorithms-934a444fca47

Gao, Z., Xu, C., Zhang, H., Li, S., & de Albuquerque, V. H. C. (2020). Trustful internet of surveillance things based on deeply represented visual co-saliency detection. *IEEE Internet of Things Journal, 7*(5), 4092–4100.

Garg, S., Kaur, K., Kumar, N., Kaddoum, G., Zomaya, A. Y., & Ranjan, R. (2019). A hybrid deep learning-based model for anomaly detection in cloud datacenter networks. *IEEE eTransactions on Network and Service Management, 16*(3), 924–935.

Gavrilescu, R., Zet, C., Foşalău, C., Skoczylas, M., & Cotovanu, D. (2018, October). Faster R-CNN: an approach to real-time object detection. In *2018 International Conference and Exposition on Electrical And Power Engineering (EPE)* (pp. 165-168). IEEE.

Ghosh. (2021). *AI transforming the Insurance industry in India.* https://www.expresscomputer.in/artificial-intelligence-ai/ai-transforming-the-insurance-industry-in-india/74458/

Giebel, G., Draxl, C., Brownsword, R., Kariniotakis, G., & Denhard, M. (2011). *The state-of-the-art in short-term prediction of wind power. A literature overview.* Academic Press.

Giffinger, R. (2008). The role of rankings in growing city competition. European Urban Research Association.

Gill, H. (2014). Smart Cities and Affordable Housing in India. *The International Journal of Business & Management, 2*(8), 67-73.

Girshick, R. (2015). Fast r-cnn. In *Proceedings of the IEEE international conference on computer vision* (pp. 1440-1448). IEEE.

Girshick, R., Donahue, J., Darrell, T., & Malik, J. (2014). Rich feature hierarchies for accurate object detection and semantic segmentation. In *Proceedings of the IEEE conference on computer vision and pattern recognition* (pp. 580-587).

Girshick, R., Donahue, J., Darrell, T., & Malik, J. (2015). Region-based convolutional networks for accurate object detection and segmentation. *IEEE Transactions on Pattern Analysis and Machine Intelligence, 38*(1), 142–158.

Global Wind Energy Council. (2015). Global wind report: Annual market update. Global Wind Energy Council.

Guennouni, S., Ahaitouf, A., & Mansouri, A. (2014, October). Multiple object detection using OpenCV on an embedded platform. In *2014 Third IEEE International Colloquium in Information Science and Technology (CIST)* (pp. 374-377). IEEE.

Guha, J. (2021). *Street Lighting.* Retrieved from https://www.neevenergy.com/street-lighting/: http://neevenergy.com/wp-content/uploads/2019/04/Intellivolt-Solution-Version-2.pdf

Gulati, S., Sousa, S., & Lamas, D. (2017). Modelling Trust: An Empirical Assessment. In *IFIP Conference on HumanComputer Interaction* (pp. 40–61). Springer.

Gulati, S., Sousa, S., & Lamas, D. (2018). Modelling Trust in Human-Like Technologies. In *Proceedings of the 9th Indian Conference on Human Computer Interaction* (pp. 1–10). ACM.

Guo, Y., Li, Y., & Wang, S. (2020, May). Cs-r-fcn: Cross-supervised learning for large-scale object detection. In *ICASSP 2020-2020 IEEE International Conference on Acoustics, Speech and Signal Processing (ICASSP)* (pp. 2553-2557). IEEE.

Guo, S., Dai, Y., Xu, S., Qiu, X., & Qi, F. (2019). Trusted cloud-edge network resource management: DRL-driven service function chain orchestration for IoT. *IEEE Internet of Things Journal, 7*(7), 6010–6022.

Gupta, M., Abdelsalam, M., & Mittal, S. (2020). *Enabling and enforcing social distancing measures using smart city and its infrastructures: A COVID-19 Use case.* arXiv preprint arXiv:2004.09246.

Gupta, A., Anpalagan, A., Guan, L., & Khwaja, A. S. (2021). Deep learning for object detection and scene perception in self-driving cars: Survey, challenges, and open issues. *Array, 10*, 100057. doi:10.1016/j.array.2021.100057

Hall, J., Tritton, T., Rowe, A., Pipe, A., Melhuish, C., & Leonards, U. (2014). Perception of own and robot engagement in human–robot interactions and their dependence on robotics knowledge. *Robotics and Autonomous Systems, 62*(3), 392–399. doi:10.1016/j.robot.2013.09.012

Hall, R. (2000). The Vision of A Smart City. *2nd International Life Extension Technology Workshop*, 1-7.

Hancock, P. A. (2017). Imposing limits on autonomous systems. *Ergonomics*, *60*(2), 284–291. doi:10.1080/00140139.2016.1190035 PMID:27409152

Hang, C., Khan, S., Kou, B., Nazir, S., Liu, W., & Anwar, H. (2020). A Smart Machine Learning Model for the Detection of Brain Hemorrhage Diagnosis Based Internet of Things in Smart Cities. *Complexity*, *2020*, 1–10. doi:10.1155/2020/3047869

Harbach, M., De Luca, A., Malkin, N., & Egelman, S. 2016. Keep on Lockin' in the Free World: A Multi-National Comparison of Smartphone Locking. In *Proc. of the 2016 CHI Conference on Human Factors in Computing Systems (CHI '16)*. ACM. 10.1145/2858036.2858273

Haris, M., & Hou, J. (2020). Obstacle detection and safely navigate the autonomous vehicle from unexpected obstacles on the driving lane. *Sensors (Basel)*, *20*(17), 4719. doi:10.339020174719 PMID:32825601

Harrison, O. (2018). *Machine Learning Basics with the K-Nearest Neighbors Algorithm*. Towards Data Science. https://towardsdatascience.com/machine-learning-basics-with-the-k-nearest-neighbors-algorithm-6a6e71d01761

Harrison, C., Eckman, B., Hamilton, R., Hartswick, P., Kalagnanam, J., Paraszczak, J., & Williams, P. (2010). Foundations for smarter cities. *IBM Journal of Research and Development*, *54*(4), 1–16.

Hasanin, T., Khoshgoftaar, T. M., Leevy, J. L., & Bauder, R. A. (2019). Severely imbalanced big data challenges: Investigating data sampling approaches. *Journal of Big Data*, *6*(1), 107. Advance online publication. doi:10.118640537-019-0274-4

Hashimoto, D. A., Ward, T., & Meireles, O. R. (2020). The Role of Artificial Intelligence in Surgery. *Advances in Surgery*, *54*, 89–101. doi:10.1016/j.yasu.2020.05.010 PMID:32713441

He, N., Cao, J., & Song, L. (2008, December). Scale space histogram of oriented gradients for human detection. In *2008 International Symposium on Information Science and Engineering* (Vol. 2, pp. 167-170). IEEE.

Hernandez-Aceituno, J., Arnay, R., Toledo, J., & Acosta, L. (2016). Using kinect on an autonomous vehicle for outdoors obstacle detection. *IEEE Sensors Journal*, *16*(10), 3603–3610. doi:10.1109/JSEN.2016.2531122

He, X., Jin, R., & Dai, H. (2018). Deep PDS-learning for privacy-aware offloading in MEC-enabled IoT. *IEEE Internet of Things Journal*, *6*(3), 4547–4555.

Hinton, G. E., Osindero, S., & Teh, Y. W. (2006). A fast learning algorithm for deep belief nets. *Neural Computation*, *18*(7), 1527–1554. doi:10.1162/neco.2006.18.7.1527 PMID:16764513

Hoang, V. D., Kurnianggoro, L., & Jo, K. H. (2014, November). Scalable histogram of oriented gradients for multi-size car detection. In *2014 10th France-Japan/8th Europe-Asia Congress on Mecatronics (MECATRONICS2014-Tokyo)* (pp. 228-231). IEEE.

Hosny, A., Parmar, C., Quackenbush, J., Schwartz, L. H., & Aerts, H. (2018). Artificial intelligence in radiology. *Nature Reviews. Cancer*, *18*(8), 500–510. doi:10.103841568-018-0016-5 PMID:29777175

Howarth, A., Quesada, J., Silva, J., Judycki, S., & Mills, P. R. (2018). The impact of digital health interventions on health-related outcomes in the workplace: A systematic review. *Digital Health*, *4*, 2055207618770861. doi:10.1177/2055207618770861 PMID:29942631

Huenerfauth, M., van Rossum, G., & Muller, R. P. (2009). *Introduction to python*. Harvard University.

Hussain, T., Muhammad, K., Del Ser, J., Baik, S. W., & de Albuquerque, V. H. C. (2019). Intelligent embedded vision for summarization of multiview videos in IIoT. *IEEE Transactions on Industrial Informatics*, *16*(4), 2592–2602.

Hussain, T., Muhammad, K., Ullah, A., Cao, Z., Baik, S. W., & de Albuquerque, V. H. C. (2019). Cloud-assisted multiview video summarization using CNN and bidirectional LSTM. *IEEE Transactions on Industrial Informatics*, *16*(1), 77–86.

Hu, X., Wu, J., Wu, M., & Hu, J. (2021). Recent developments of infrared photodetectors with low-dimensional inorganic nanostructures. *Nano Research*, 1–13. https://www.businesswire.com/news/home/20210126005806/en/Global-Cybersecurity-Market-2020-to-2025---Growth-Trends-and-Forecasts---ResearchAndMarkets.com

Infosys. (2019). *AI and Insurance: Everything Changes*. https://www.infosys.com/about/knowledge-institute/insights/Documents/ai-insurance.pdf

Integra, N. (n.d.). *Smart Cities, how they work and how to tackle change*. https://nexusintegra.io/smart-city-what-is-how-it-works/

Jiang, D., Wu, K., Chen, D., Tu, G., Zhou, T., Garg, A., & Gao, L. (2020). A probability and integrated learning based classification algorithm for high-level human emotion recognition problems. *Measurement*, *150*, 107049. doi:10.1016/j.measurement.2019.107049

Jia, X. (2012). RFID technology and its applications in Internet of Things (IoT). In *2nd International Conference on Consumer Electronics Communications and Networks (CECNet)* (pp. 1282-1285). Yichang, China: IEEE.

Jin, C., Chen, W., Cao, Y., Xu, Z., Tan, Z., Zhang, X., ... Feng, J. (2020). Development and evaluation of an artificial intelligence system for COVID-19 diagnosis. *Nature Communications*, *11*(1), 1–14.

Joshi, N. (2019). *4 challenges faced by smart cities*. https://www.allerin.com/blog/4-challenges-faced-by-smart-cities

Kagermann, H., Wahlster, W., & Helbig, J. (2013). Acatech–National Academy of Science and Engineering. *Recommendations for implementing the strategic initiative INDUSTRIE, 4*.

Kang, M. S., & Lim, Y. C. (2017, November). High performance and fast object detection in road environments. In *2017 seventh international conference on image processing theory, tools and applications (IPTA)* (pp. 1-6). IEEE.

Kanimozhi, S., Gayathri, G., & Mala, T. (2019, February). Multiple Real-time object identification using Single shot Multi-Box detection. In *2019 International Conference on Computational Intelligence in Data Science (ICCIDS)* (pp. 1-5). IEEE.

Kapur, R. (2021). *What is MIOTY Technology?* Retrieved from www.everythingrf.com: https://www.everythingrf.com/community/what-is-mioty-technology

Karmakar, G., Chowdhury, A., Das, R., Kamruzzaman, J., & Islam, S. (2021). Assessing trust level of a driverless car using deep learning. *IEEE Transactions on Intelligent Transportation Systems*, *22*(7), 4457–4466. doi:10.1109/TITS.2021.3059261

KCT. (2021). *BLE (Bluetooth Low Energy).* Retrieved from www.blog.kct.ac.in: https://blog.kct.ac.in/ble-bluetooth-low-energy/

Kezer, M., Sevi, B., Cemalcilar, Z., & Baruh, L. (2016). Age differences in privacy attitudes, literacy and privacy management on Facebook. *Cyberpsychology (Brno)*, *10*(1), 1. doi:10.5817/CP2016-1-2

Khan, Z., Anjum, A., & Kiani, S. L. (2013). Cloud based big data analytics for smart future cities. *2013 IEEE/ACM 6th International Conference on Utility and Cloud Computing*. 10.1109/UCC.2013.77

Khrais. (2020). Role of Artificial Intelligence in Shaping Consumer Demand in E-Commerce Future Internet. *MDPI*, 1-14.

Kierkegaard, S. (2008). Cybering, online grooming and ageplay. *Computer Law & Security Review*, *24*(1), 41–55. doi:10.1016/j.clsr.2007.11.004

Kilts, S. (2021). *LoRa Sensors*. Retrieved from www.radiobridge.com: https://radiobridge.com/wireless-technologies/lora

Kitchin, R. (2013). The real-time city? Big data and smart urbanism. *GeoJournal*, *79*(1), 1–14. doi:10.100710708-013-9516-8

Knight, H., & Simmons, R. (2016, May). Laban head-motions convey robot state: A call for robot body language. In *2016 IEEE international conference on robotics and automation (ICRA)* (pp. 2881-2888). IEEE.

Koçi, J., Topal, A. O., & Ali, M. (2020, December). Threat Object Detection in X-ray Images Using SSD, R-FCN and Faster R-CNN. In *2020 International Conference on Computing, Networking, Telecommunications & Engineering Sciences Applications (CoNTESA)* (pp. 10-15). IEEE.

Kohl, C., Knigge, M., Baader, G., Böhm, M., & Krcmar, H. (2018). Anticipating acceptance of emerging technologies using twitter: The case of self-driving cars. *Journal of Business Economics*, *88*(5), 617–642. doi:10.100711573-018-0897-5

Konezny, R. (2021). *Zigbee Wireless Mesh Networking*. Retrieved from www.digi.com: https://www.digi.com/solutions/by-technology/zigbee-wireless-standard

Kopf, D. (2020). *The countries with the worst traffic congestion - and ways to reduce it.* World Economic Forum. https://www.weforum.org/agenda/2020/07/cities-congestion-brazil-colombia-united-kingdom/#:~:text=Traffic%20analytics%20company%20Inrix%20puts,to%20traffic%20each%20year%2C%20respectively

Kota City. (n.d.). *What is a Smart City?* https://smartcitykota.com/about/

Krishna Chaitanya, G., & Maragatham, G. (2021). Object and obstacle detection for self-driving cars using GoogLeNet and deep learning. In *Artificial intelligence techniques for advanced computing applications* (pp. 315–322). Springer. doi:10.1007/978-981-15-5329-5_30

Kshirsagar, S. (2002). A multilayer personality model. In *Proceedings of the 2nd international symposium on Smart graphics* (pp. 107-115). ACM. 10.1145/569005.569021

Kumar, A., Priyanka, S., Dhanashree, K., Praveen, V., Rekha, R. (2022). Efficient binary grasshopper optimization based neural network algorithm for bitcoin value prediction. *International Journal of Nonlinear Analysis and Applications*, *13*, 53-60. . doi:10.22075/ijnaa.2022.6330

Kumar, J., & kumar, J. (2016). Affective modelling of users in HCI using EEG. *Procedia Computer Science*, *84*, 107–114. doi:10.1016/j.procs.2016.04.073

Kundu, D. (2019). Blockchain and Trust in a Smart City. *Environment & Urbanization Asia*, *10*(1), 31–43. doi:10.1177/0975425319832392

Kursa, M. B., & Rudnicki, W. R. (2010). Feature selection with the Boruta package. *Journal of Statistical Software*, *36*(11), 1–13.

Kuznetsova, A., Rom, H., Alldrin, N., Uijlings, J., Krasin, I., Pont-Tuset, J., ... Ferrari, V. (2020). The open images dataset v4. *International Journal of Computer Vision*, *128*(7), 1956–1981.

Lai, Y. K., Ho, C. Y., Huang, Y. H., Huang, C. W., Kuo, Y. X., & Chung, Y. C. (2018, October). Intelligent vehicle collision-avoidance system with deep learning. In *2018 IEEE Asia Pacific Conference on Circuits and Systems (APCCAS)* (pp. 123-126). IEEE. 10.1109/APCCAS.2018.8605622

Lee, K., Jang, J., Yu, Y., Heo, J. S., Han, H., Yoon, Y., Kang, C. M., Hwang, H. K., & Kang, S. (2021). Usefulness of artificial intelligence for predicting recurrence following surgery for pancreatic cancer: Retrospective cohort study. *International Journal of Surgery*, *93*(106050), 393–401. doi:10.1016/j.ijsu.2021.106050 PMID:34388677

Leng, J., Liu, Y., Du, D., Zhang, T., & Quan, P. (2019). Robust obstacle detection and recognition for driver assistance systems. *IEEE Transactions on Intelligent Transportation Systems*, *21*(4), 1560–1571. doi:10.1109/TITS.2019.2909275

Letaifa, S. (2015). How to strategize smart cities: Revealing the Smart model. *Journal of Business Research, 68*, 1414–1419.

Li, K., & Cao, L. (2020, May). A Review of Object Detection Techniques. In *2020 5th International Conference on Electromechanical Control Technology and Transportation (ICECTT)* (pp. 385-390). IEEE.

Liao, R. F., Wen, H., Chen, S., Xie, F., Pan, F., Tang, J., & Song, H. (2019). Multiuser physical layer authentication in Internet of Things with data augmentation. *IEEE Internet of Things Journal*, *7*(3), 2077–2088.

Libelium. (2020). *50 Sensor Applications for a Smarter World.* https://www.libelium.com/libeliumworld/top_50_iot_sensor_applications_ranking/

Li, C., Zhang, Y., & Qu, Y. (2018, March). Object detection based on deep learning of small samples. In *2018 Tenth International Conference on Advanced Computational Intelligence (ICACI)* (pp. 449-454). IEEE.

Li, F., Shinde, A., Shi, Y., Ye, J., Li, X. Y., & Song, W. (2019). System statistics learning-based IoT security: Feasibility and suitability. *IEEE Internet of Things Journal*, *6*(4), 6396–6403.

Li, F., Shi, Y., Shinde, A., Ye, J., & Song, W. (2019). Enhanced cyber-physical security in internet of things through energy auditing. *IEEE Internet of Things Journal*, *6*(3), 5224–5231.

Liu, X., Xie, L., Wang, Z., & Fu, D. (2012). Cognitive-affective regulation process for micro-expressions in active field state space. In *2012 IEEE 2nd International Conference on Cloud Computing and Intelligence Systems* (Vol. 2, pp. 831-835). IEEE. 10.1109/CCIS.2012.6664292

Liu, B., Zhao, W., & Sun, Q. (2017, October). Study of object detection based on Faster R-CNN. In *2017 Chinese Automation Congress (CAC)* (pp. 6233-6236). IEEE.

Liu, G., Bao, H., & Han, B. (2018). A stacked autoencoder-based deep neural network for achieving gearbox fault diagnosis. *Mathematical Problems in Engineering*, *2018*, 2018. doi:10.1155/2018/5105709

Liu, Y., Shi, J., & Jian, M. (2017). Understanding visitors' responses to intelligent transportation system in a tourist city with a mixed ranked Logit model. *Journal of Advanced Transportation*, *2017*, 1–13. doi:10.1155/2017/2854895

Liu, Y., Wang, H., Peng, M., Guan, J., Xu, J., & Wang, Y. (2019). DeePGA: A privacy-preserving data aggregation game in crowdsensing via deep reinforcement learning. *IEEE Internet of Things Journal*, *7*(5), 4113–4127.

Lloyds. (2019). *Taking control Artificial intelligence and insurance*. Emerging Risk Report 2019 Technology. https://www.lloyds.com/~/media/files/news-and-insight/risk-insight/2019/aireport_2019_final_pdf.pdf

Lodgaard, E., & Dransfeld, S. (2020). Organizational aspects for successful integration of human-machine interaction in the industry 4.0 era. *Procedia CIRP*, *88*, 218–222. doi:10.1016/j.procir.2020.05.039

Lombardia, P. (2012). Modelling the smart city performance. *Innovation - The European Journal of Social Science Research, 25*(2), 137 - 149.

LoRa, S. (2021). *Understanding the LoRaWAN Architecture.* Retrieved from www.tech-journal.semtech.com: https://tech-journal.semtech.com/understanding-the-lorawan-architecture

Lou, Y., Caruana, R., Gehrke, J., & Hooker, G. (2013, August). Accurate intelligible models with pairwise interactions. In *Proceedings of the 19th ACM SIGKDD international conference on Knowledge discovery and data mining* (pp. 623-631). 10.1145/2487575.2487579

Lu, L., Jiang, C., Hu, G., Liu, J., & Yang, B. (2021). Flexible Noncontact Sensing for Human–Machine Interaction. *Advanced Materials, 33*(16), 2100218. doi:10.1002/adma.202100218 PMID:33683745

Lundervold, A.S., & Lundervold, A. (2019). An overview of deep learning in medical imaging focusing on MRI. *Zeitschrift für Medizinische Physik, 29*(2), 102-127. doi:10.1016/j.zemedi.2018.11.002

Luong, N. C., Hoang, D. T., Gong, S., Niyato, D., Wang, P., Liang, Y. C., & Kim, D. I. (2019). Applications of deep reinforcement learning in communications and networking: A survey. *IEEE Communications Surveys and Tutorials, 21*(4), 3133–3174.

Lynggaard, P. (2020). Using neural networks to reduce sensor cluster interferences and power consumption in smart cities. *International Journal of Sensor Networks, 32*(1), 25–33.

Madakam, S., & Ramaswamy, R. (2014). Smart Cities - Six Dimensions. *International Conference on Advances in Computing and Information Technology - ACIT*, 38-41.

Madhavan, P., & Wiegmann, D. A. (2004, September). A new look at the dynamics of human-automation trust: Is trust in humans comparable to trust in machines? *Proceedings of the Human Factors and Ergonomics Society Annual Meeting, 48*(3), 581–585. doi:10.1177/154193120404800365

Madhavan, P., & Wiegmann, D. A. (2007). Similarities and differences between human–human and human–automation trust: An integrative review. *Theoretical Issues in Ergonomics Science, 8*(4), 277–301. doi:10.1080/14639220500337708

Magaia, N., Fonseca, R., Muhammad, K., Segundo, A. H. F. N., Neto, A. V. L., & de Albuquerque, V. H. C. (2020). Industrial Internet-of-Things Security Enhanced With Deep Learning Approaches for Smart Cities. *IEEE Internet of Things Journal, 8*(8), 6393–6405.

Manimuthu, A., & Dharshini, V. (2020). *Framework for load power consumption in Han using machine learning and IoT assistance. IEEE Design & Test.*

Maran, P. S., Reddy, B. S., & Saiharshavardhan, C. (2021). Air Quality Prediction (IoT) Using Machine Learning. In *Advances in Electronics, Communication and Computing* (pp. 583–591). Springer.

Mayer, R. C., Davis, J. H., & Schoorman, F. D. (1995). An Integrative Model of Organizational Trust. *Academy of Management Review*, *20*(3), 709–734. doi:10.2307/258792

McAllister, D. J. (1995). Affect- and cognition-based trust as foundations for interpersonal cooperation in organizations. *Journal of The Academy of Management*, *38*(1), 24–59.

Mcknight, D. H., Carter, M., Thatcher, J. B., & Clay, P. F. (2011). Trust in a Specific Technology: An Investigation of Its Components and Measures. *ACM Transactions on Management Information Systems*, *2*(2), 12. doi:10.1145/1985347.1985353

Meenakshi, K., & Maragatham, G. (2019, March). A review on security attacks and protective strategies of machine learning. In *International Conference on Emerging Current Trends in Computing and Expert Technology* (pp. 1076-1087). Springer.

Metaxiotis, K., Kagiannas, A., Askounis, D., & Psarras, J. (2003). Artificial intelligence in short term electric load forecasting: A state-of-the-art survey for the researcher. *Energy Conversion and Management*, *44*(9), 1525–1534. doi:10.1016/S0196-8904(02)00148-6

Misbahuddin, S., Zubairi, J. A., Saggaf, A., Basuni, J., Sulaiman, A., & Al-Sofi, A. (2015, December). IoT based dynamic road traffic management for smart cities. In *2015 12th International conference on high-capacity optical networks and enabling/emerging technologies (HONET)* (pp. 1-5). IEEE.

Moallem, A. (Ed.). (2018). *Human-Computer Interaction and cybersecurity handbook.* CRC Press. doi:10.1201/b22142

Mobile air quality monitoring with IoT. (2021). *Manx Technology Group.* https://www.manxtechgroup.com/mobile-air-quality-monitoring-with-iot/

Mohamed, Y. (2018). *What is the role of IoT in Smart Cities?* Academic Press.

Morales, J. M., Minguez, R., & Conejo, A. J. (2010). A methodology to generate statistically dependent wind speed scenarios. *Applied Energy*, *87*(3), 843–855.

Mosannenzadeh F., & Vettorato D. (2014). Defining smart city: a conceptual framework based on keyword analysis. *Tema - Journal of Land Use, Mobility and Environment*, 683 - 694.

Mower, E., Mataric, M. J., & Narayanan, S. (2009). Human perception of audio-visual synthetic character emotion expression in the presence of ambiguous and conflicting information. *IEEE Transactions on Multimedia*, *11*(5), 843–855. doi:10.1109/TMM.2009.2021722

Mrinal. (2021). *Top 10 Open-Source Datasets for Object Detection In 2021.* Analytics Vidhya. https://www.analyticsvidhya.com/blog/2021/05/top-10-open-source-datasets-for-object-detection-machine-learning-in-2021/

Muhammad, K., Hussain, T., Del Ser, J., Palade, V., & De Albuquerque, V. H. C. (2019). DeepReS: A deep learning-based video summarization strategy for resource-constrained industrial surveillance scenarios. *IEEE Transactions on Industrial Informatics*, *16*(9), 5938–5947.

Muhammad, K., Hussain, T., Tanveer, M., Sannino, G., & de Albuquerque, V. H. C. (2019). Cost-effective video summarization using deep CNN with hierarchical weighted fusion for IoT surveillance networks. *IEEE Internet of Things Journal*, *7*(5), 4455–4463.

NAIC. (2020). *Artificial Intelligence*. https://content.naic.org/cipr_topics/topic_artificial_intelligence.htm

Najafabadi, M. M., Villanustre, F., Khoshgoftaar, T. M., Seliya, N., Wald, R., & Muharemagic, E. (2015). Deep learning applications and challenges in big data analytics. *Journal of Big Data*, *2*(1), 1. Advance online publication. doi:10.118640537-014-0007-7

Nam, T., & Pardo, T. A. (2011, June). Conceptualizing smart city with dimensions of technology, people, and institutions. In *Proceedings of the 12th annual international digital government research conference: digital government innovation in challenging times* (pp. 282-291). Academic Press.

Nardo, M., Forino, D., & Murino, T. (2020). The evolution of man–machine interaction: The role of human in Industry 4.0 paradigm. *Production & Manufacturing Research*, *8*(1), 20–34. doi:10.1080/21693277.2020.1737592

Nass, C., Steuer, J., & Tauber, E. R. (1994, April). Computers are social actors. In *Proceedings of the SIGCHI conference on Human factors in computing systems* (pp. 72-78). ACM.

Nguyen, V. D., Van Nguyen, H., Tran, D. T., Lee, S. J., & Jeon, J. W. (2016). Learning framework for robust obstacle detection, recognition, and tracking. *IEEE Transactions on Intelligent Transportation Systems*, *18*(6), 1633–1646. doi:10.1109/TITS.2016.2614818

Nicole, M., Ewald, S., Weiss, A., Kühnlenz, B., Dirk, W., & Manfred, T. (2015). Can You Read My Face? *International Journal of Social Robotics*, *7*(1), 63–76. doi:10.100712369-014-0261-z

Norman, D. A. (1981b) A psychologist views human processing: Human errors and other phenomena suggest processing mechanisms. *Proceedings of the International Joint Conference on Artificial Intelligence*.

Norman, D. A. (1983a). Design principles for human-computer interfaces. *Proceedings of the CHI 1983 Conference on Human Factors in Computing Systems*.

Norman, D. A. (1984). Four stages of user activities. In *INTERACT '84, First Conference on Human-Computer Interaction*. North Holland.

Norman, D. A. (1981a). Categorization of action slips. *Psychological Review*, *88*(1), 1–15. doi:10.1037/0033-295X.88.1.1

Nott, G. (2017, June 23). *Google's research chief questions value of "Explainable AI."* Computerworld. https://www2.computerworld.com.au/article/621059/google-research-chief-questions-value-explainable-ai/

Obinikpo, A.A., Kantarci, B. (2017). Big Sensed Data Meets Deep Learning for Smarter Healthcare in Smart Cities. *Journal of Sensor and Actuator Networks*, *6*(4), 1-17. doi:10.3390/jsan6040026

Oizom. (2019). *4 Pillars of Smart Cities of India and the Role of Smart Environment*. Academic Press.

Oliveira, D., Rocha, H., Yang, H., Ellis, D., Dommaraju, S., Muradoglu, M., Weir, D., Soliman, A., Lin, T., & Ebner, N. 2017. Dissecting spear phishing emails for older vs young adults: On the interplay of weapons of influence and life domains in predicting susceptibility to phishing. In *Proc. of the 2017 CHI Conference on Human Factors in Computing Systems*. ACM. 10.1145/3025453.3025831

Oltramari, A., Henshel, D. S., Cains, M., & Hoffman, B. (2015). Towards a Human Factors Ontology for Cyber Security. *Stids*, *2015*, 26–33.

Paessler. (2021). *What is LPWA?* Retrieved from www.paessler.com: https://www.paessler.com/it-explained/lpwa

Pai, A. (2020). *Your Social Distancing Detection Tool: How to Build One using your Deep Learning Skills*. Analytics Vidhya. https://www.analyticsvidhya.com/blog/2020/05/social-distancing-detection-tool-deep-learning/

Pandl, K. D., Thiebes, S., Schmidt-Kraepelin, M., & Sunyaev, A. (2020). On the convergence of artificial intelligence and distributed ledger technology: A scoping review and future research agenda. *IEEE Access : Practical Innovations, Open Solutions*, *8*, 57075–57095.

Pantelis, K., & Aija, L. (2013). Understanding the value of (big) data. *2013 IEEE International Conference on Big Data*. 10.1109/BigData.2013.6691691

Patel, D. (2021). *Low Power Wide Area Networks (LPWAN): Technology Review And Experimental Study on Mobility Effect*. South Dakota State University. Retrieved from www.openprairie.sdstate.edu: https://openprairie.sdstate.edu/etd/2667/

Patient Exposure and the Air Quality Index. (2021). *United States Environmental Protection Agency (US EPA)*. https://www.epa.gov/pmcourse/patient-exposure-and-air-quality-index

Paul, D., Sanap, G., Shenoy, S., Kalyane, D., Kalia, K., & Tekade, R. K. (2021). Artificial intelligence in drug discovery and development. *Drug Discovery Today*, *26*(1), 80–93. doi:10.1016/j.drudis.2020.10.010 PMID:33099022

Pawłowicz, B., Salach, M., & Trybus, B. (2018, September). Smart city traffic monitoring system based on 5G cellular network, RFID and machine learning. In *KKIO Software Engineering Conference* (pp. 151-165). Springer.

Paz, A. (n.d.). *New Technologies in Smart Cities*. https://www.e-zigurat.com/innovation-school/blog/new-technologies-smart-cities/

Peiser, C. (2018). *Role of Artificial Intelligence in Commercial Insurance Underwriting and Renewals*. Capgemini. https://www.capgemini.com/wp-content/uploads/2018/06/Role-of-Artificial-Intelligence-in-Commercial-Insurance-Underwriting-and-Renewals.pdf

Phillips, Repperger, Kinsler, Bharwani, & Kender. (2007). *A quantitative model of the human–machine interaction and multi-task performance: A strategy function and the unity model paradigm.* doi:10.1016/j.compbiomed.2006.12.007

Pierce, Fox, Merrill, & Wong. (2018). Differential Vulnerabilities and a Diversity of Tactics: What Toolkits Teach Us About Cybersecurity. *Proc. ACM Hum.-Comput. Interact., 2.* doi:10.1145/3274408

Pinquart, M., & Silbereisen, R. K. (2004). Human development in times of social change: Theoretical considerations and research needs. *International Journal of Behavioral Development, 28*(4), 289–298. doi:10.1080/01650250344000406

Prabhakar, G., Kailath, B., Natarajan, S., & Kumar, R. (2017, July). Obstacle detection and classification using deep learning for tracking in high-speed autonomous driving. In 2017 IEEE region 10 symposium (TENSYMP) (pp. 1-6). IEEE. doi:10.1109/TENCONSpring.2017.8069972

Punn, N. S., Sonbhadra, S. K., Agarwal, S., & Rai, G. (2020). *Monitoring COVID-19 social distancing with person detection and tracking via fine-tuned YOLO v3 and Deepsort techniques.* arXiv preprint arXiv:2005.01385.

Qian, H., Xu, J., & Zhou, J. (2018). Object detection using deep convolutional neural networks. In *2018 Chinese Automation Congress (CAC)* (pp. 1151-1156). IEEE.

Qian, Y., Zheng, H., He, D., Zhang, Z., & Zhang, Z. (2018, August). R-CNN Object Detection Inference With Deep Learning Accelerator. In *2018 IEEE/CIC International Conference on Communications in China (ICCC Workshops)* (pp. 297-302). IEEE.

Raj. (2019). *How Are Insurance Companies Implementing Artificial Intelligence (AI)?* Towards Data Science. https://towardsdatascience.com/how-are-insurance-companies-implementing-artificial-intelligence-ai-aaf845fce6a7

Raparthi, K. (2014). Assessing Smart-Growth Strategies in Indian Cities: Grounded Theory Approach to Planning Practice. *ASCE - Journal of Urban Planning and Development*, 1-10.

Rezaei, M., & Azarmi, M. (2020). Deepsocial: Social distancing monitoring and infection risk assessment in covid-19 pandemic. *Applied Sciences (Basel, Switzerland), 10*(21), 7514.

Riley, V. (1989, October). A general model of mixed-initiative human-machine systems. In *Proceedings of the Human Factors Society Annual Meeting* (Vol. 33, No. 2, pp. 124-128). SAGE Publications. 10.1177/154193128903300227

Robertson, D. I., & Bretherton, R. D. (1991). Optimizing networks of traffic signals in real time-the SCOOT method. *IEEE Transactions on Vehicular Technology, 40*(1), 11–15.

Rocha, N. P., Dias, A., Santinha, G., Rodrigues, M., Queirós, A., & Rodrigues, C. (2019). Smart Cities and Healthcare: A Systematic Review. *Technologies, 7*(3), 1-8. doi:10.3390/technologies7030058

Roh, M. C., & Lee, J. Y. (2017, May). Refining faster-RCNN for accurate object detection. In *2017 fifteenth IAPR international conference on machine vision applications (MVA)* (pp. 514-517). IEEE.

Rowe, D. C., Lunt, B. M., & Ekstrom, J. J. (2011, October). The role of cyber-security in information technology education. In *Proceedings of the 2011 conference on Information technology education* (pp. 113-122). 10.1145/2047594.2047628

Russakovsky, O., Deng, J., Su, H., Krause, J., Satheesh, S., Ma, S., ... Fei-Fei, L. (2015). Imagenet large scale visual recognition challenge. *International Journal of Computer Vision*, *115*(3), 211–252.

Salama, A. S., Saleh, B. K., & Eassa, M. M. (2010, November). Intelligent cross road traffic management system (ICRTMS). In *2010 2nd International conference on computer technology and development* (pp. 27-31). IEEE.

Sanil, N., Rakesh, V., Mallapur, R., & Ahmed, M. R. (2020, January). Deep learning techniques for obstacle detection and avoidance in driverless cars. In *2020 International Conference on Artificial Intelligence and Signal Processing (AISP)* (pp. 1-4). IEEE. 10.1109/AISP48273.2020.9073155

Santana, M. C., Passos, L. A., Moreira, T. P., Colombo, D., de Albuquerque, V. H. C., & Papa, J. P. (2019). A novel siamese-based approach for scene change detection with applications to obstructed routes in hazardous environments. *IEEE Intelligent Systems*, *35*(1), 44–53.

Sarda, A., Dixit, S., & Bhan, A. (2021, February). Object Detection for Autonomous Driving using YOLO [You Only Look Once] algorithm. In *2021 Third International Conference on Intelligent Communication Technologies and Virtual Mobile Networks (ICICV)* (pp. 1370-1374). IEEE.

Savjani, R. (2019). *Understanding firmware updates: what, why and when*. Einfochips. https://www.einfochips.com/blog/understanding-firmware-updates-the-whats-whys-and-hows/

SEMTECH. (2021). *LoRa White Papers, Smart Lighting*. Retrieved from www.semtech.com: https://www.semtech.com/lora/resources/lora-white-papers

Shearer, C. (2000). The CRISP-DM Model: The New Blueprint for Data Mining. *Journal of Data Warehousing*, *5*, 13–22.

Sheikhi, A., Rayati, M., Bahrami, S., Ranjbar, A. M., & Sattari, S. (2015). A cloud computing framework on demand side management game in smart energy hubs. *International Journal of Electrical Power & Energy Systems*, *64*, 1007–1016. doi:10.1016/j.ijepes.2014.08.020

Sheng, S., Holbrook, M., Kumaraguru, P., Cranor, L. F., & Downs, J. (2010). Who Falls for Phish?: A Demographic Analysis of Phishing Susceptibility and Effectiveness of Interventions. In *Proc. of the SIGCHI Conference on Human Factors in Computing Systems (CHI '10)*. ACM. 10.1145/1753326.1753383

Shu, C., Ding, X., & Fang, C. (2011, April). Histogram of the oriented gradient for face recognition. *Tsinghua Science and Technology*, *16*(2), 216–224. doi:10.1016/S1007-0214(11)70032-3

Silva, B., Khan, M., Jung, C., Seo, J., Muhammad, D., Han, J., Yoon, Y., & Han, K. (2018). Urban planning and smart city decision management empowered by real-time data processing using big data analytics. *Sensors (Basel), 18*(9), 2994. doi:10.339018092994 PMID:30205499

Singh, R. K., Dalal, S., Chauhan, V. K., & Kumar, D. (2019, March). Optimization of FAR in intrusion detection system by using random forest algorithm. *Proceedings of 2nd International Conference on Advanced Computing and Software Engineering (ICACSE).*

Sodhro, A. H., Pirbhulal, S., & De Albuquerque, V. H. C. (2019). Artificial intelligence-driven mechanism for edge computing-based industrial applications. *IEEE Transactions on Industrial Informatics, 15*(7), 4235–4243.

Söllner, M., Hoffmann, A., Hoffmann, H., Wacker, A., & Leimeister, J. M. (2012). Understanding the Formation of Trust in IT Artifacts. *Proceedings of the International Conference on Information Systems (ICIS 2012)*, 2, 1038–1055.

Song, M., Cen, L., Zheng, Z., Fisher, R., Liang, X., Wang, Y., & Huisingh, D. (2017). How would big data support societal development and environmental sustainability? Insights and practices. *Journal of Cleaner Production, 142*, 489–500. doi:10.1016/j.jclepro.2016.10.091

Sousa, S., Lamas, D., & Dias, P. (2014). A Model for HumanComputer Trust. *International Conference on Learning and Collaboration Technologies*, 128–137.

Souza, J. T., Francisco, A. C., Piekarski, C. M., & Prado, G. F. (2019). Data Mining and Machine Learning to Promote Smart Cities: A Systematic Review from 2000 to 2018. *Sustainability, 11*(4), 1–11, 1077. doi:10.3390u11041077

Spectrum Enterprise. (n.d.). *What is the aim of smart cities?* https://enterprise.spectrum.com/support/faq/smart-cities/what-is-the-aim-of-smart-cities.html

Springett, M., Law, E. L. C., & Coulson, M. (2015). Integrating the strengths of cognitive emotion models with traditional HCI analysis tools. *Universal Access in the Information Society, 14*(2), 203–214. doi:10.100710209-013-0335-y

Sreedhar, B. K., & Shunmugam, N. (2020, July). Deep Learning for Hardware-Constrained Driverless Cars. In *2020 IEEE 44th Annual Computers, Software, and Applications Conference (COMPSAC)* (pp. 26-29). IEEE.

Stefanini. (2020). *Top 5 Applications of IoT in Building Smart Cities.* https://stefanini.com/en/trends/news/top-5-applications-of-iot-in-building-smart-cities

Stone, S. (2018). *Key Challenges of Smart Cities & How to Overcome Them.* https://ubidots.com/blog/the-key-challenges-for-smart-cities/

Stryker & Nancy. (1988). The society of mind. *Systems Research, 5*(3), 266–266.

Sun, P., Aljeri, N., & Boukerche, A. (2020). Machine learning-based models for real-time traffic flow prediction in vehicular networks. *IEEE Network, 34*(3), 178–185.

Surasak, T., Takahiro, I., Cheng, C. H., Wang, C. E., & Sheng, P. Y. (2018, May). Histogram of oriented gradients for human detection in video. In *2018 5th International conference on business and industrial research (ICBIR)* (pp. 172-176). IEEE.

Surya, L., & Yarlagadda, R. T. (2020). AI Economical Smart Device to Identify Covid-19 Pandemic, and Alert on Social Distancing WHO Measures. *International Journal of Creative Research Thoughts.*

Syam, N., & Sharma, A. (2018). Waiting for a sales renaissance in the fourth industrial revolution: Machine learning and artificial intelligence in sales research and practice. *Industrial Marketing Management, 69*, 135–146. doi:10.1016/j.indmarman.2017.12.019

Tamayo-Sarver, J. H., Hinze, S. W., Cydulka, R. K., & Baker, D. W. (2003). Racial and ethnic disparities in emergency department analgesic prescription. *American Journal of Public Health, 93*(12), 2067–2073. doi:10.2105/AJPH.93.12.2067 PMID:14652336

Tang, C., Feng, Y., Yang, X., Zheng, C., & Zhou, Y. (2017, July). The object detection based on deep learning. In *2017 4th international conference on information science and control engineering (ICISCE)* (pp. 723-728). IEEE.

Tascikaraoglu, A. (2018). Evaluation of spatio-temporal forecasting methods in various smart city applications. *Renewable & Sustainable Energy Reviews, 82*, 424–435.

Tewari, N., & Joshi, M. (2020). Water quality prediction system (WQPS) and method using fog of thing (FoT). *2020 9th International Conference System Modeling and Advancement in Research Trends (SMART).* 10.1109/SMART50582.2020.9337109

Tewari, N., & Datt, G. (2020). Towards FoT (fog-of-Things) enabled architecture in governance: Transforming E-Governance to smart governance. *2020 International Conference on Intelligent Engineering and Management (ICIEM).* 10.1109/ICIEM48762.2020.9160037

Thales. (n.d.). *Secure, sustainable smart cities and the IoT.* https://www.thalesgroup.com/en/markets/digital-identity-and-security/iot/inspired/smart-cities

Tian, Z., Luo, C., Qiu, J., Du, X., & Guizani, M. (2019). A distributed deep learning system for web attack detection on edge devices. *IEEE Transactions on Industrial Informatics, 16*(3), 1963–1971.

Ting, D. S. W., Carin, L., Dzau, V., & Wong, T. Y. (2020). Digital technology and COVID-19. *Nature Medicine, 26*(4), 459–461. doi:10.103841591-020-0824-5 PMID:32284618

Tiwari, A., & Jain, K. (2014). GIS Steering Smart Future for Smart Indian Cities. *International Journal of Scientific and Research Publications, 4*(8), 1-5.

Turton, W., & Mehrotra, K. (2021). *Hackers Breached Colonial Pipeline Using Compromised Password.* Bloomberg.https://www.bloomberg.com/news/articles/2021-06-04/hackers-breached-colonial-pipeline-using-compromised-password

TWI Ltd. (n.d.). *What is a Smart City? Definition and Examples.* https://www.twi-global.com/technical-knowledge/faqs/what-is-a-smart-city

Ullah, F., Naeem, H., Jabbar, S., Khalid, S., Latif, M. A., Al-Turjman, F., & Mostarda, L. (2019). Cyber security threats detection in internet of things using deep learning approach. *IEEE Access: Practical Innovations, Open Solutions*, *7*, 124379–124389.

Umer, M. A., Jilani, M. T., Junejo, K. N., Naz, S. A., & D'Silva, C. W. (2021). Role of Machine Learning in Weather Related Event Predictions for a Smart City. In *Machine Intelligence and Data Analytics for Sustainable Future Smart Cities* (pp. 49–63). Springer.

University, T. O. (n.d.). *Introduction to smart cities.* https://www.open.edu/openlearn/ocw/mod/oucontent/view.php?id=67877

VE, S., Shin, C., & Cho, Y. (2021). Efficient energy consumption prediction model for a data analytic-enabled industry building in a smart city. *Building Research and Information*, *49*(1), 127–143.

Vilajosana, I., Llosa, J., Martinez, B., Domingo-Prieto, M., Angles, A., & Vilajosana, X. (2013). Bootstrapping smart cities through a self-sustainable model based on big data flows. *IEEE Communications Magazine*, *51*(6), 128–134. doi:10.1109/MCOM.2013.6525605

Vles, M. D., Terng, N., Zijlstra, K., Mureau, M., & Corten, E. (2020). Virtual and augmented reality for preoperative planning in plastic surgical procedures: A systematic review. *Journal of Plastic, Reconstructive & Aesthetic Surgery*, *73*(11), 1951–1959. doi:10.1016/j.bjps.2020.05.081 PMID:32622713

Wade, M. R. (2016). *What is a Smart City anyways?* https://www.imd.org/research-knowledge/articles/what-is-a-smart-city-anyways/

Wang, C., & Peng, Z. (2019, June). Design and implementation of an object detection system using faster R-CNN. In *2019 International Conference on Robots & Intelligent System (ICRIS)* (pp. 204-206). IEEE.

Wang, L., Fan, X., Chen, J., Cheng, J., Tan, J., & Ma, X. (2020). 3D object detection based on sparse convolution neural network and feature fusion for autonomous driving in smart cities. *Sustainable Cities and Society*, *54*, 1–5. doi:10.1016/j.scs.2019.102002

Wang, Z. (2009). *Artificial Emotion.* Machine Press.

West, R., Mayhorn, C., Hardee, J., & Mendel, J. (2009). The weakest link: A psychological perspective on why users make poor security decisions. In *Social and Human elements of information security: Emerging Trends and countermeasures* (pp. 43–60). IGI Global. doi:10.4018/978-1-60566-036-3.ch004

Whitty, M., Doodson, J., Creese, S., & Hodges, D. (2015). Individual differences in cyber security behaviors: An examination of who is sharing passwords. *Cyberpsychology, Behavior, and Social Networking*, *18*(1), 3–7. doi:10.1089/cyber.2014.0179 PMID:25517697

Witeck, C. (2021). *Smart Streetlight*. Retrieved from https://internetofthingsagenda.techtarget.com/definition/smart-streetlight: https://internetofthingsagenda.techtarget.com/definition/smart-streetlight

Witeck, C. (2021). *What is internet of things (IoT)?* Retrieved from www.internetofthingsagenda.techtarget.com: https://internetofthingsagenda.techtarget.com/definition/Internet-of-Things-IoT

Wu, C. N., Luo, S. D., Lin, H. C., Huang, J. T., Lee, C. H., Liu, S. Y., Tsai, M. H., Wang, C. C., Fan, S., Wang, P. S., & Lan, K. C. (2021). Eligibility for live, interactive otolaryngology telemedicine: 19-month experience before and during the COVID-19 pandemic in Taiwan. *Biomedical Journal*. Advance online publication. doi:10.1016/j.bj.2021.07.012

Xiao, G., Ma, Y., Liu, C., & Jiang, D. (2020). A machine emotion transfer model for intelligent human-machine interaction based on group division. *Mechanical Systems and Signal Processing*, *142*, 106736. doi:10.1016/j.ymssp.2020.106736

Xu, C., Fan, J., & Liu, L. (2020, December). A novel object detection algorithm based on enhanced R-FCN and SVM. In *2020 IEEE Conference on Telecommunications, Optics and Computer Science (TOCS)* (pp. 363-367). IEEE.

Yang, D., Ye, Z., Lim, L. H. I., & Dong, Z. (2015). Very short-term irradiance forecasting using the lasso. *Solar Energy*, *114*, 314–326.

Yang, D., Yurtsever, E., Renganathan, V., Redmill, K. A., & Özgüner, Ü. (2021). A vision-based social distancing and critical density detection system for covid-19. *Sensors (Basel)*, *21*(13), 4608.

Yan, L., Zhong, B., & Song, W. (2017, November). Region-based fully convolutional networks for vertical corner line detection. In *2017 International Symposium on Intelligent Signal Processing and Communication Systems (ISPACS)* (pp. 159-163). IEEE.

Yao, S., Ren, P., Song, R., Liu, Y., Huang, Q., Dong, J., O'Connor, B. T., & Zhu, Y. (2020). Nanomaterial-Enabled Flexible and Stretchable Sensing Systems: Processing, Integration, and Applications. *Advanced Materials*, *32*(15), 1902343. doi:10.1002/adma.201902343

Yin, H., & Berger, C. (2017, October). When to use what data set for your self-driving car algorithm: An overview of publicly available driving datasets. In *2017 IEEE 20th International Conference on Intelligent Transportation Systems (ITSC)* (pp. 1-8). IEEE.

Yin, X., Yang, Y., Xu, H., Li, W., & Deng, J. (2020, December). Enhanced Faster-RCNN Algorithm for Object Detection in Aerial Images. In *2020 IEEE 9th Joint International Information Technology and Artificial Intelligence Conference (ITAIC)* (Vol. 9, pp. 2355-2358). IEEE.

Yustiawati, R., Husni, N. L., Evelina, E., Rasyad, S., Lutfi, I., Silvia, A., Alfarizal, N., & Rialita, A. (2018, October). Analyzing of Different Features Using Haar Cascade Classifier. In *2018 International Conference on Electrical Engineering and Computer Science (ICECOS)* (pp. 129-134). IEEE.

Zanatto, D., Chattington, M., & Noyes, J. (2021, July). Sense of Agency in Human-Machine Interaction. In *International Conference on Applied Human Factors and Ergonomics* (pp. 353-360). Springer.

Zhang, J., Chen, M., Zhao, S., Hu, S., Shi, Z., & Cao, Y. (2016). ReliefF-based EEG sensor selection methods for emotion recognition. *Sensors (Basel)*, *16*(10), 1558. doi:10.339016101558 PMID:27669247

Zhang, J., Chen, S., & Hou, Y. (2020, November). Accurate Object Detection with Relation Module on Improved R-FCN. In *2020 Chinese Automation Congress (CAC)* (pp. 7131-7135). IEEE.

Zhang, L., Zhou, W., Li, J., Li, J., & Lou, X. (2020, December). Histogram of Oriented Gradients Feature Extraction Without Normalization. In *2020 IEEE Asia Pacific Conference on Circuits and Systems (APCCAS)* (pp. 252-255). IEEE.

Zhang, X., Chen, X., Liu, J. K., & Xiang, Y. (2019). DeepPAR and DeepDPA: Privacy preserving and asynchronous deep learning for industrial IoT. *IEEE Transactions on Industrial Informatics*, *16*(3), 2081–2090.

Zhang, Y., & Li, L. (2014). A Personality Model based on NEO PI-R for emotion simulation. *IEICE Transactions on Information and Systems*, *97*(8), 2000–2007. doi:10.1587/transinf.E97.D.2000

Zhang, Z., Xu, H., Chao, Z., Li, X., & Wang, C. (2014). A novel vehicle reversing speed control based on obstacle detection and sparse representation. *IEEE Transactions on Intelligent Transportation Systems*, *16*(3), 1321–1334. doi:10.1109/TITS.2014.2360337

Zhao, X., Wu, H., Xu, Z., & Min, H. (2020). Omni-Directional Obstacle Detection for Vehicles Based on Depth Camera. *IEEE Access: Practical Innovations, Open Solutions*, *8*, 93733–93748. doi:10.1109/ACCESS.2020.2993934

Zhao, Y., Zhang, Y., Cheng, R., Wei, D., & Li, G. (2015). An enhanced histogram of oriented gradients for pedestrian detection. *IEEE Intelligent Transportation Systems Magazine*, *7*(3), 29–38.

Zhao, Z. Q., Zheng, P., Xu, S. T., & Wu, X. (2019). Object detection with deep learning: A review. *IEEE Transactions on Neural Networks and Learning Systems*, *30*(11), 3212–3232.

Zhou, W., Gao, S., Zhang, L., & Lou, X. (2020). Histogram of oriented gradients feature extraction from raw bayer pattern images. *IEEE Transactions on Circuits and Wystems. II, Express Briefs*, *67*(5), 946–950.

Zhou, X., Guo, Y., Shen, M., & Yang, G. (2020). Artificial Intelligence in Surgery. *Frontiers of Medicine*, 1–28. doi:10.100711684-019-0733-5 PMID:32705406

Zhu, Y., Zhao, C., Wang, J., Zhao, X., Wu, Y., & Lu, H. (2017). Couplenet: Coupling global structure with local parts for object detection. In *Proceedings of the IEEE international conference on computer vision* (pp. 4126-4134).

Zubizarreta, I. (2015). Smart City Concept: What It Is and What It Should Be. *ASCE - Journal of Urban Planning and Development,* 1-8.

About the Contributors

Rajeev Kumar is Professor and Research Head, Computer Science and Engineering, Chandigarh University, Mohali Punjab, India. He is Visited international countries London, United Kingdom and Mauritius as professional activities. Under his supervision awarded 5 scholar their Ph.D. and 5 Scholar is Ongoing their research work. He is awarded 3 times Best project supervisor award; He is delivered a guest lecture, keynote speaker, chaired a session in many National and International conferences, Faculty Development program and workshop; He is completed one Research Project from DST in Bhimtal; He is organized many IEEE international conferences and workshop. He is authored and coauthored more than 85 papers in refereed international journal (SCI and Non- SCI) and conferences like IEEE, Springer, American Institute of Physics, New York Science international Journal New York City (USA), American Science Journal, BioInfo science Journal, Academic science of international journal (USA), International Journal of researcher, American Journal of Physics (USA) and many international Conferences and National Conferences, like IIT Roorkee (International Conference), etc.

Rakesh Kumar Dwivedi is Professor and Director, CCSIT, Teerthanker Mahaveer University, Moradabad, Uttar Pradesh, India. Under his supervision awarded 8 scholar their Ph.D. and 4 Scholar is Ongoing their research work. He is delivered a guest lecture, keynote speaker, chaired a session in many National and International conferences, Faculty Development program and workshop; He is completed one Research Project from DST in Bhimtal; He is organized many IEEE international conferences and workshop. He is authored and coauthored more than 85 papers in refereed international journal (SCI and Non- SCI) and conferences like IEEE, Springer, American Institute of Physics, New York Science international Journal New York City (USA), American Science Journal, BioInfo science Journal, Academic science of international journal (USA), International Journal of researcher, American Journal of Physics (USA) and many international Conferences and National Conferences, like IIT Roorkee (International Conference), etc.

* * *

Abeda Begum is currently pursuing Ph.D. in Computer Science and Engineering at Bhagwant University, Ajmer, India and working as Consultant Oracle Database Administrator. She has pursed Master of Computer Applications from Sikkim & Manipal University, India. She has pursed Project Management Professional and ITIL Standards Certification along with Oracle DBA Certifications. She has worked for more than 12yrs as Oracle Database Administrator in Dubai at Emirates NBD Bank & Ingram Micro, during her career she has worked on various IT projects and supported terabytes of Production databases. With her IT experience and her personal journey towards health issues, gave her the courage and focus to research on health problems in women and youth using Computer Applications. Her research work focuses on Artificial Intelligence, Machine Learning, Deep Learning algorithms, Blockchain technology, Big Data, Data Mining, Data Analytics and Visualization. She has 15yrs of IT industry experience in banking and supply chain management domains.

Ashutosh Kumar Bhatt is working in School of Computer Science and Information Technology, Uttarakhand Open University. Dr. Bhatt has completed Ph. D. in Computer Science in 2009. His work area of research was Artificial Neural Network. He has more than Seventeen years of teaching and research experience in various organizations of repute for PG & UG courses of Computer Science & IT. He is also associated with many renowned National/International Journals as Lead Guest Editor /reviewer / editorial board member. More than 12 SCIE/ESCI and more than 55 Scopus indexed research publications are credited to him in reputed National/ International Journals. He is awarded by the significant contribution award by CSI for the year 2012-13 and Distinguished Professor award by CSI Mumbai Chapter. He is also awarded by Adarsh Vidya Saraswati Rastriya Purasker. He has published and edited number of books e.g. C# Programming using Dot Net Framework, Technology Enabled Learning, Introduction to Computers with basics of Programming etc. He had worked in a various research under the Scheme for Young Scientists and Professional DST Govt. of India, Uttarakhand Science Education and Research Center (U-SERC), SWYAM PRABHA CHANNEL-20 and under TEQIP AICTE. He has published 4 patents. He has also sanctioned one copyright.

Sandeep Kumar Budhani is currently working as Head, School of Computing at Graphic Era Hill University, Bhimtal Campus, Uttarakhand, India. He has more than 20 years of academic and administrative experience. His area of expertise includes machine learning, data mining and artificial neural network.

Menal Dahiya is working as an Associate Professor at the Department of Computer Science, Maharaja Surajmal Institute (Affiliated to GGSIPU, Dwarka, Delhi), Janakpuri, New Delhi. She carries 14 years of experience in academics. Apart from academics she also has a stronghold on administrative skills. She has a Ph.D. from Maharshi Dayanand University (MDU), Rohtak, Haryana in Computer Science from the Department of Computer Science and Applications in 2017. She received her M.Phil. in Computer Science from Chaudhary Devi Lal University (CDLU), Sirsa, Haryana, India in 2007. Before she had studied at Guru Jambheshwar University of Science & Technology (GJU), Hisar and Kurukshetra University (KU), Kurukshetra, Haryana, India. Her main research interest areas are Neural networks, Wireless Security, Soft Computing, Artificial Intelligence and Wireless Communication. More than 40 of her research papers have been published in international peer-reviewed journals indexed in Scopus, Copernicus, ISI, ESCI and others.

Gopal Datt is currently working as an Assistant Professor in Uttarakhand Open University, Haldwani, Uttarakhand, India and has a total of 12 years of teaching experience in higher education. His area of interest is E-Learning, Learner support services in Open and Distance Learning (ODL), and Web programming.

Rupal Gupta is currently working as Assistant Professor, FOE&CS, Teerthanker Mahaveer University, Moradabad. He is parallel pursuing Ph.D in information technology from University School of Information, Communication and Technology, GGSIPU, New Delhi. He holds Master degree in Information Technology, Maters in Computer Applications and M.Phil in Computer Science. He is UGC-NET (Computer Science) qualified and also a Wipro certified faculty. He has a rich experience of 15 years in teaching and published multiple research papers in various international and national conferences and journals of repute. His research interest is in the field of Semantic Web, SPARQL Query Processing, Machine Learning and Data Science.

Ruchika Gupta is currently working as APT and Pro Instructor at WhiteHat Jr, a global platform of online learning. she has completed her M.Tech from University School of Information, Communication and Technology, GGSIPU, New Delhi. She also holds Master degree in Computer Applications from MDU, Rohtak. She has a rich experience of 14 years in teaching and published multiple research papers in various international and national conferences and journals. Her research interests include GPS, GIS, Semantic Web and Machine Learning.

Mukesh Joshi is working as an Associate Professor in the Department of SOC, Graphic Era Hill University, Bhimtal. Dr. Mukesh Joshi, having more than 17 years of Teaching experience in Post Graduate and Under Graduate Program. He did his

Ph.D.(Computer Science) in 2018. Currently, he has 7 patents published and many research papers in SCI, ESCI, SCOPUS, IEEE & UGC Approved Journals. He has attended several conferences, seminars, and workshops in various renowned Institutions. His areas of interest are Cloud Computing, Database, Bigdata etc.

Preety Khatri has more than 16 years of teaching experience as an Assistant Professor in various institutes. Published various research papers in national as well as international journals. Participated and presented various research papers in national as well as international conferences. Published various books and patents also. Apart from teaching, worked aa NAAC Coordinator at institute level. Having responsibilities like event coordinator, research committee member, anti-ragging committee member, placement committee member in previous institutes. Guiding and mentoring graduates and post-graduate students in their projects.

Rohit Kumar is the Program Coordinator BCA/MCA Cum Assistant Professor at Institute of Management Studies, Noida.

Nikita Malik is presently working as an Assistant Professor in Department of Computer Applications at Maharaja Surajmal Institute, New Delhi. She is parallelly pursuing her Ph.D in IT from University School of Information, Communication and Technology, GGSIPU. Nikita completed her M.Tech from Ambedkar Institute of Advanced Communication Technologies and Research, GGSIPU (presently East campus-NSUT) in 2017, where she was the Gold Medalist in Information Security branch. She has over 2 years of teaching experience and has published multiple research papers in various international and national journals and conferences of repute. Her research interests include Semantic Web Technologies, IoT, WSN, MANETs, Data Analytics and Information Security aspects.

Ghanshyam Parmar is working as an Assistant Professor since 2006 in Natubhai V. Patel College of Pure and Applied Sciences, a constituent college of The CVM University. Completed PhD from Rai University, Ahmedabad, Gujarat in 2017, in area of Cloud Computing.

Monika Patel is working as an Assistant Professor since 2005 in Natubhai V. Patel College of Pure and Applied Sciences, a constituent college of The CVM University. Completed PhD from Sardar Patel University, Gujarat in 2017, in area of Watermark extraction using Artificial Intelligence.

Priyadarsini Patnaik is a knowledgeable, experienced, and seasoned marketing professional with a huge experience in both industry and academics. Currently pursuing Ph.D. in Birla Global University, Bhubaneswar, Odisha, India. Her research area of interest are Artificial Intelligence, Consumer Behaviour, Advertising, Retail Management, Digital Marketing.

Arun Kumar Rai is currently working as an Assistant Professor in CSE at Graphic Era Hill University, Bhimtal, Uttarakhand, India and has a total of 12 years of teaching experience in higher education. His area of interest are Automata, Web Security, Soft Computing, etc.

Abhay Saxena is Dean of School of Technology, Management and Communication and Professor, Computer Science at Dev Sanskriti Vishwavidyalya, Hardwar, Uttrakhand, India. He holds Doctoral Degree of Computer Science in the area of Artificial Neural Networks (ANN). He is a visiting Professor for Budapest Business School, Hungary, National Formosa University Taiwan. He had been appointed as a Honorary Research fellow for two years by University of Malaysia (UNIMAS). He had also delivered Technical talks in Swinburne University, Malaysia, University of Malaysia, Yunlin University, Taiwan along with his contribution and participation in national and international workshop, conferences across India and abroad. In his 28 years of Teaching and Industry experience, he had authored twelve books on discrete areas of Core computing i.e. Computer Networks, DBMS, E-commerce, Web technology, .Net Framework, Life management and Human values, Internet of things for futuristic computing (2018) and Vedic Informatics (2109). His upcoming two books are Artificial Intelligence for the societal development and Global well-being (IGI global publication May 2022) and Block Chain (2021 December).

Anu Sharma is working as an Assistant professor in CCSIT, TMU, Moradabad, UP, India. She is pursuing Ph.D in CSE from Uttrakhand Technical University, Dehradaun. She has done M.Tech (CSE) from MMU, Mullana, Ambala. Her current research interest includes data Mining, Neural Network, Machine Learning. She is having experience of more than 12 years. She has published more than 27 research papers in International Journals/Conferences.

Naveen Tewari, having more than 13 years of Teaching experience in Post Graduate and Under Graduate Program. He did his Ph.D. (Computer Science) in 2015. He has done MCA in 2008. He is also a Developer in Multimedia (Certified by Arena Multimedia). Currently, he has 7 patents published and many research papers in SCI, ESCI, SCOPUS, IEEE & UGC Approved Journals. He has attended several conferences, seminars, and workshops in various renowned Institutions. He is also on the Reviewer board of many reputed International journals. 2 research scholars are doing Ph.D. under his guidance. His areas of interest are Internet of

Things, Fog Computing, Web programming, Programming Methodologies, Cloud computing, and Object-Oriented Concepts.

Nipun Walia is a third-year student pursuing Bachelor's of Computer Application from Maharaja Surajmal Institute, affiliated to GGSIPU. With a keen interest in the field of Machine Learning and Development, he is currently working on honing his skills in these respected fields.

Index

E

F

G

H

I

L

M

N

P

Q

R

S

T

V

W

Y

Z

www.ingramcontent.com/pod-product-compliance
Lightning Source LLC
LaVergne TN
LVHW080557300926
849162LV00005B/975